The Diseased Brain and the Failing Mind

Explorations in Science and Literature

Series Editors:
John Holmes, Anton Kirchhofer and Janine Rogers

Explorations in Science and Literature considers the significance of literature from within a scientific worldview and brings the insights of literary study to bear on current science. Ranging across scientific disciplines, literary concepts, and different times and cultures, volumes in this series will show how literature and science, including medicine and technology, are intricately connected, and how they are indispensable to one another in building up our understanding of ourselves and of the world around us.

Published Titles
Biofictions, Josie Gill
Imagining Solar Energy, Gregory Lynall

Forthcoming Titles
Narrative in the Age of the Genome, Lara Choksey
Fictions of Prevention, Benedetta Liorsi
The Social Dinosaur, Will Tattersdill
Writing Remains, Josie Gill

The Diseased Brain and the Failing Mind

Dementia in Science, Medicine and Literature of the Long Twentieth Century

Martina Zimmermann

BLOOMSBURY ACADEMIC
LONDON • NEW YORK • OXFORD • NEW DELHI • SYDNEY

BLOOMSBURY ACADEMIC
Bloomsbury Publishing Plc
50 Bedford Square, London, WC1B 3DP, UK
1385 Broadway, New York, NY 10018, USA
29 Earlsfort Terrace, Dublin 2, Ireland

BLOOMSBURY, BLOOMSBURY ACADEMIC and the Diana logo are trademarks of
Bloomsbury Publishing Plc

First published in Great Britain 2020
This paperback edition published in 2022

Copyright © Martina Zimmermann, 2020

Martina Zimmermann has asserted her right under the Copyright, Designs and Patents Act, 1988, to be identified as Author of this work.

For legal purposes the Acknowledgements on p. ix–xii constitute an extension of this copyright page.

Cover design by Toby Way
Cover image © Getty Images

This work is published subject to a Creative Commons Attribution Non-commercial No Derivatives Licence. You may share this work for non-commercial purposes only, provided you give attribution to the copyright holder and the publisher.

Bloomsbury Publishing Plc does not have any control over, or responsibility for, any third-party websites referred to or in this book. All internet addresses given in this book were correct at the time of going to press. The author and publisher regret any inconvenience caused if addresses have changed or sites have ceased to exist, but can accept no responsibility for any such changes.

A catalogue record for this book is available from the British Library.

Library of Congress Cataloging-in-Publication Data
Names: Zimmermann, Martina (Researcher in health humanities), author.
Title: The diseased brain and the failing mind : dementia in science, medicine and literature of the long twentieth century / Martina Zimmermann.
Identifiers: LCCN 2020010911 (print) | LCCN 2020010912 (ebook) | ISBN 9781350121805 (hardback) | ISBN 9781350121812 (ebook) | ISBN 9781350121829 (epub)
Subjects: LCSH: Dementia in literature. | Literature and science–History–20th century. | Literature, Modern–20th century–History and criticism.
Classification: LCC PN56.D4645 Z56 2020 (print) | LCC PN56.D4645 ebook) | DDC 809/.933561–dc23
LC record available at https://lccn.loc.gov/2020010911
LC ebook record available at https://lccn.loc.gov/2020010912

ISBN: HB: 978-1-3501-2180-5
PB: 978-1-3502-4936-3
ePDF: 978-1-3501-2181-2
eBook: 978-1-3501-2182-9

Series: Explorations in Science and Literature

Typeset by Newgen KnowledgeWorks Pvt. Ltd., Chennai, India

To find out more about our authors and books visit www.bloomsbury.com and sign up for our newsletters.

Meinem Mann

Contents

Acknowledgements	ix
Series Preface	xiii

1 Introduction   1
   Alzheimer's disease: A twenty-first-century first-world scare   1
   Dementia in history   6
   Methodology: Literature and science   9
   Overview   15

Part I The organic paradigm   21

2 From brain inspection to cell death   23
   The Forsyte Saga: The cultural image of dementia in the
      fin-de-siècle family novel   23
   Dementia and memory loss in science, medicine and literature
      before 1880   29
   Auguste D. and Johann F.: Alzheimer's clinical cases and
      histological research   34
   Degeneration: The old and new narrative of loss and decline in
      medico-scientific literature on dementia and Alzheimer's disease 41
   There Were No Windows: The patient's illness experience in the
      modernist novel   45

Part II The ageing perspective   53

3 Culture shapes politics shapes science   55
   Researching old age: From medical science to old-age psychiatry 55
   At The Jerusalem: Dementia defines the elderly in 1960s new
      realist fiction   60

4	The loss of self in healthcare and cultural discourse	67
	Caregiver guides: Helpers in the face of loss and decline	67
	Out of Mind: The postmodern novel delves into the mind of the patient	74

Part III The cognitive picture — 85

5	The narrative of loss in a growing biomedical and literary marketplace of Alzheimer's disease	87
	Neurodegeneration: The biochemical narrative of lost molecules, pathways and communication	89
	On genes and genealogy: The patient as specimen, carrier and type in research and popular science	94
	Death in Slow Motion: Past identities, lost plots and old age in caregiver life-writing	100
6	Neurotechnologies and narrative examine the failing mind	109
	The visual exploration of the brain and fascination with the mind	109
	The Dying of the Light: Detective fiction claims back patient authority	115
	Who Will I Be When I Die? Patient life-writing around the year 2000	123

Part IV The whole-person prospects — 131

7	The dichotomy of Alzheimer's disease	133
	Immunization hope and hype: The patient as non-responder	135
	La guardiana di Ulisse: The patient beyond forgetting in children's literature and adult fiction of the new century	142
	Alzheimer mon amour: Healthcare changes and patient personality in contemporary caregiver memoirs	148
	We Are Not Ourselves: The cultural image of Alzheimer's disease in the twenty-first-century bildungsroman	154
8	Conclusion	163

Glossary	169
Notes	177
Bibliography	223
Index	257

Acknowledgements

This book, like my own training, crosses several disciplines. It immeasurably benefitted from my exchanges with literary and film scholars, scholars of ageing and the health humanities, cultural historians and historians of science as well as colleagues working in medicine and pharmaceutical sciences, within and beyond King's College London. I am particularly grateful to Brian Hurwitz and James Whitehead. A practicing clinician in his first life, who enacts narrative medicine in each and every conversation, Brian's critical evaluation and analytical observations have opened my eyes to the challenges of discipline-crossing articulation and argument. James Whitehead rose to the challenge of initiating a non-native speaker to the infinite thesaurus of literary scholarship. His perspective has significantly influenced how my 'neuromolecular style of thought' shifted towards more literary reasoning.

I am also indebted to Neil Vickers and Jonathan Day for inspirational conversations about my research. Their comments were instrumental in giving the project a firmer grounding in Ian Hacking's outlook on the role of science in discourse developments, and in deepening the study's focus on how literary trends have contributed to shaping current thinking about dementia.

My research also benefited from engaging with Helen Small and Sander L. Gilman. Helen's philosophy-centred perspective on ageing in literature and culture pushed me more fully to consider the impact on the patient's unrecoverable self of a century-long discourse of dispossession, decline and loss. Sander's reflections on the cultural politics of neurodiversity added further dimensions to my analysis of literary writings. Through conversation with them both, I considered in more detail the challenges faced in the medical clinic of having to engage with patients and caregivers while having precious little to offer in terms of remedial intervention.

I greatly appreciated the support of Richard Kirkland, Josephine McDonagh, Mark Turner and Jane Elliott throughout my time at the English Department of King's. Thanks are also due to Ludmilla Jordanova and Katherine Foxall for helpful comments on the original Wellcome Trust grant application (099351/Z/12/Z); their observations helped to frame this project more clearly as a historically situated literary study. I am grateful to Susan Greenfield for having endorsed this application, and I thank Julia Howse for administrative support during the application process and Sabrina Beck for helping with organizational tasks during this research period. Members of staff at the British Library, the Radcliffe Science Library and the Wellcome Library have been enormously supportive throughout.

This project has also benefitted from my conversations with many national and international scholars, too numerous to mention individually, at health humanities and literature and science meetings as well as workshops and round tables concerned with ageing, dementia and literary writing more broadly – including the conference A Narrative Future for Health Care (London, June 2013), a workshop on Medical Case Histories as Genre (London, July 2013), the first Narrative Medicine Summer Meeting (Berlin, August 2013), a Medical Humanities Colloquium at UNC (Chapel Hill, September 2013), a round table about Science and Storytelling at the University of Reading (November 2013), a workshop on The Writer's Diary (London, April 2014), the Boundaries of Illness meeting (London, June/July 2014), two CHCI Health and Medical Humanities meetings (Dartmouth College, NH, July 2015; King's College London, June 2016), the twelfth annual conference of the British Society for Literature and Science (Bristol, April 2017), the first Dementia and Cultural Narrative Symposium at Aston University (December 2017), a symposium on Living a Good Life in Older Age at the University of Warwick (July 2018), a workshop on Dementia, Violence, and the Politics of Memory at Freie Universität Berlin (September 2018), a workshop on Working Together beyond the Academy in Research in Dementia and Culture (London, November 2018) and a Centre for the Humanities and Health seminar at King's College London (January 2019).

Several scholars directed me to specific texts or helped me explore research materials in particular ways. First and foremost, I feel deeply indebted to Sally Shuttleworth for pointing me to Timothy Forsyte's 'losing his mind'. With Victoria Coulson, I had a wonderfully inspirational discussion about 'wicked' scientists; with Tania Gergel, I deliberated on ageing in Ancient Greece and Rome; Eileen Gillooly mentioned Barbara Pym's Marcia Ivory to me; to John Holmes, I owe my attention to the influence on discourse developments of film; Keisuke Maruyoshi answered my questions regarding Japanese culture and language in relation to two dementia narratives; with Gordon McMullan, I discussed late-life creativity and redemptive turns in caregiver accounts; with Ulrich Metzner, I had an enthusiastic exchange about the correlation between character traits of musicians and the role of their instruments within a string quartet; I thank Giancarlo Pepeu for his prompt assistance in obtaining original language material; Ruth Richardson pointed me to an early literary dementia narrative centred on the condition's early-onset form; discussions with Max Saunders and Julia Watson added nuances to my readings of life-writing; and a conversation with Claudia Stein, long before the inception of this project, inspired my focus on imaging methodologies. I owe thanks to David Avital and Ben Doyle at Bloomsbury, to the anonymous reader for a perceptive and thoughtful reflection on this discipline-crossing research and to the series editors for their careful reading, considered observations and generous support. Finally, conversations with Jochen Klein continue to inspire me in my discipline-crossing undertakings, and I am immensely grateful for his observations on my work.

Throughout this research activity, I audited several MA modules at the English Department of King's, which shaped how I approached my research material. I am grateful to Ian Henderson for welcoming me into his 2013/2014 seminar series on 'Victorian Sensations', and to Madeleine Wood for her stimulating discussion of Wilkie Collins's work. I also wish to thank Brian Glasser and Michael Clark for encouraging my participation in their 2014/2015 seminar series 'Medicine on Screen', which introduced me, among others, to identity-affirming documentaries like those by Nicolas Philibert. I also had the great fortune of following Neil Vickers's 2015/2016 module on 'Biopower: The Fate of an Idea'. The course readings and seminar discussions have influenced my perspective on Alzheimer's disease as a case study of how the power pertaining to disciplinary and institutional knowledge shapes the wider societal and cultural thinking about illness.

It is several years now that I have had the privilege of teaching illness narratives in a pharmaceutical sciences as well as a health humanities context. I feel especially indebted to all undergraduate students who entered into exploring a field outside textbook physiology and pharmacology, both at the Faculty of Biochemistry, Chemistry and Pharmacy at Goethe University Frankfurt and at the School of Biomedical Sciences at King's. Teaching at graduate level in the English Department at King's and discussing my research in a series of postgraduate seminars in Frankfurt especially enriched my reading of current life-writing. The continued interest in my research endeavours by science colleagues, especially Jochen Klein, Michael Karas and Paul Layer, is further encouragement to persevere at the boundary between science, medicine and literature.

Finally, I wish to thank Peter and Elizabeth Howarth for having hosted me on so many occasions, and all those who have provided support and encouragement (while mercilessly lengthening my list of research materials with TV recordings, newspaper clippings and the yet latest bestseller).

The Wellcome Trust generously enabled this research and its open access publication.

Series Preface

In spite of the myth of the 'Two Cultures', science and literature have always been shaped by one another. Many of our most powerful scientific concepts, from natural selection to artificial intelligence, from germ theory to chaos theory, have been formed through the careful – and sometimes careless – use of written language. Poets, novelists, playwrights and journalists have taken up scientific ideas, medical research and new technologies, exploring them, reworking them, at times distorting or misjudging them, but always shaping profoundly the wider culture's understanding of what they mean. This intimate and productive relationship between literature and science generated a steady stream of insightful scholarship and commentary throughout the twentieth century and has grown into a substantial field of study in its own right since the turn of the millennium. Where the idea of 'Two Cultures' does still have a hold, however, is in academic disciplines themselves. In schools and universities, we study science and arts subjects in different classrooms, taught by different people with different expectations. Literature and science studies has, so far, been largely a sub-discipline of literature, with only rare contributions from or addressed to scientific experts. In a world of ever-increasing specialization, failure to communicate across these disciplinary divides risks failing to appreciate the contribution that the study of literature can make to our understanding of science, medicine and technology, the uses that science makes of images, narratives and fictions, and the insights that scientists can bring to bear on literature and on culture at large.

Explorations in Science and Literature aims to speak across this divide. It has a particular mandate to bring the insights of literary study to bear on science itself; to consider the significance of literature from a scientific point of view; and to explore the role of literature within the history of science. The books therefore examine the complex interrelations between science and literature in cross-disciplinary ways. They are written equally for scholars and students of literature and for scientists and science students, but also for historians and sociologists of science, as well as general readers interested in science and its place in culture and society. By showing how each field can be enhanced by a knowledge of the others, we hope to enrich scientific as well as literary research, and to cultivate a new cross-disciplinary approach to fundamental questions in both fields.

The series will encompass topics from across the physical, biological and social sciences, medicine and technology, wherever literature can inform our understanding of the science, its origins and its implications. It will also include books on literary forms and techniques that are informed by science, as well as studies that consider how science itself has been articulated. Along with literature in the broad sense of written texts, books in the series will also consider other cultural forms including drama, film, television, and other arts and media.

John Holmes, Anton Kirchhofer and Janine Rogers

1

Introduction

Alzheimer's disease: A twenty-first-century first-world scare

> She was not Inés any more, she was only a body in front of the doctor, a handful of inner organs, two lungs, a brain that was destroying itself.
> Andrés Barba, *Ahora tocad música de baile*[1]

> We are, above all, human beings with strengths and weaknesses, highs and lows, and with a disease that touches an organ of our body: the brain.
> Claude Couturier, *Puzzle, Journal d'une Alzheimer*[2]

When we think about dementia or Alzheimer's disease, we think about it in terms of loss: the loss of mental capacity, the loss of skills and agency, the loss of memory, identity and personhood.[3] This book charts the development of this discourse from its origins in the nineteenth century to the present day and specifically considers concepts that have shaped the neurological and psychological understanding of dementia. It explores the extent to which the societal dementia discourse, which induces Andrés Barba to remove Inés's personhood and reduce her to mere bodily parts, is rooted in a science-led and medicine-adopted discourse. This discourse relates the broad-spectrum concept of loss of cognition, personhood and identity to the actual loss of cells in specific brain areas, as illustrated by the following extract from *Brain Facts*, a public information initiative of the Society for Neuroscience launched in 2012:

> Initially, people experience memory loss and confusion ... symptoms of AD [(Alzheimer's disease)] gradually lead to behavior and personality changes, a decline in cognitive abilities such as decision-making and language skills ... AD ultimately leads to a severe loss of mental function. These losses are related to the worsening breakdown of the connections between certain neurons in the brain and their eventual death.[4]

I worked in experimental research related to Alzheimer's disease and other forms of dementia for fifteen years and have participated in this scientific discourse. Underlying it is what Nikolas Rose and Joelle M. Abi-Rached have termed a 'neuromolecular style of thought'. This way of reasoning argues that mental states express themselves in material processes in the brain, and that these processes can be anatomized at a molecular level.[5] Applying this style of thought to reading literary texts could lead me to argue that a focus on unravelling molecular mechanisms of cell death in dementia has reduced patients to mere carriers of symptoms; and that this reductionism, as we hear it in the voice of Inés's husband Pablo, underestimates patients' need for and right to dignified care. But how well can Inés's objectification be explained by the appropriation within caregiver discourse of scientifically and medically sanctioned concepts of loss and reduction? Or is Pablo's objectification of his wife the result of overwhelming burdens of care placed on the caregiver? These questions bridge material-scientific and discursive-cultural understandings of Alzheimer's disease.

Concern with this nexus of issues has pushed my research interests more and more towards the intersection of literature, science and medicine. I began to focus on the context in which specific scientific and medical texts have been produced. I had to change my view of scientific work as being merely an isolated discipline of knowledge acquisition: science and its published products are clearly a form of culture.[6] The process in part has turned me into a subject of my own enquiry, as I have come to understand my scientific research as historically situated. My initial work developed in the context of the early 2000s' fascination with biomarkers (which this book covers in Chapter 5), while later studies, in the framework of my *Habilitation*, evolved in response to growing doubts about the validity of specific disease hypotheses (discussed in Chapter 7). But this is not to say that this study is an auto-ethnographic work by a neuropharmacologist-turned-health-humanities researcher; nor is it evidence of lived two-culture splits that tell of the frustrations of a scientist healed by turning to literature. Science is neither weak nor wicked. It studies aspects of health and disease which the health humanities have no access to; and, vice versa, the medical humanities open additional perspectives onto health and illness.

But medical systems, especially in rich societies, have increasingly failed the patient. This is a phenomenon not unique to dementia: patients are objectified in the process of diagnosis and functional assessment of many diseases, and healthcare programmes continue to develop around the notion of the patient as burden and as having no voice.[7] The patient's diminished capabilities, dependence and passivity are central to the cultural dementia narrative (i.e.

how the condition's nature and patient's identity are perceived and understood in Western societies today); and this narrative has become core to beliefs about dementia in the wider cultural and societal discourse. As this book will show, health humanities work, including literary scholarship, has greatly assisted the patient on the journey back to centre stage. Yet, this has come at the price of what I increasingly perceive to be an antagonistic mode that sets up the sciences against the humanities. This happens because the biomedical approach has become a hold-all for processes that remove the personhood from the patient, where the effects of the scientific and the medical are insufficiently separated out, either in terms of discourse developments or in terms of their preferred methodologies, goals and interest groups.[8] This study, although it strongly aligns itself with the medical humanities, is not a priori intended as a 'challenge and corrective to the hierarchies of evidence that have come to define theoretical, practice-based and policy-oriented instantiations of the biomedical'.[9] Rather, it will illustrate how current criticism of the biomedical approach could be more nuanced by drawing attention to the crucial difference between the process of methodological reductionism in science research and notions of objectification in medical and clinical practice.

This book explores to what extent the current cultural dementia narrative is grounded in historical scientific and medical language related to brain diseases, beginning with its origins in the nineteenth century, when the condition was first described as an organic disease, to the present day, when it is viewed as a disorder of cognition. I pursue a historical approach, because I share the view of medical historian Jesse F. Ballenger that 'without a sense of history, without the ability to construct a coherent narrative linking the present to the past as well as the future, public discourse on Alzheimer's will itself be confused, disoriented'.[10] I pinpoint changes in the scientific and medical understandings of dementia throughout the long twentieth century, and examine how these changes have influenced the wider cultural thinking about dementia over time. I take Ballenger's groundbreaking study as an invitation to historicize the literary approaches to dementia and Alzheimer's disease. Marlene Goldman has thought along these lines in her recent work on narratives about Alzheimer's disease in Canada. She turns to literature 'because fiction often explicitly or implicitly adopts a non-biomedical, historical approach to dementia', although she largely focuses on how current literary productions explore historically inflected notions of the condition.[11]

The present study, by comparison, takes literary writing as an active contributor to discourse developments throughout the entire period under investigation; in doing so, it critically negotiates the possibility that literary writings perpetuate and enhance biomedical discourse. It contends that whole-person approaches – approaches that acknowledge the patient as a feeling, thinking, communicating and engaging individual, with a mind and world of experience of their own – can integrate an understanding of Alzheimer's disease as a local pathological entity – an organic condition of the brain. Conversely, I argue that considering Alzheimer's disease as a whole-person phenomenon can equally subject the patient to a process of objectification ascribed to biomedical discourse. My overarching aim is to understand where the cultural path taken by the presentation and perception of dementia – since its respective definitions as an organic disease around 1900 and a cognitive, performance-limiting condition around 2000 – leads the illness and its sufferers to in the twenty-first century. Two processes interest me in particular: how have literary renderings of dementia been shaped by an evolving medico-scientific dementia discourse, and how has this cultural narrative fed back into the scientific and medical gaze at the condition? To address this, I will attend to four different kinds of text: fiction (including film), caregiver accounts, patient life-writing, and scientific and medical texts as well as their popular echoes in mass media reports and generally accessible scientific book publications.

Fictional representations can be understood in part as an expression of societal and cultural developments, and fictional depictions such as Andrés Barba's can be taken as both a product and an illustration of the current cultural dementia narrative. Caregiver life-writing, in turn, gives us a glimpse into how family members make sense of the illness; but the existence of these texts itself appears to objectify the patient. Such objectification has motivated individuals like Claude Couturier to reclaim identity and selfhood through their writing. Patient accounts convince me that dementia life-writing is well positioned to instigate a counter-discourse that focuses attention on patient experience and makes this experience central to both: the medico-scientific understanding of the condition and sociopolitical approaches to the illness.[12] Cultural renderings of dementia and ageing, indeed, do feature in more and more specialist clinical journals.[13] These papers evince a clear interest by the medical and scientific professions in cultural representations of dementia. They also suggest that cultural images as well as the patient's and caregiver's own writing can reshape medico-scientific dementia discourse: I will explore this through analysis of a selection of scientific and medical research papers.

These different types of text already point to the fact that the condition is not only a scientific concept caught up in what Oliver Sacks termed neurology's 'deficit' language: language that denotes 'loss of speech, loss of language, loss of memory, loss of vision, loss of dexterity, loss of identity and myriad other lacks and losses of specific functions (or faculties)'.[14] It is also an experience. And as such, the narrative of loss in relation to Alzheimer's disease reaches beyond what the biomedical model of the condition can explain. With the rise during the second half of the nineteenth century of cellular theories and the notion of mental life residing in a granular substance, the cell came to be understood as the locus of the disease. Wider cultural discourse began to link the integrity of the cell to that of identity.[15] As William R. Clark puts it in his account of cell death as a creative principle, 'Human beings seem to have decided that cells of the brain are more important in defining life than other cells.'[16] Once brain had acquired mind, wider sociocultural concepts equated the loss of cognitive skills and memory with the patient's loss of awareness and self; and concepts that link a sense of self to a sound mind (hence functioning brain) and intact memory link this sense of self also to the ability to narrate.[17]

In addition, there is a growing preoccupation with dementia as the synthesis of old age and illness.[18] Ageing itself is equated with experiencing loss – the loss of family and friends, the loss of opportunities, power and prowess.[19] Celebratory perspectives on ageing – notions of gain, growth and progress – hardly seem to feed into the current cultural dementia narrative. Instead, biological understandings of ageing appear to reinforce a dementia narrative that centres on loss; and a lot of gerontological research and geriatric practice integrates 'social values responsible for ageism and the declining status of old people'.[20] On top of all this, the perception of dementia has become yoked to 'loss of wholeness, loss of certainty, loss of control, loss of freedom to act, and loss of the familiar world' that is fundamental to any experience of severe illness.[21]

The cultural profile of the condition is partly related to significant demographic shifts in Western societies and the politicization of these shifts. Alzheimer's disease has become the most common form of dementia among the elderly. And, representing the most feared neurological condition today, it has also turned into a trope for all forms of dementia. This anxiety is played on by the popular press in relation to the condition's epidemic proportions. Cultural figures about dementia include images of a 'rising tide' and 'millennium demon', and the condition has become a metaphor for all kinds of sociopolitical ills.[22] This is also the case because plague-like terminology abounds which, as Susan Sontag explicated in relation to a quite different disease, has long been used metaphorically as the highest standard of collective calamity, evil, scourge ... Although the disease to which the word is permanently affixed produced the most lethal of recorded epidemics, being experienced as a pitiless slayer is not necessary for a disease to be regarded as plague-like. [An illness] has been regarded as a plague ... not because it killed often, but because it was disgracing, disempowering, disgusting.[23]

This book will delve into the historical grounding of objectifying patient presentations and explore how specific metaphorical language became affixed to condition and patient.

Dementia in history

Dementia has been conceived of as both an abstract and isolated molecular event and a condition related to an individual as a whole. Recent sociological work elaborates on these continued tensions: first, whether behavioural changes are the manifestation of a distinct pathology or are inherently related to ageing; second, whether specific organic changes in the brain cause behavioural changes or whether the process of ageing itself impacts on the mind and leads to changes that are pathological; and third, whether dementia is socially conditioned or genetically programmed.[24]

While the second and third such tensions emerge successively from medico-scientific work at the end of the nineteenth and twentieth centuries, the first issue resonates with longstanding ideas about memory loss and ageing. Homer associated ageing with physical change, Juvenal coupled it with ideas of second childhood and foolishness and Horace described it as anticipating death.[25] In ancient Greece, figures of public life, such as Xenophon, 'hesitated to live a longer time lest he be forced to pay the penalties of old age – to ... fail in intellectual capacities'; and literary representations feature characters like Strepsiades in Aristophanes's comedy *The Clouds*, who 'has a bad memory, is dull of comprehension, and is too old to learn'.[26] I agree with Pat Thane that we should not draw simplified cultural perceptions from classical Greek comedy.[27] But such literary stereotypes could not exist in a conceptual void, even if modern scholarship of classical texts can find almost no reference 'specific to mental illness in old age'.[28]

I do not claim that the present-day cultural dementia narrative has its roots in the Ancient World, nor do I read currently recognized symptoms of dementia into the cultural expressions of a society that thrived more than two thousand years ago.[29] However, I agree with Helen Small 'that what philosophers and non-philosophers have had to say about old age has, in essence, changed very little since classical antiquity'.[30] Shakespeare's *King Lear* is perhaps the most frequently mentioned post-classical portrayal of ageing as a state characterized by memory loss and 'second childhood'.[31] In common with Nina Taunton, I identify Lear as 'driven by anger and shame, two emotions associated with aged impotence', as acting under 'a misguided belief that he continues to wield power over his children and over nature itself'.[32] Yet, Lear's behaviour is frequently linked to Alzheimer's disease. Work claiming parallels between Lear and the clinical case of a 'frail elder' and probable Alzheimer's, that 'Lear's increasing isolation results from a tragic combination of cognitive impairment, caregiver stress, and generational conflict', is one such example.[33] Shakespeare's interest in the development of character and familiarity with the human experience have been observed to invite continuous reinterpretation for a contemporary audience.[34] Still, I read such contemporary keenness to identify Alzheimer's disease in cultural representations of the Antiquities, Middle Ages and Early Modern Times first and foremost as a manifestation of the current societal concern with the condition.

From the end of the sixteenth century until well into the eighteenth, the term 'dementia' was employed as synonymous with 'melancholia' and 'madness'. By the late eighteenth century, the term held the meaning 'out of one's mind', around the time when the condition began to be treated as a mental illness.[35] This idea of the condition built on a growing interest in brain pathology, which emerged from the biological psychologies of Erasmus Darwin, Franz Josef Gall and other brain scientists, from the 1790s onwards, and played a crucial role in establishing the brain 'as the organ of thought'.[36] Methodological progress and technical advancements which became available towards the end of the nineteenth century supported the idea that the various morphological subdivisions of the brain consisted of discrete areas of particular function, which enabled development of a 'materialist science of the self'.[37] By the 1880s, biological explanations of physiological phenomena also included a changing conceptualization of dementia.

The structure of the modern sciences of memory became established between 1874 and 1886.[38] In particular, lesion theories began to fuel enquiries into potential links between bodily signs (symptoms related to a failing mind) and organic (cerebral) changes. The changing methods of clinical observation led, as Michel Foucault argued, to the 'projection of illness onto the plane of absolute visibility' and a 'gaze [that] dominat[ed] the entire field of possible knowledge'.[39]

Alois Alzheimer's case of Auguste D. in 1906 is widely considered a landmark.[40] But it was part of a much larger shift in the discipline's approach to dementia, driven by new methodologies that created different conceptual perspectives.[41] In this instance, the new methodologies included an expanding array of visualizing techniques, among them a nerve cell staining method developed in 1872 by Camillo Golgi (1843–1926).[42] This staining technique enabled the study of the morphology of the brain, including pathological changes. And this, in turn, fuelled in the wake of Alzheimer's work a neuroscientific redefinition of the condition as an organic disease of the brain.

The nineteenth-century fin de siècle has attracted scholarly attention not only because of the 'new technologies that became available … the new sites of scientific enquiry thereby opened, and the new forms of knowledge they facilitated'.[43] It was also a time when 'biological explanations of psychological states held sway', and 'ideas about the functioning of the mind and body seeped from science into broader cultural discourses'.[44] In the context of the memory sciences, literary notions of memory impairment throughout the nineteenth century became increasingly related to concepts of lesion as a form of loss. Nicholas Dames, for example, establishes a clear link between the researched understanding of memory and the literary representation thereof, and Sally Shuttleworth shows how fictional representations of memory changed during the second half of the nineteenth century.[45] Also from a wider cultural perspective the fin de siècle makes a suitable beginning for this study. The concept of identity in dementia has shifted in response to changes in the meaning of terms such as 'idiocy', 'retardation', 'senility' and 'madness'.[46] These shifts happened, towards the end of the 1800s, in response to changing societal preoccupations, scientific understandings and wider cultural concerns about health and disease.[47] And related to this, socio-economic, political and legal language at the fin de siècle assumed a decidedly degenerative aspect.[48] This suggests there existed, at the end of the nineteenth century, a range of dimensions to the cultural counterparts of the evolving medico-scientific discourse of dementia.

This study is not meant to foreground ageing, but there is consensus that 'the quality of our ideas about the meaning of aging has considerable implications for medical ethics and practice, psychotherapy and the education of older people, research on the biology of aging and the prolongation of human life, public policy, and religious and spiritual life'.[49] These insights directly feed into social, cultural, political and economic implications of dementia. In addition, attending to the close connections between ageing and memory loss will help to explore questions of identity within dementia. 'When we think about old age', emphasizes Helen Small, 'our thinking rests on larger, but usually tacit, assumptions about what a life is, what a person is, what a *good* life is, what social justice is, and much else besides'. Small explores questions of identity and agency within Alzheimer's disease, arguing that the experience of decline and ageing depends on personal, political as well as historical circumstance.[50] The same is true for how ageing is negatively viewed in Western culture and how ageism is gendered, that is, how old men and women are treated differently within a wider presumption of ageism. Discourses of ageing and gender both influence how individuals with dementia are stereotyped negatively – in the wider societal discourse as well as in cultural productions.

Methodology: Literature and science

The core procedure of my research is to read literature alongside science and medicine in a way that takes into account the emphasis placed by scientific as well as medical research on dementia as being, first, mental, behavioural and psychiatric; second, bodily, organic and biochemical; and third, cognitive and performance-limiting in nature.

I distinguish science from medicine based on how their practices relate to the patient. For the purpose of this study, I take science to denote medical science: an applied discipline, whose 'goals are interventions to produce health'.[51] Medical science research takes place in the laboratory and uses, for instance, tissue samples; in its procedures and approaches it comes close to what Bruno Latour terms 'technoscience'.[52] It relies on simplified model systems or focuses on individual molecular events. This practice is necessary so that the conclusions drawn can remain objectively true within the methodological framework employed. Where this practice directs understanding that the whole system can be explained by causalities and connections between the individual molecular events investigated, it may become problematic and reductionist. It overlooks that, with each hierarchic level, 'entirely new laws, concepts, and generalizations are necessary, requiring inspiration and creativity to just as great a degree as in the previous one'.[53] Such reasoning is especially important where medical science involves the patient, for example, for the collection of imaging data. Here, I see an overlap with medicine. This area of overlap technically concerns the sphere of scientific medicine. It is in this area of overlap that scientific research procedures have been applied to studying, for instance, molecular pathways in the living patient; and from here approaches 'generated by the microperspectives of biomedicine' make their way into the area of medicine.[54] My use of the term 'medico-scientific' in this book acknowledges the blurred nature of this area of overlap. But medicine is much more than applied medical science.[55]

I use the term 'medicine' in the context of clinical work, a practice in which the healing of, and the care for, the individual patient is the primary concern. It is in these settings that cognitive assessment takes place, behavioural observations are made and psychiatric research is carried out. Where thinking within the framework of reductionist scientific research procedures filters into such clinical encounters, objectification of the patient may ensue as the practitioner aspires to intervene in the molecular process gone awry – without embracing the further dimensions relevant to any illness experience, including considering the patient as the ultimate authority of and within their condition. What my investigations will not aim for is to describe how knowledge articulated in a single case report becomes part of a body of scientific learning. The process of experimental work itself and how research results become part or not of a specialist journal publication is not my focus. Nor am I primarily interested in what Bruno Latour termed 'the disorderly mixture revealed by science in action' in relation to Alzheimer's disease.[56]

Popular science contributes to the public understanding of what is going on inside research laboratories. But especially where popular science is concerned with medical science, it is a genre prone to blurring medical science, scientific medicine and clinical and healthcare practice, for example, by telling its story through the prism of medical cases. Although not always possible, my intention is to separate these entities, which is reflected in my choice of primary research materials, as I will read scientific research papers and popular scientific publications side by side. The research papers included relate to core scientific questions and medical concepts that were developed during the long twentieth century. I specifically address the central early papers related to Alzheimer's disease, between 1907 and the 1960s (Chapters 2 and 3), and discuss the seminal publications appearing in the high-impact journals *Science* and *Nature* once research picked up from the 1980s onwards (Chapter 5). From the 1990s onwards, biomedical research accelerated with the number and diversity of scientific journals drastically increasing and publication in high-impact journals becoming more and more competitive. For covering the final three decades (Chapters 6 and 7), I therefore choose representative research articles, whose citation frequency is suggestive of their scientific relevance as much as the impact factor of the journal in which they appeared. My reading of the medico-scientific literature, lay press and popular science texts will attend to the figurative language used in the context of patients and illness. But it also includes the language related to biomolecular and cellular events as well as behavioural and psychiatric symptoms and descriptions.

Scientific specialization since the 1950s does complicate this procedure. Ever more compartmentalized expert knowledge and professional language have made biomedical research nearly inaccessible to the general reader. My analysis puts these publications not only into their historical but also their scientific context. But such scientific contextualization is not intended as a running commentary on recent research projects or an assessment of state-of-the-art treatment strategies. Similarly, popular scientific texts cannot serve as a more easily accessible substitute for primary scientific papers. I use popular scientific outputs as primary research materials in themselves rather than as an objective source of information on scientific findings, because the selective emphasis of popular scientific texts reflects fascination with particular methodologies or insights symptomatic of a specific cultural context. I have made every attempt to explain scientific ideas in generally accessible ways in the running text, without introducing new metaphorical concepts. For clarity, I follow the abbreviations used in each research paper, but since these abbreviations vary from paper to paper, I do not employ abbreviated terminology throughout the running text; I also offer lay definitions of current medical and scientific understandings in the glossary at the end of this book, although these definitions cannot be free of the metaphorical understandings that this study aims to investigate.

Metaphoric language is key to this study. Metaphors are, as Andrew S. Reynolds puts it so well, 'a central element of the scientific process, every bit as important as the material instruments and microscopes with which [scientists] investigate and create understanding of the world'.[57] But they also represent a challenge to interdisciplinary conversation.[58] I employ the relevant terms as they are used in each context: metaphoric or non-metaphoric. More generally, I take images and metaphors as 'rhetorical tools in the construction of a public meaning', but also as 'intermediaries, establishing a link between nature and society, science and culture, reality and representation'.[59] I want to understand why specific images related to Alzheimer's disease are so persistent, and why certain research concepts maintain a dominant position in both the medical sciences and health humanities, at the expense of the patient's individuality and personhood. Nearly all the contributions on patient life-writing published to date concentrate on how these accounts capture changing notions of selfhood and identity.[60]

This means that I am not invested in drawing 'arrows of influence', as G. S. Rousseau would have it.[61] With the exception of a few cases of recent life-writing, I do not claim their authors were directly engaged with Alzheimer's disease research. Rather, I take literary texts to be representative examples of certain moods and discourses or phenomena at a given time. I do not contend that literary writing serves as an access point of information about the condition per se, not least since a fictional plot might require 'a measure of idealization or vilification' of character or condition.[62] But I agree with writer and literary critic Stefan Merrill Block, who, with reference to dementia fiction, intimates: 'I'm left with urgent questions that only fiction can answer: What do those late stages *feel* like? What is it like to lose oneself and still live? Could there be some essential kernel of selfhood that survives until the end? Mid- to late-stage sufferers, lost in their aphasia, can't explain it to us.'[63] Therefore, I ask what models, and perhaps alternatives, fictional accounts develop. More generally, I argue that the fascination with specific narrative themes and strategies coincides with selected research interests or medical concerns, that certain narrative trends have given us the terms in which we have come to understand dementia, consciousness and the mind.

I especially focus on prose, mindful of 'the essentially *narrative* nature of human consciousness'.[64] I am interested in how, for example, modernist and postmodernist approaches to consciousness developed ideas of dementia, and what conceptual overlaps there are with wider medico-scientific and sociocultural concerns during specific periods. Like David Herman, I take such approaches to be 'a window onto changes and innovations in the wider sociocultural context'.[65] But this is not to say that I exclude either poetry or drama. The latter offers additional perspectives on interactional challenges (to which I return in Chapters 6 and 7); a specific example of the former allows for an analysis of the density of nouns linked to the condition during the initial popularization of Alzheimer's disease (Chapter 4). I also include film, taking entertainment culture to be a key indicator of broader sociocultural concerns especially since the last third of the twentieth century.

In thinking with different narrative themes and trends, this book relies on close readings across a range of genres, including the family novel, detective fiction and children's literature, each of them emerging and dominant at different times. Following John Frow, I take genre, albeit dynamic and elastic, as given by three structural dimensions, namely formal organization, rhetorical structure and thematic content.[66] Genres can change in response to cultural shifts, and, given the possibility for change, I am particularly interested in two things: which narratives were told at particular times, and how these narratives were conditioned by the genre through which they were told. In addition, I explore how genre predetermines representational expectations and what we learn about shifts in cultural climate and discourse developments where such expectations remain unmet.

Different genres produce (but also delimit) different kinds of representations of clinical encounters, agency and subjectivity, depending on cultural and historical context.[67] There also is an enormous difference in purpose and tone of life-writing as compared to fiction. This is why I read patient and caregiver life-writing – memoirs, autobiographical accounts and diaries – against contemporary illness narrative theory, which to date has only to a limited extent touched upon cognitive and neurological disorders. I will focus on questions of the different concerns of the patient as compared to those of the carer, and those of the spouse as compared to those of the adult child. But in comparison to my earlier work on life-writing which attended to agency and independence, I particularly concentrate on the author-narrator's emphasis on scientific evidence, treatment strategies and attention to the loss of performance as compared to remaining abilities. Aware that life-writing is limited in how much it can depict the dissolution of the self, I will read these narratives against memory-related concepts of identity, and specifically focus on how representations depend on cultural and historical connotations of memory, cognitive performance and self-sufficiency. Dementia life-writings are limited in number. I expand my reading by including life-writings, inter alia, by Huntington's and Parkinson's disease dementia patients and even fictional accounts that centre on them. My premise is that comparing these narratives to those on Alzheimer's disease will help me address how memory-related brain diseases are considered to threaten identity, not least since motor neuron diseases like amyotrophic lateral sclerosis, and Parkinson's disease, by comparison, predominantly feature impaired motor function.

In view of the central role taken by German, French, Italian and British scientific research during the nineteenth century and the first two decades of the twentieth century, to begin with, this study centres on European developments. It will emphasize how a dementia discourse – how the medico-scientific and clinical establishment defined, and terminologically referred to, the condition – developed across Europe. The impact of Alois Alzheimer's research on twentieth-century science and medicine will become clear from including transatlantic research and politics: sociopolitical awareness of ageing and related research into old-age diseases arose much earlier in the United States than in Europe, during the first decade of the twentieth century.[68] My literary and cultural consideration therefore echoes this pattern.

The choice of narratives discussed is the fruit of broad reading across British and other European, North American and Australian literatures, and the resulting selection is representative of the key themes of the century-long cultural dementia discourse. I take early-twentieth-century narratives from the British context, while my selection of 1980s and 1990s fiction focuses (though not exclusively) on American and German contexts. I also look at life-writing of the 2000s and take into account the French, Italian and Spanish situation. This approach is complicated by the fact that a country's cultural inheritance produces specific images. But the narratives chosen reflect themes and cultural developments that I have identified in parallel works of other national contexts, and I will position them in relation to primary texts that already have a strong readership.

This approach is supported, for instance, by Raquel Medina's recent work. In an intersectional study of non-mainstream filmic portrayals of dementia, the Spanish scholar reveals how Alzheimer's disease is approached metaphorically to explore issues of gender and ageing across world cinema. In other work, Medina has studied how biomedical discourses of ageing and dementia are negotiated within Spanish documentary; I find largely the same principles of medicalization and politics of memory at work in Dutch fiction (which I read in English translation) and American life-writing. Intergenerational challenges of caregiving feature in Spanish cinema in a way that is comparable to my reading of Italian narratives and German films.[69] Similarly, dementia caregiver and patient mobilization in France has been identified as echoing the American movement.[70] By engaging with material from different cultural and national backgrounds within their respective contemporary medico-scientific and sociopolitical contexts, I suggest that cultural concerns about Alzheimer's disease and dementia have varied only in degree, not kind – that today there exists a common global concern about the condition. That said, as my archive suggests, this study focuses on the global North. For the global South, the tale of the meshing of science and culture has not been told in the same depth and detail. In addition, unlike in the global North where 'aging followed industrialization', in Africa, India and other Asian countries the timelines of development, urban migration and improving standards of living (at least for a growing proportion of society) have been compressed together.[71] As such, I would not want to extrapolate my findings indiscriminately to developing nations, although I will point to different conceptualizations of ageing and illness in countries especially of the global South.

Introduction

As this work seeks to trace cross-cultural discourse developments, I have relied, wherever possible, on original language texts, and offer my own working translations in citations. This approach also avoids reliance on present-day translations of earlier, key scientific papers, because these translations carry the intellectual assumptions of a century-long discourse it is the purpose of this work to examine.

Following the contemporary scientific and/or medical discourse, I employ the term 'dementia' in relation both to a diseased brain and a failing mind, a local pathological entity or a condition related to the whole person. The terms 'Alzheimer's' and 'Alzheimer's disease' are interchangeable, and I adopt 'Alzheimer' (disease) only where original work employs this form. I use the term 'Alzheimer's disease' where the condition is referred to as such in the medico-scientific context; where it relates to an experience; and where the sociocultural or popular discourses employ it, whether as a synecdoche or construct for types of dementia other than Alzheimer's disease, such as vascular dementia, dementia with Lewy bodies, frontal lobe disorder and Parkinson's and Huntington's disease. I prefer the terms 'sufferer' and 'patient' where the individual with dementia is the object created by the scientific, medical, healthcare, popular or societal discourse of the 'healthy'.[72] I avoid such terminology when exploring the life-affirming presentations of individuals living with dementia and where changes in discourse take place.[73] I define the caregiver as the person who writes about the patient, although her or his role might not primarily be to look after the patient.[74]

Overview

This book is divided into four parts. Part I (Chapter 2) provides an account of how the work of clinicians before 1920 changed the definition of dementia from a psychiatric condition to an organic disease, which sanctioned sociocultural perceptions of dementia as a form of degeneration. Part II (Chapters 3 and 4) analyses the dynamics between literary images of the elderly with memory loss and the socio-medical awareness of the ageing population since the 1940s, which eventually led to an exponential increase in research funding for Alzheimer's disease. Part III (Chapters 5 and 6) investigates how, during the 1980s and 1990s, reductionist research methodologies began to be used in scientific medicine, and how this impacted on the objectification of patients and reformulated the illness as one of limited cognitive performance. It also analyses how detective fiction, while reflecting the period's fascination with cognitive performance, helped to claim back authority for patients. Part IV (Chapter 7) illustrates how recent medico-scientific research has focused on disease prediction and prevention, while concurrent patient advocacy has led to the redrafting of a biopsychosocial approach to patient care. Fiction and life-writing has begun to show a growing sociocultural and healthcare appreciation of the continued personhood of people with dementia. Such literature challenges the concepts of loss and degeneration that continue to dominate socio-economic thinking.

Part I (Chapter 2) explains how the conversation between science, medicine and literature related to dementia first arose in the second half of the nineteenth century when psychology began to take shape as an academic discipline organized around the science of memory. This instigated the systematic neurological study of the dementias at the fin de siècle. The microscopic observations of tissue loss by Alois Alzheimer and other researching clinicians during the first two decades of the twentieth century were taken to corroborate a psychiatric definition of dementia as a form of degeneration. Reflections of this can be found in literary works of the period: the ailment of John Galsworthy's Timothy in *The Forsyte Saga* (1906–21) – one of ageing, decline and loss – is a micro-cultural image of macro-cultural concepts of degeneration at the fin de siècle.

Between the 1910s and 1930s, brain inspection was carried out in younger as compared to old-age patients, who increasingly were all conceived to be in stages of involution. By the 1940s, this way of looking at the brain had narrowed dementia down to a biomolecular process of shrinkage and loss. In *There Were No Windows* (1944), the Irish novelist Norah Hoult explores notions of disorientation in space and time as well as issues of care. The fate of the central character Claire Temple illustrates the growing general awareness of ageing mid-century, and it brings out the link between ageing and memory loss, a societal challenge that by the 1960s would be recognized internationally. Hoult's narrative, like William Faulkner's *The Sound and the Fury* (1929), also explores the uncontrolled intrusion of memory. I take Hoult's inward turn as an example of how modernist mimetic approaches developed during a period marked by post-Freudian psychiatric interest in the patient's illness experience, which became increasingly important to sociological research from the 1950s onwards.

Part II focuses on conceptions of dementia as pertaining to old age. Research interest in ageing and old-age diseases first arose during the first third of the twentieth century, with gerontology and geriatrics established as disciplines by mid-century. Chapter 3 covers the socio-medical studies that followed from these developments. These studies played an important role in the conversation between science, medicine and literature related to dementia, because by using interviews, their focus was the patient's experience and inner world. But comprehending dementia as a mental condition did not aid the patient: it underwrote a healthcare discourse centred on images of the patient as dependent, passive and failing. This discourse began to take shape in the 1960s. By that time, as I will also explore in Chapter 3, socio-medical concerns about population ageing were reflected in cultural productions which established the elderly as sufferers of memory loss and as a task for care management. One example is British writer Paul Bailey's lonely and isolated old woman Faith Gadny in the short novel *At The Jerusalem* (1967).

In Chapter 4, I show how the first caregiver narratives tended to emphasize caregiver burden, such as American Rosalie Walsh Honel's *Journey with Grandpa* (1988), and perpetuated the idea of the patient as passive, something also seen in public healthcare campaigns. And rather than developing counter-narratives to these representations of dependence, fiction of the 1970s and 1980s propagated the idea that a functional self relies on intact memory and narrative capabilities. As part of a surge of post-war memory discourses, these narratives also enmeshed dementia with notions of extermination and the Holocaust, as in Dutch writer J. Bernlef's *Out of Mind* (1984), laying the ground for Alzheimer's as metonym for all types of memory loss and forgetting and as underwriting pessimism about ageing more generally.

Part III explores the dominance of biomedical research in establishing a shared discourse of dementia as cognitive loss. This dominance can be traced back to the mid-1970s, as sociopolitical awareness of demographic changes pushed Alzheimer's disease into research laboratories. This happened through a sharp increase in funding from the National Institute on Aging, which was founded as an agency of the US Department of Health, Education and Welfare in 1974. In Chapter 5, I attend to salient research papers from the 1980s and 1990s and their neurochemical and genetics focus detached from the patient's experience – which additionally endorsed a healthcare discourse that dehumanized patients. By the end of the 1980s, the scientific rush for cause and cure as well as caregiver-focused healthcare campaigns induced readers to overlook the patient's plight in a growing literary marketplace for caregiver accounts that told of the patient's social death. Literary scholarship played its part in this discourse. This was the case, because, until recently, the Oxford don John Bayley's account about his wife Iris Murdoch from 1998 has dominated literary studies about dementia. In this account, Murdoch's previously fulfilled and youthful life is shown as destroyed by dementia.

A growing fascination with intellectual performance and healthy cognitive ageing towards the end of the twentieth century further demoted the standing of the patient, as I explore in Chapter 6. In exchange with a growing cultural interest in the visual, both imaging studies of atrophying brains and pharmaceutical efforts to achieve cognitive enhancement additionally stressed the patient's losses, also furthered by popular scientific accounts, like neurogeneticist Rudolph E. Tanzi's *Decoding Darkness* (2000). Yet, despite this century-long conceptual adversity, the patient finally began to be credited with authority towards the turn of the twenty-first century. As I also bring out in Chapter 6, the fascination with cognition, together with many other elements that contributed to this conversation between science, medicine and literature since the 1970s, eventually supported the releasing of the patient from being typecast so negatively. Linguists studied the patient narrative; and patients began to hold centre stage in popular detective novels, such as British crime novelist Michael Dibdin's satirical *The Dying of the Light* (1993), and eventually contributed to the conversation through their own writing and institutionalized advocacy.

The final part of this study (Chapter 7) covers discourse developments since the turn of the century. Immunological approaches to Alzheimer's disease introduced notions of flexibility and spontaneity into the medico-scientific writing about the condition. These scientific ideas strengthened budding cultural concepts of the patient as active and resistant. Disappointing drug trial outcomes related to these immunological approaches, however, left patients all the more vulnerable, pushing them back into a position of passive objects of pity waiting for a cure. Given that enterprise societies assume cure to be superior to care, with access to care frequently being conditioned on a formal diagnosis, the dementia patient inevitably takes on victim status in these societies. As such, appreciation of the patient's individuality appears to be tragically short-lived in today's prosperous societies. Such societies don't have space or place for the ageing and cognitively challenged; and, paradoxically, as patients finally took measures to assert their continued agency, their condition became the object of economic interests. The market made them cardboard figures yet again. Since the turn of the century, the condition has dominated bestseller lists and made for controversial hit movies like German director Til Schweiger's *Honig im Kopf* (2014). But it is not only a market for accounts of dementia that has arisen; there is also a market of people afflicted by dementia, and their condition guarantees the profit of the cognitive health industry. At present, patient individuality remains confined to niche children's books, like Italian primary school teacher Alessandro Borio's *La guardiana di Ulisse* (2006), and caregiver texts like French psychologist Cécile Huguenin's *Alzheimer mon amour* (2011). Such depictions challenge the concepts of degeneration, decline and loss that continue to dominate socio-economic thinking in rich societies, as my reading of Matthew Thomas's *We Are Not Ourselves* (2014) in the final section of this book will bring out.

Over the long twentieth century, two approaches to understanding and representing dementia have crystallized: on the one hand, as a disease of the brain, and on the other hand, as a condition of the mind. Both evolved in parallel, not always in tension, and intertwined with literary representations; and both perpetuated notions of the patient as elderly and in decline – with far-reaching consequences for how, today, Alzheimer's disease signifies the synthesis of illness, ageing and dying. Literary explorations have developed important counterweights to a more fatalistic view of Alzheimer's disease as a condition that strips patients of their humanity. But contrary to the often invoked dualism between the psycho-sciences and the neurosciences, understanding the condition as an illness of the mind did not aid the patient. It underwrote images of dependence, dispossession and failure, and this conditioned a remarkable characteristic of literary writing about Alzheimer's disease: compassion focused on the caregiver rather than the sufferer. This prepared the ground for how today dementia is the most prominent condition for representing the burden of care, the passivity of the patient and the death of the self. I begin this study with an early and perceptive literary representation of dementia as a form of material and social dispossession in old age, as living death.

Part I

The organic paradigm

2

From brain inspection to cell death

The Forsyte Saga: The cultural image of dementia in the fin-de-siècle family novel

Timothy Forsyte's ailment in *The Forsyte Saga* (1906–21) is one of ageing, decline and loss, the micro-cultural image of macro-cultural concepts of degeneration at the turn from the nineteenth to the twentieth century. It is the representation of an illness that would exert its presence over the entire twentieth century and that continues to be part and parcel of the societal perception of what, today, is termed 'Alzheimer's disease'. John Galsworthy's saga has been described as having a special place in English fiction for its 'quality of human truth, for its intimate portrait of a culture in transition, its understanding of people as creations of both family and history, and for its imaginative development of the scope of the saga as a narrative form'.[1] In addition to its intricate portrait of a particular culture over a period of decline, I argue that John Galsworthy's saga holds a special place for its early and perceptive literary representation of dementia.

The story of three generations of an upper-middle-class British family of ten siblings traces Victorian fin-de-siècle fears of decline and degeneration through the prism of the lives of James and Old Jolyon, their respective sons Soames and Young Jolyon and in turn their children. Timothy is the youngest male of the first generation of siblings. Surviving his brothers and sisters, he dies in 1920, at the age of 101, afflicted by severe memory loss. Existing critical scholarship of *The Forsyte Saga* passes over Timothy's condition. Earlier readings focused on social change, while current criticism, at least ostensibly, refers to the saga in the context of studies on Victorian society more broadly, rather than on ageing in particular.[2] The 2002 televised *Forsyte Saga* miniseries entirely cuts out Timothy as a character, let alone his dementia.[3] Yet, I consider Timothy Forsyte's faltering cognitive state to be Galsworthy's ultimate and most powerful symbol in the characterization of the family clan's decline and loss. Not only does it represent the dissolution of a member of the Forsyte family, but it is, as I will argue throughout this section, *the* synecdoche of the entire clan's fragmentation, the Forsytes' slow dying during the forty years spanning the saga's three novels. Timothy's condition in the story suggests an account of late-life dementia as a further form of degeneration at the end of the nineteenth century in a culture that was becoming increasingly concerned with social deviance and aberration from the norm.[4]

The youngest of the first-generation Forsyte brothers, Timothy is 'supposed to resemble their father' (504), the stronghold of the family. Timothy's house embodies stability as a home for the clan, as 'one of hundreds of such homes in the City of London – the homes of neutral persons of the secure classes, who are out of the battle themselves, and must find their reason for existing, in the battles of others' (132).[5] Propelled forward by family members gossiping about family matters as well as wider political issues at Timothy's house, the narrative circles around the Forsytes trying to hear Timothy's opinion. Yet for most of the time he remains invisible: 'He had become almost a myth – a kind of incarnation of security haunting the background of the Forsyte universe. He had never committed the imprudence of marrying, or encumbering himself in any way with children' (22). Where Timothy actually appears in the novel, his behaviour symbolizes always yet another stage in the Forsytes' dissolution. Timothy first anticipates, then perceives and, eventually, embodies the Forsyte family's decline. This decline unfolds in the light of the supposed social threats stoking contemporary fears of loss and decline from the 1880s onwards. Central to Timothy's appearances, these threats could be summarized in William Greenslade's words as anxieties 'about poverty and crime, about public health and national and imperial fitness, about decadent artists, "new women" and homosexuals'.[6]

The depiction of how Timothy's brother James sees him is telling:

> The figure of a thickset man, with the ruddy brown face of robust health … He spoke in a grudging voice. 'Well, James', he said; 'I can't – I can't stop.' And turning round, he walked out. It was Timothy. James rose from his chair. 'There!' he said; 'there! I knew there was something wro-' He checked himself, and was silent, staring before him, as though he had seen a portent. (267)

James's premonition becomes tangible in Timothy's second appearance, in the second part of the saga. Young Jolyon's son has died in the Boer War, which, to the father, 'marked the decline of the Forsyte type' (537). The story of the family's breaking up becomes ever more intently interwoven with an exploration of the perception of the nation's decline at the beginning of the twentieth century. Reference to the decline of a 'type' suggests that Galsworthy depicts dementia as one of the hereditary conditions supposed to threaten the health of Victorian society. Next to health concerns, war and feminism play a major role in the Forsytes' premonition of decline, again explored through Timothy. Timothy's second appearance in the flesh 'on the Sunday after the evacuation of Spion Kop' was again 'in the nature of a portent' (505). It makes the war, and the Empire's loss of identity through this war, tangible to all Forsytes. Galsworthy lets Timothy concentrate on what the upper-middle classes perceived as one of its ultimate threats, namely the 'new woman'. On hearing about Soames's divorce, he asserts: ' "Don't let's have anything of that sort in the family," he said. "All this enlistin's bad enough. The country's breakin' up; I don't know what we're comin' to." He shook a thick finger at the room: "Too many women nowadays, and they don't know what they want" ' (509).

The overarching symbol for the country's breaking up, the nation's perceived loss of unity and cohesion, is Queen Victoria's death, which Soames links to the impending death of the Forsyte clan as a dynasty of old values.[7] This event, 'supremely symbolical', impresses Soames's 'fancy', as it sums up 'a long rich period': 'Morals had changed, manners had changed, men had become monkeys twice-removed ... Sixty-four years that favoured property, and had made the upper middle class; buttressed, chiselled, polished it, till it was almost indistinguishable in manners, morals, speech, appearance, habit, and soul from the nobility' (567). Galsworthy explicitly invokes reference to Darwin's evolutionary theories in Soames's thought process. In this way, he underlines how perceptions of the nation's decline and loss were interwoven with post-Darwinian myths of degeneration.[8] Perceptions of decline and loss are pushed further during the post-war period of 1920 and 1921, covered by the saga's third novel, as 'manners, flavour, quality, all gone, engulfed in one vast, ugly, shoulder-rubbing, petrol-smelling Cheerio' (622). As the saga draws out the Forsytes' final decline against the background of Queen Victoria's as well as the Empire's ageing and ultimate death, the feeling of change is enmeshed with the Forsyte family's failure: their ageing and dying.

Galsworthy depicts a culture in transition at the turn of the nineteenth and twentieth centuries. Queen Victoria's ageing and death played a significant role in increasing the visibility of age and ageing during the final two decades of the nineteenth century.[9] The awareness of ageing and old age in society – at least as a societal challenge – grew only in the latter third of the nineteenth century, when ever larger parts of the population, which had moved into the cities for work and life, began to grow old. These elderly citizens were, as Charles Booth pointed out in 1892, at the margins of productivity and society:

> Old age fares hardly in our times. Life runs more intensely than it did, and the old tend to be thrown out. Not only does work on the whole go faster and require more perfect nerve, but it changes its character more frequently, and new men – young men – are needed to take hold of the new machines or new methods employed. The community gains by this, but the old suffer. They suffer beyond any measures of actual incapacity, for the fact that a man is old is often in itself enough to debar him from obtaining work, and it is in vain he makes pretence by dyeing his hair and wearing false teeth.[10]

As Thomas R. Cole fittingly points out, 'Old age emerged as the most poignant – and most loathsome – symbol of the decline of bourgeois self-reliance' during the Victorian Age.[11] Predestined to explore the changing of times through the protagonists' ageing, the saga as a narrative form relies on degenerational concepts.[12] Ageing anticipates dying, and death threatens what is core to a Forsyte's identity: the ability to possess. Since, as Geoffrey Harvey points out, death is the 'ultimate check on [the Forsytes'] acquisition of property' and its administration, continuously altering their wills is their only remaining way to exert influence beyond their death.[13] Here again, Timothy's condition exemplifies the ageing Forsytes' situation most acutely. It is a condition of dispossession, and as part of the cultural dementia narrative this notion explains, at least to some extent, the fears and anxieties of a materialist society in relation to dementia. Formerly an expert in making money and managing inherited wealth – he held shares in consolidated annuities (abbreviated as 'Consols') and 'gilt-edged securities' (161) – Timothy loses the ability to look after his property. As Timothy's servant informs Soames, 'Every now and then he asks the price of Consols, and I write it on a slate for him – very large. Of course', she adds, 'I always write the same, what they were when he last took notice, in 1914' (653). Timothy is unable to continue to tinker with his will, only 'turn[ing] it over and over, not to read it, of course' (653). His condition prophesizes that of the Forsytes and parallels that of the nation: 'The attitude of the nation was typified by Timothy's map [which he used to follow the nation's movements in the Boer War by inserting flags], whose animation was suspended – for Timothy no longer moved the flags, and they could not move themselves' (575). The map's stillness mirrors Timothy's apathy – a term I use to denote not mere unwillingness or indolence, but a disease-related inability to act.

The narrative reaches its climax, the Forsytes' anticlimax, when we behold Timothy the third time, during Soames's visit to the centenarian's 'Mausoleum' (647). Strolling round the house, Soames appears to dip into his memories. Yet, the Forsytes' conviction that Timothy was preserving the old Victorian values, cherished and nourished throughout almost forty years, is about to be destroyed. Both confirming and anticipating the clan's final dissolution, Soames discovers in his uncle's study that Timothy's intellect had been mere façade, when the 'third wall he approached with more excitement. Here, surely, Timothy's own taste would be found. It was. The books were dummies' (650). These dummies are the image of what Timothy had been throughout the narrative: never there, never a stronghold, just imagination. A myth – as Galsworthy had put it early on (22). Dementia becomes a metaphor for emptiness, for 'Nothing – nothing!' (655). Galsworthy had declared one of the core aims of writing *The Forsyte Saga* to divest the idea of property as a social value, 'to leave property as *an empty shell*'.[14] In the context of what Timothy embodies in the novel, Galsworthy's statement enhances the concept of dementia as emptiness all the more. This is also true because Timothy's memory loss becomes a symbol for the nation's condition. Memory loss afflicts the nation and its people, who do not any longer remember the old values: 'And old Timothy – what could *he* not tell them, if he had kept his memory! Things were unsettled, people in a funk or in a hurry, but here were London and the Thames, and out there the British Empire, and the ends of the earth' (782; emphasis original).

The role of Timothy's memory loss becomes even clearer when considered in the context of Victorian interest in centenarianism. Conceding that the 'perception that the old bear a special responsibility for the preservation and transmission of knowledge from and about the past is probably true for all cultures', Helen Small emphasizes that this interest particularly recognized that in 'bearing the "dead past" with them, [the elderly] sometimes troubled a progressive society's assurance that the past was, indeed, past'.[15] Timothy's age makes him the embodiment of the (past) nineteenth century, and his death the embodiment of the death of that century. This is especially the case, because memory helped to explore 'the relation between the individual and society, the relation between the mental and the physical ... the roles of heredity and environment in shaping the individual'.[16] It is therefore fitting that Soames hears the 'sound, as of a child slowly dragging a hobby-horse about' (653), as Galsworthy depicts Timothy as 'resum[ing] his babyhood' (653) in his final appearance. I offer this description here in full, because it will become central to images of dementia patients throughout this study:

> The last of the old Forsytes was on his feet, moving with the most impressive slowness, and an air of perfect concentration on his own affairs, backward and forward between the foot of his bed and the window, a distance of some twelve feet. The lower part of his square face, no longer clean-shaven, was covered with snowy beard clipped as short as it could be, and his chin looked as broad as his brow where the hair was also quite white, while nose and cheeks and brow were a good yellow. One hand held a stout stick, and the other grasped the skirt of his Jaeger dressing-gown, from under which could be seen his bed-socked ankles and feet thrust into Jaeger slippers. The expression on his face was that of a crossed child, intent on something that he has not got. Each time he turned he stumped the stick, and then dragged it, as if to show that he could do without it. (653)

The reduction of space represents a culturally pervasive metaphor for an ageing individual's involution. Elaine Scarry notes that 'in very very (*sic*) old and sick people, the world may exist only in a circle two feet out from themselves'; and this world eventually 'shrinks ... to a casket'.[17] Considered rather a place and institution, Timothy's so described involution and end are the consequence of a novel-long depersonalization.[18]

Eventually and significantly, Galsworthy directly links ageing, old age and dementia to decline and loss, and the depression perceived during the post-war era. Ageing and, linked to this, dementia *are* degeneration: 'The world was in its second childhood for the moment, like old Timothy' (783). Underlying this idea of second childhood was a nineteenth-century 'materialist perspective on the life course' which dismissed both childhood and old age as burdensome and unproductive.[19] This discourse had a scientific counterpart. With the increasing spread of Ernst Haeckel's (1834–1919) hypothesis of the recapitulation of phylogeny in the foetus's ontogeny (that the foetus, over the gestation period, passes through the different evolutionary developmental stages of the human species), the child quickly came to be represented by a state of underachievement. Among others, the physician Havelock Ellis (1859–1939) placed the child 'naturally, by his organisation, nearer to the animal, to the savage, to the criminal, than the adult'.[20] Knowledge of arrested brain development during ontogeny added further meaning to cerebral alterations in individuals with dementia that were more and more often described. What I am more interested in, though, is that there existed a scientific counterpart to Galsworthy's depiction of dementia as a further form of degeneration, and I will show this in the next two sections.

Already Galsworthy's depiction of memory loss as the ultimate culmination of ageing, which anticipates death, brought together what the nineteenth century had elaborated in its medical, scientific and cultural domains. Galsworthy built his narrative around concepts of ageing. Therefore, it is possible to claim that dementia, for him, represented a tool for depicting Timothy – as the centre of the Forsytes' belief system – as exaggeratedly old. This seems particularly likely since, as the historian F. B. Smith stresses, 'The years between 1880 and 1914 are the crucial ones in the great transition from the age-old pattern of mass morbidity and mortality occasioned by infectious diseases, poor nutrition and heavy labour, to the contemporary assemblage of functional disorders, viral diseases and bodily decay associated with old age.'[21] Even the saga's literary style would support this – firm nineteenth-century social realism, and therefore a lacking exploration of Timothy's fragmented mind in narrative.

A detailed look at medico-scientific developments throughout the nineteenth century will reveal that Timothy's portrayal anticipates much of the twentieth-century sociocultural dementia discourse. I would even argue that most societal concerns about dementia, as well as the core medical and scientific questions, were established by the early 1920s if not before the First World War. They merely were revived and intensified after the Second World War. I will thus explore the conceptual overlaps between Galsworthy's disease representation (he completed his saga twenty years into the new century) and the medico-scientific discourse that developed around Alois Alzheimer's research. Before addressing this, however, I will take a step back in time to show how Galsworthy's depiction of memory loss in old age reflects conceptual changes in physiological psychology as they developed from the mid-1800s.

Dementia and memory loss in science, medicine and literature before 1880

Nicholas Dames has offered a detailed analysis of the meaning of memory and its loss in the Victorian novel: memories were only seen as retained inasmuch as they continued to be beneficial to the individual's well-being in the present. Dames specifically argues: 'Over and throughout both mid-century psychology and mid-century fiction plays a mind that functions, perhaps, like a machine: winnowing and sorting, eliminating and arranging, turning the past into the story that always only confirms our sense of the present.'[22] Timothy's sense of the present, if his memory were still functioning, would be one of loss and fin-de-siècle fears. His increasing lack of awareness may literally be read as his strategy for surviving, as the last of the old Forsytes, into the new century and post-Great War Britain. In Galsworthy's words: 'That terrific symbol Timothy Forsyte ... the only man who hadn't heard of the Great War – they found him wonderful – not even death had undermined his soundness' (841). In the aftermath of the Great War, earlier fin-de-siècle fears of degeneration became renewed and enhanced. Society became anew 'worried about poverty and social deprivation, the power of organised labour, the democratisation of culture, the shifting in manners and gender roles'.[23] In closing his eyes to the Great War, Timothy could remain unchanged. Reflected against Dames's deliberations, we can read Timothy's condition as a desperate attempt to remain linked to and embedded in the past, and thus escape change.

Galsworthy's representation of memory loss has much in common with this Victorian perspective. But, as Dames acknowledges, since the mid-nineteenth century the search for material explanations for mental function had begun to occupy physiological psychology. Such studies were fuelled by work with soldiers returning from the Crimean and Boer Wars. Their conditions raised awareness of the close relation between emotional trauma and bodily ailment.[24] Galsworthy would probably have been aware of this, since, by the time he wrote volumes two and three of the saga, he had worked for two months in a hospital for shell-shocked French soldiers.[25] In fact, Timothy's memory loss reflects the changes in how science understood memory loss towards the end of the nineteenth century: a form of degeneration deployed in relation to irreversible cognitive impairment of the elderly. To make this point, I will briefly sketch out how material explanations for mental functions evolved since mid-nineteenth century and shaped physiological understanding of memory, against a background of post-Darwinian degenerationism.

Throughout the eighteenth century, dementia was considered to be both a medical and legal state, based on the idea of the individual lacking competence to perform independently in society. Timothy Forsyte belongs to a long line of fictional feeble-minded male characters, who are typified as having lost the ability to keep control of money and maintain paternal lineage; in this line, Patrick McDonagh particularly identifies John Galt's Walter Walkinshaw in *The Entail* (1822), Charles Dickens's helpless Smike in *Nicholas Nickleby* (1839) and Elizabeth Gaskell's Willie in the short story 'Half a Life-Time Ago' (1855).[26] But even though Timothy Forsyte fits their dependence and incapacity, he is nonetheless exceptional as a depiction of this type who is also identifiable as a person with dementia.

For much of the nineteenth century the scientific and medical understandings of dementia exclusively built on clinical observations. Such observations, by the

French physician Philippe Pinel (1745–1826) and his pupil Jean Étienne Esquirol (1772–1840), connected the condition to age-related decline and attempted to separate it from imbecility and idiotism.[27] The following description of 'senile dementia', by British physician James Cowles Prichard (1786–1848), strongly relied on Pinel and resonates in many ways with Timothy's final appearance: 'Rapid succession or uninterrupted alternation of insulated ideas, and evanescent and unconnected emotions; continually repeated acts of extravagance; complete forgetfulness of every previous state; diminished sensibility to external impressions; abolition of the faculty of judgement; perpetual activity'.[28]

Notions of age-related decline or 'senile decay' placed the condition with ideas of decay and loss amid theories of degeneration.[29] These were elaborated mid-century by the French psychiatrist Bénédict-Auguste Morel (1809–1873) and were to shape psychiatric and neurological discourses for decades to come.[30] Coinciding with Morel's work, Charles Darwin (1809–1882) published his theory of evolution according to natural selection in 1859. This work posited that progressive development of species could still involve simplification and loss. Yet, as Gillian Beer points out, 'many of Darwin's first readers favoured the counter-form of evolutionary myth: that of growth, ascent, and development towards complexity'.[31] Morel's theory of degeneration, by comparison, explicitly stated that evolution could produce errors. This perspective was central to the confluence of social Darwinist and fin-de-siècle degenerationist discourses. These were represented in *Entartung* (Degeneration), published by the journalist and Zionist Max Nordau in 1892, and influential all over Europe and the United States.[32]

Morel's treatises provided a background for the classification of dementias and a basis for pathological descriptions. The British neurologist John Hughlings Jackson (1835–1911), for instance, claimed that dementia was 'the behavioural reflection of successive stages of the dismantling of cerebral structures'.[33] And brain dysfunction, for him, represented a process of dissolution bearing a threat of degeneration.[34] So, by the end of the nineteenth century, dementia as a form of insanity had become a form of degeneration that described irreversible behavioural changes linked to cognitive impairment – a concept that overlaps with Galsworthy's rendering of Timothy's condition.[35]

These developments were paralleled by changing understandings of brain pathology, which also incorporated concepts of degeneration. This happened as localization and lesion theories were gaining ground throughout the nineteenth century. The combined approaches of clinical observations and experimental scientific work were developing an understanding that physiological changes, namely morphological alterations or lesions, could mark out disease as an aberration. First and foremost, language began to be conceived of as a product originating from specific regions in the brain, and aphasia (the inability to understand or produce speech), in turn, comprehended as symptomatic of the lesion of a clearly circumscribed brain area.[36] Such insights also influenced how dementia began to be viewed: a form of memory loss explained by a lesion related to brain cell death. These developments received additional support from the rise of neurology as a discipline, which held that, once its organic cause was known, a brain disease would eventually be treatable.[37]

In Victorian Britain this perspective was championed by Hughlings Jackson. He separated brain from mind in his hypothesis of strict psychophysical parallelism. For him, there was 'no physiology of the mind any more than there [was] psychology of the nervous system'.[38] He believed that mental and behavioural symptoms could be diagnosed, but they could not be interpreted psychologically 'as immediate manifestations of abnormal states of mind'. Moreover, as Michael J. Clark illustrates, professional conceptions of the physician's moral-pastoral responsibilities explained the period's objections to the treatment of the mentally disordered in psychoanalytical terms. Freudian depth-psychological approaches elicited serious distrust, because they seemed to exaggerate the moral depravity of patients.[39]

Sigmund Freud (1856–1939) was actually one of those who questioned localization theories as early as 1891 in his work *On Aphasia*.[40] But what he became most invested in was looking for a physiological model of the nervous system that could give insight into mental processes: 'a physiological account of associationism'. In 'Project for a Scientific Psychology' (1895) Freud elaborated that associations did not correlate with locations of cells; they were determined by 'differing structures of nervous connections, spread across the brain surface'. In direct consequence, mental disorders were not to be explained by a specific lesion. Rather they could arise because 'unhelpful associations had been formed', with lived experiences leaving memory traces.[41] Yet, during the final third of the nineteenth century, organic explanations significantly overshadowed psychological approaches to problems of mental disorder.[42] This polarity between brain physiology and mind psychology would permeate neuropsychological thinking in relation to dementia and Alzheimer's disease throughout the twentieth century.[43]

Alois Alzheimer (1864–1915) and his work hold a central place in the evolution of this materialist discourse. As a student, Alzheimer was influenced by German psychiatrist Wilhelm Griesinger (1817–1868), the figurehead of the school proposing a materialistic theory of mental diseases, and neurologist Carl Westphal (1833–1890).[44] He argued early on that psychiatric diseases were diseases of the brain and soon would support this view with scientific data obtained from detailed microscopic studies. These methodological and related scientific innovations gained importance in the practice of medicine towards the end of the century and led to a mutual enhancement and eventual merging of science and medicine. This would provide a fruitful working environment for Alzheimer's generation of psychiatrists.

Of particular importance were the installation of hospitals and the growing emphasis on laboratory research which enhanced and supplemented diagnoses made in the clinical setting. By the 1870s, coherent professional psychiatric groups had emerged in Britain, France and Germany. The political dimensions of science and related technologies also became more and more evident, especially in Germany, where optical, pharmaceutical and chemical industries were particular features of its economic growth.[45] At the same time, the emerging research platform – specialized journals, societies and conferences – helped Alzheimer's findings quickly gain notice on the international stage. This development was enhanced by the fact that throughout the nineteenth century the highly esteemed German university research tradition continued to attract students as well as guest researchers from abroad.[46] These constellations placed Alzheimer in an influential position in the critical psychiatric discussions at the beginning of the twentieth century.[47] His work was destined to reach the attention of an audience much broader and better equipped to extend his insights than had been the case for Pinel's or Esquirol's contributions sixty years earlier.

It may appear far-fetched to link Timothy Forsyte's condition to a twenty-first-century diagnosis of dementia of the Alzheimer's type. Yet, the cultural connotations of degeneration and decline reflected in Galsworthy's saga resemble present-day images of the affliction which developed early in the medical discourse around the disease. This position can be substantiated by turning to Alois Alzheimer's case study of Auguste D. in 1906. This case has suggestive similarities to Timothy's state, almost making Galsworthy's Forsyte centenarian a case study in itself. Turning from the British fin-de-siècle literary context to scientific research in *Kaiserreich* Munich around the same period may invite questions. Are we looking at shared cultural concerns regarding memory loss, ageing and dementia; or are we dealing with parallel developments during a period marked by post-industrial societal and cultural challenges and turn-of-the-century fears? There is evidence for the pan-European nature of these concerns. Scholars such as Laura Otis have discussed other novels that portray family drama through a lens of disintegration in the eighteenth and nineteenth centuries, among them Thomas Mann's *Buddenbrooks* (1901). Otis explores Mann's preoccupation with heredity, degeneration and memory among the Hanseatic bourgeoisie, and identifies similar themes in Spanish and French literatures, comparable to Silvia Acocella's observations on Italian texts.[48]

These illustrations suggest that a dementia discourse developed in similar, connected ways across Europe. Galsworthy had created Timothy in 1906, before he could have heard of Alois Alzheimer's research. By the time he had published the second (1920) and third (1921) volumes, so completing the full version of his saga, over a decade had passed since the publication of Alzheimer's cases. But these cases were discussed in specialist journals rather than generalist periodicals so that it is unlikely that Galsworthy built his character's decline into dementia on the scientific knowledge of the day. Rather, these scientific investigations developed within a sociocultural environment whose reflections Galsworthy traced in his saga. As such, I argue that the look into the microscope and subsequent descriptions of tissue by Alzheimer and his colleagues scientifically strengthened an already existing societal degenerationist dementia discourse.

Auguste D. and Johann F.: Alzheimer's clinical cases and histological research

The first clinical descriptions of brain atrophy associated with dementia and senility date back to 1864, when the British physician Samuel Wilks (1824–1911) observed 'wasted convolutions' in the brains of 'those long demented'.[49] Likewise, the medical psychologist James Crichton-Browne (1840–1938) had lectured on clinical cases of 'senile dementia' already in 1874, describing the condition as 'a gradual degeneration and decay of the mental faculties' and linking it to 'cerebral atrophy'.[50] Alzheimer himself had been interested in and published on *altersbedingten Schwachsinn* (senile dementia) before the turn of the century, relating this condition to both nerve cell death and an altered blood supply to the brain.[51] In this line of investigation, the case of Auguste Deter (abbreviated as Auguste D.) represented just one case among many, and it is not surprising that Alzheimer received no significant attention when communicating his findings to the 37th Meeting of the South-West German Psychiatrists in Tübingen in 1906. But despite this initial neglect, the case, published in print one year later, was to become seminal for the discourse that developed around dementia and Alzheimer's disease later in the twentieth century. This is at least partly the case, because it bore, in Lauren Berlant's words, 'the weight of an explanation worthy of attending to and taking a lesson from'.[52]

The case

Although overlooked at the time, the case of Auguste D. was striking for the fact that she was comparatively young for a dementia patient. Only 51 years old on admission to the Frankfurt asylum, she showed symptoms and behaviour that hitherto had only been described in elderly patients. (Then and now senile dementia or late-onset Alzheimer's disease is defined as afflicting individuals aged above 65. Presenile dementia or early-onset Alzheimer's disease concerns individuals younger than 65 years of age.[53]) But Auguste D.'s symptoms were very intense and progressed comparatively rapidly. At admission, Alzheimer noted Auguste D.'s ideas of jealousy, auditory hallucinations, feebleness of memory and lack of orientation. Alzheimer emphasized absolute perplexity, disorientation in time and space, and occasional pronouncements of lacking understanding. He closed his clinical observations with a description that echoes the visual representation of a dementia patient by Esquirol: 'general stultification (*allgemeine Verblödung*) progresses. Death occurs after 4 ½ years of illness. The patient was at the end totally lethargic, lying in bed with her legs tucked up' (147).

After having detailed these clinical observations, Alzheimer turned to macroscopic and, thereafter, microscopic descriptions of brain tissue, which he had analyzed after Auguste D.'s death. He reported a 'uniformly atrophic brain' and larger blood vessels as 'arteriosclerotically changed', that is, their inner diameter narrowed due to the build-up of deposits. Then, he moved on to describing individual cell components. With an improved Golgi silver stain (still in use today) Alzheimer could perceive 'peculiar alterations of neurofibrils', assembled in bundles (147–8). 'Finally', Alzheimer concluded, 'nucleus and cell disintegrate' (148), a perspective that anticipates how the look through the microscope would strengthen concepts of decay.

The paper dedicates similar amounts of text to clinical as compared to histological aspects (gathered by observing tissue through the microscope). This formal choice underscored Alzheimer's conviction that histological examinations complemented and clarified clinical observations. It also lets us appreciate how carefully Alzheimer separated the patient's behaviour (and its description) from his histological findings. I emphasize Alzheimer's caution so much because, as Michael Lynch puts it, 'visibility in science', in the form of data displays, '*constitute[s]* the material form of scientific phenomena'. Specific cellular structures only become visible, measurable and reportable once they are labelled; and microscopic visibility (and, linked to this, the production of *camera lucida* drawings, for which Alzheimer was renowned) relies on the specimen's thinness, histological treatment and microscopic magnification.[54] In summary, cellular phenomena visible with a particular staining method were described with terminology hitherto employed in the cultural discourse of degeneration.

In the case of Auguste D., this terminology, that is, *Zerfall* (disintegration), *Untergang* (demise) and *Verschwinden* (disappearance) of cells, is still clearly set apart from the vocabulary employed for the patient who experiences *Gedächtnisschwäche* (feebleness of memory) and, eventually, reaches *allgemeine Verblödung* (general stultification). These latter expressions are exactly those used in nineteenth-century clinical descriptions of dementia. I note that Alzheimer does not recount these events in terms of degeneration in this case study. By the 1920s, Alzheimer's work was internationally known and referenced, and the condition, by then carrying his name, had also been linked to cases of senile dementia. In what follows I explore the contexts in which Alzheimer and his close co-workers actually used the term 'degeneration' and, related to this, whether they were consistent in their separation of clinico-medical and scientific-histological terminology. It will become ever clearer how Alois Alzheimer's look into the microscope and subsequent descriptions of tissue loss by him and other researching clinicians scientifically underpinned a societal dementia discourse reliant on degenerationist concepts.

The discourse of degeneration

Alzheimer's biographers repeatedly emphasize how strongly the neurologist resisted the medico-scientific developments of the time, which Melissa M. Littlefield describes as follows: 'Between 1890 and 1930, in both psychology and neurology, the body was undergoing a transition ... Studies of the body were split between delineating the body's fragments – its structure – and describing the complex interactions – the functional relationships – among its respective parts.'[55] Alzheimer was a practising clinician and psychiatrist as well as a highly skilled scientist with neurological training. As such, he saw these two disciplines as complementing each other.[56] He consistently placed patients and their interview at the centre of his dedication as a clinician and of his research philosophy, and the case study remained the tool with which he approached the classification of diseases.[57] This may explain Alzheimer's motivation to request of his guest researcher Gaetano Perusini (1879–1915) to review Auguste D.'s case and three further cases, one of which Francesco Bonfiglio (1883–1966) had described in 1908. I briefly touch on these two publications here, because, compared to Alzheimer's first case, they manifest three changes that anticipate the increasingly marginalized place of the patient in the cultural dementia discourse. The first of these changes – a shift in focus from patient interview to tissue analysis – at least in part, is representative of a continuous evolution of the case study as a form of writing.[58] In addition, I note a related emphasis on differential diagnosis (meaning post-mortem histological analysis of tissue to separate the condition from senile dementia); and a further evolution of vocabulary, as dementia as a form of degeneration becomes knitted to the notion of ageing as involution.

In Bonfiglio's case of a 60-year-old patient, microscopic observations amount to a third of the paper.[59] His emphasis on cellular morphology gains an additional dimension when he writes that 'the nucleus of the nerve cell often looks degenerate (*degenerato*)' (201), and summarizes the microscopic picture as showing 'regressive changes' (202) and 'degenerative, destructive events' (205). We should not over-interpret Bonfiglio's use of the term 'degeneration' in the cellular context, even though it distinguishes his paper from Alzheimer's case of Auguste D. Since mid-century, the term had been used in relation to necrotic cell death and fatty alterations of cerebral arteries.[60] In addition, as Michel Foucault has emphasized, the description of visual evidence has frequently necessitated abstract language, which, in turn, results in a 'nominalist reduction on the essence of the disease'.[61] All the same, at this moment in time the term entered the work of medical scientists whose studies became points of reference for further research on what would shortly be termed 'Alzheimer's disease'.

Perusini's paper, in turn, witnessed to the rising reliance on microscopic study.[62] Clinical notes remained limited to no more than a fifth of the entire article. Its main focus rested on methodological approaches and anatomo-pathological observations, especially the origin and nature of plaques (i.e. deposits around nerve cells). Anticipating a nearly century-long debate, Perusini believed presenile forms of the condition were the same illness as senile dementia – a position in opposition to Alzheimer's mentor Emil Kraepelin (1856–1926).[63]

The definition of the disease

Kraepelin was particularly known for his classification of psychiatric conditions. In the 1910 edition of his psychiatry textbook, he introduced, in relation to the cases described by Alzheimer, Bonfiglio and Perusini, the term 'Alzheimer's disease' within the section on 'Der Altersblödsinn' (senile dementia; 593–632).[64] Concurrently, he treated presenile dementia in a separate section on 'Das präsenile Irresein' (presenile dementia; 534–54).[65] But considering the comparatively young age of the patients described as having Alzheimer's disease, he suggested a disease entity separate from senile dementia, namely *senium praecox* (627). The clinical characterization of these Alzheimer's disease cases echoed cases with senile dementia: Kraepelin described them as weak in memory, poor in thought and disturbed (624), before the highest degree of stultification develops (626). On a microscopic and anatomic level, Kraepelin gave a detailed account of loss of components of nervous tissue (621). Behavioural descriptions included the patient's 'playing like children' (600), 'nocturnal restlessness' (601) and 'lack of judgement in relation to time' (608). These descriptions overlap with how Galsworthy describes Timothy as 'too old for anything but baby's slumber' (704). They also evoke notions of the Forsyte's involution: Timothy 'had been their baby, getting younger and younger every day, till at last he had been too young to live' (842).

Also important for further discourse developments, Kraepelin used clinical vocabulary for ageing in the context of brain inspection. This means that he knitted the concept of dementia as a form of degeneration to that of ageing as regression and involution. Specifically, he opened the chapter with observations on 'events of regression (*Rückbildungsvorgänge*), which we call human ageing' (533).[66] Yet, further on, he observed *Rückbildungsvorgänge* in the context of brain inspection, namely 'strong pigmentation, vacuolization, shrinking of the nucleus' (619). This observation brings us full circle: morphological events of cell loss occurring in dementia (in the brains of younger or older individuals) became linked to concepts of degeneration (on a cellular level) and regression (both on a cellular and individual level). The look into the microscope both confirmed and expanded dementia as a condition of decline and loss, degeneration and involution.

As the first decade of the new century came to a close, Kraepelin had – in giving the condition his assistant's name and promulgating concepts of disease pathology in a widely read textbook – pushed the illness and its eponym into the limelight.[67] He christened the disease at a time when only five cases had been reported, in this way introducing the term with only limited evidence for a disease pattern. But once 'new patterns are suggested', Mary Boyle argues, 'we are apt to talk of "a new disease being discovered"'; further research consequently focuses on details rather than questioning the patient classification in the first place.[68] With his classification, Kraepelin accentuated three specific characteristics that determined how Alzheimer's disease would be researched and conceptualized over the ensuing decades. First, by introducing the term *senium praecox*, he separated Alzheimer's disease from senile dementia (even though clinical terminology was similar in both categories); second, he privileged histopathological features (i.e. microscopically observed tissue alterations), in particular plaques and tangles, while treating vascular changes (i.e. changes in blood vessels) separately; and third, he emphasized cognitive symptoms at the expense of affective and delusional symptomatology (what Alzheimer had described as ideas of jealousy and auditory hallucinations).[69] Kraepelin's move selectively to emphasize specific disease characteristics evokes Foucauldian notions of how specialist knowledge creates a subject as well as the parameters according to which this specialist knowledge wants to see that subject defined.[70]

The reaction to the definition of the disease

I will consider shortly how Alzheimer reacted to his mentor's move to introduce the disease pattern and define its subjects in these terms. First, I want to reflect on the far-reaching implications for discourse developments of Kraepelin's step to introduce a new disease category with these specifications.[71] Related to the three above disease characteristics, the following points may explain Kraepelin's selective emphases to some degree. First, one could argue that Kraepelin in fact anticipated the problems of defining the disease by means of age of onset: placing the Alzheimer's disease cases within the section on senile dementia foreshadowed political moves in the 1970s that eliminated age of onset as the criterion separating senile dementia from Alzheimer's disease. Second, during a period of fascination with the steadily improving microscopic visibility of neuronal cells, a focus on the histopathological features of plaques and tangles can be explained, and their study will dominate subsequent biomedical research. The issue I consider worthy of speculation touches on the third characteristic – Kraepelin's removing affective and delusional symptomatology. To my mind, this choice cemented the organic paradigm in twentieth-century dementia research and discourse developments. Kraepelin defended materialistic explanations of mental diseases, and sought the connection between clinical observation and established somatic sciences based on localization theories and the lesion model.

His classification of diseases dominated early-twentieth-century psychiatry, including the rise of psychoanalytical approaches.[72] But given that, as already stated, Freud aimed at developing a 'thoroughgoing physiology of association to explain normal and pathological states', it is unlikely that Freud was opposed to 'an organic basis for dementia'.[73] Rather, I would argue that Kraepelin sought to establish Alzheimer's disease with a stress on cognitive symptoms at the expense of delusional phenomena because he had the explanatory possibilities of neurology as compared to psychiatry in mind.

Affective and delusional symptoms had featured in nineteenth-century descriptions as well as Alois Alzheimer's own report of Auguste D. They were disease dimensions that a psychiatric approach could embrace in its emphasis on patient experience, and which lesion theories were hard pressed to explain. Notions of 'excess', as the neurologist Oliver Sacks calls them, challenge the principal mechanistic concepts of neurology; they only receive attention in psychiatry.[74] Seen in this light, Kraepelin's move predestined how Alzheimer's disease would be researched and defined. Histopathological details and the question of age of onset became key concerns to medico-scientific research, while affective and delusional phenomena would emerge in literary writing significantly earlier than in medico-scientific research.

Alois Alzheimer himself must have sensed these ambiguities and uncertainties, feeling the need to summarize the current knowledge of dementia as he understood it. So he produced two further publications on the condition before his early death in 1915. It comes as no surprise that, in his 1910 programmatic article on 'diagnostic difficulties in psychiatry', Alzheimer mainly wanted to devise a clear means by which to separate organic brain diseases from functional ones (including manic depression, paranoia and hysteria).[75] Reading this paper brings us back to our original enquiry into notions of degeneration in the writing of Alois Alzheimer and his immediate colleagues.

In relation to functional mental diseases, Alzheimer was in tune with contemporary views of the hereditary nature of these conditions. Most importantly, however, his deliberations moved from the individual to the histopathological. For him, there were 'other conditions, which we must relate to degeneration (*Entartung*), since they are ... familial conditions. They partially even lead to degeneration (*Degeneration*) of nervous tissue' (12). I note the distinction between the German medico-scientific term *Degeneration* and *Entartung*, the latter being the German word employed in socio-economic, political and legal contexts into which Max Nordau's concepts of degeneration had been assimilated. This distinction underpins why Alzheimer was not one of the psychiatrists from whose works Nordau had borrowed freely.[76] But more importantly, in a single line of thought, this statement confirms the observations made on Bonfiglio's, Perusini's and Kraepelin's writings: there is a distinct shift in the use of degenerationist vocabulary in the context of dementia and Alzheimer's disease. Previously employed in general social terms for an individual's condition, the term 'degeneration' was now applied to describe a specimen taken from that individual and scrutinized microscopically.

A close reading of Alzheimer's only other case study in relation to dementia in a middle-aged subject, published in 1911, underlines this observation.[77] Alzheimer opens his study of the 56-year-old Johann F. with explicit reference to the 1906 'case of illness of presenile age' (356) and employs expressions similar to those featuring in Auguste D.'s clinical notes. Where he describes the patient's brain tissue post-mortem analysis, however, he calls on notions of degeneration: cells in dissolution are found to be with 'shrunken and degenerated nucleus' (366). Alzheimer further observes 'regressive alterations' (366) and 'degenerative appearances' (376). In a young-onset sufferer he links, like Kraepelin before him, concepts of age-related regression to those of degeneration for the purpose of brain inspection.

So, by 1911, Alzheimer's disease embodied concepts of degeneration and regression at the scientific-microscopic level. In relation to the clinical subject, aspects of feebleness of memory featured centre stage, with notions of hitherto age-related regression also gaining in importance underpinned by scientific observation. What we still need to find out is whether Alzheimer's descriptions buttressed or enhanced *clinical* conceptualizations: that is, whether cell loss was considered the confirmation of the degeneration of the patient. Such a question is not unjustified, since we see overlap between fin-de-siècle literary degenerationism and the medical and scientific dementia vocabulary. Timothy's involution encompasses his age, erratic behaviour and memory function, and his dementia constitutes Galsworthy's core metaphor for degeneration – that of the Forsytes as much as the Victorian Age.

Degeneration: The old and new narrative of loss and decline in medico-scientific literature on dementia and Alzheimer's disease

Alois Alzheimer limited his use of the term 'degeneration' to microscopically observed cellular events of loss. Today, such terminology is pervasive, in the context of both the individual patient and the condition's cellular and subcellular pathological events. In clinical practice, Alzheimer's disease did not feature as a diagnostic category until political moves in the 1970s. But its medico-scientific terminology developed during the thirty to forty years following Alois Alzheimer's own work.

Three specific changes took place. Alzheimer's strict ethos of considering histopathological observations as supplementing clinical descriptions was inverted to bring brain examination centre stage. Closely linked to this switch, vocabulary originally used in the specific context of microscopic tissue observations was eventually employed for describing the patient's clinical characteristics and behaviours. This second step was helped by the fact that research confirmed what Kraepelin's grouping of Alzheimer's disease cases with those of senile dementia had anticipated: cellular events taking place in patients suffering from Alzheimer's disease (subjects in their fifth or sixth decade) were also identified in the brains of elderly people, who found themselves with conditions such as senile dementia and living years of 'involution' and 'regression'. Vocabulary used in relation to an individual's ageing began to be employed for the description of cellular observations. Third and in parallel, a discourse of infantilization gradually came to the fore.

In one of the first research articles specifically referring to Alzheimer's disease, Gonzalo R. Lafora (1886–1971) explicitly compared the patient to the child, writing that, 'eventually, the patients become silent and idiotic (*verblödet*), are childlike in their behaviour and sometimes show restlessness or fear'.[78] In papers of the 1920s and 1930s, we find this discourse of infantilization interchangeably in descriptions of presenile and senile cases of dementia. For example, the German neurologist and psychiatrist Ernst Grünthal (1894–1972) described one of his female cases as 'sad and childishly snivelling'.[79] Similarly, the German psychiatrist Johannes Schottky (1902–1992) reported on a patient's 'childish scribbling' and described another as 'somewhat infantile'.[80] First photographic images further reinforced this concept, as they showed the patient's childlike exploratory behaviour, sucking reflex and eating habits.[81]

Along with an increasing infantilization, references to individual patients in medico-scientific publications dwindled. Grünthal's papers are a case in point. The neurologist never focused on the clinical case and individual patient in the first place. Rather, he offered a post-mortem diagnosis based on a macroscopic and microscopic study of brain sections of fourteen patients.[82] Only three pages cover the 'clinical final states' of the patients. These he separated into mild, moderate and severe cases. Mild cases were described as disoriented; moderate cases were fearful, restless and talked little; and severe cases were very irritable and fidgety, totally helpless, showed sucking reflex and communication with them was impossible. From this classification we gather that Grünthal comprehended the condition in terms of progressive clinical decline, which he saw as paralleling findings observed in brain tissue (134); in his words: 'Cortical atrophy and severity of the disease go hand in hand' (143). The narrative of loss – of brain tissue as well as memory and orientation – remained dominant. Degeneration became linked to loss in Grünthal's paper, to *Schrumpfung* (shrinking; 130, 133, 134), *Schwund* (disappearance; 130) and *Verkleinerung* (reduction; 130).

This increasing focus on brain tissue as both leading and enabling diagnosis (while telling a narrative of loss) was supported by the development of brain imaging methodologies. An early form of encephalography became available soon after the arrival of medical X-ray technology. It allowed neuropsychiatrists to begin to gaze inside the brain and study its structure while the patient was still alive. Schottky, in particular, confirmed atrophy in frontal areas of the brain by means of this investigation, further underlining the centrality of a concept of tissue loss to dementia.[83] At the same time, the narrative increasingly emphasized a clinical description of loss.[84] W. H. McMenemey's (d. 1977) publication of six cases in 1940 is a good example.[85]

McMenemey continued to develop a narrative that rested on the concept of degeneration. The pathologist used this expression in relation to macroscopic observations, referring to both peripheral organs, like the lung or heart, and the brain itself; and he employed it for microscopic descriptions at a cellular level. In this way, Grünthal's provisional assumptions (that central nervous events were paralleled by external clinical appearance and behaviour) became manifest in the vocabulary applied to patient and tissue alike. Scientific and medical concepts of degeneration and involution began to refer to an individual as a whole. Terminology used in relation to ageing, namely involution and regression, carried a meaning similar to that of degeneration: decline, decay and loss. In summary, by the early 1940s, studies correlated the bodily to the cerebral, and the clinical to the scientific, and their terminology freely oscillated between patient (as individual) and brain (as specimen). This oscillation fuelled a medico-scientific understanding of the patient as an involution-suffering carrier of morphological cerebral features, whose behavioural traits were compared to those of children.

Over the ensuing decades, researchers continued to focus on the organic cause of the condition. At the same time, their enthusiasm for the diagnostic powers of tissue examination and imaging technologies steadily grew. By the

1950s, access to, and examination of, the living brain became usual practice. Cerebral biopsy (a sample of brain tissue removed from the patient's brain under anaesthesia) was the new tool for the ex vivo diagnosis, if not confirmation, of Alzheimer's disease.[86] (I explicitly adopt the term 'ex vivo' here to set biopsy-related diagnosis as taking place outside the human body apart from in vivo diagnosis performed with imaging technologies during the 1990s.) Related to this, histological methods for the diagnosis of Alzheimer's disease became worthy of publication in their own right, separate from evidence collected by means of patient observation.[87] While attention to the individual patient decreased, the focus on the cardinal types of lesions, that is, plaques and neurofibrillary tangles, continued to tell of 'almost total loss of neurons', involving the use of such phrases as: 'marked loss of myelin', 'gross lobar atrophy' and 'diffuse loss and degeneration of neurons'.[88]

The increasing interest in research on dementia culminated in a first international conference dedicated to 'Senile Dementia – Clinical and Therapeutic Aspects' in Lausanne in 1967.[89] Attended by scientists from across Europe, this conference brought together the current clinical knowledge on 'one of the central psychopathological problems of the aging process: the psycho organical decline and its relations with many other pathological conditions of senescence'.[90] Researchers were keen to understand how the disease process of Alzheimer's disease related to that of normal ageing and how senile dementia, in turn, related to presenile events like those typical of Alzheimer's disease. Among others, investigators refuelled discussion of age-related vascular aspects.[91] Research work carried out during this period of internationalization eventually politicized dementia as a condition of ageing. In addition, the field became dominated by arguments about differences in disease pathology related to age of onset.

At the same time, the continued emphasis on the assumed parallel nature of clinical and cerebral histopathological developments increasingly devalued the subject. This happened in the reduced amount of text dedicated to the description of the clinical case. Such a move conferred decreasing importance on the information gained from patient interviews and greater importance on the scientist's description of the brain rather than the patient – a pattern we can observe in the case history's evolution more generally.[92] If any focus was to remain on the patient's well-being, at a time of increasing specialization and while the cause of the condition was still being identified, a split of disciplines was unavoidable. The rise of old-age psychiatry in the 1950s, which I cover in Part II, and the slightly later surge of work in the neurosciences in the 1960s, to which I dedicate Part III, both witnessed this split, even as both trends took Alzheimer's disease to be a condition of major concern.

There Were No Windows: The patient's illness experience in the modernist novel

Before moving further ahead in the timeline of our investigations, let us consider how microscopic inspection and the resulting enthusiasm for histopathological means of diagnosis influenced the societal awareness of dementia and the perception of these patients. Had the discussion about the distinction between presenile dementia (i.e. Alzheimer's disease) and senile dementia reached general awareness in any noticeable way? With the slowly growing international attention to Alzheimer's disease, medico-scientifically the idea predominated that this condition and senile dementia were separate entities. But although Alois Alzheimer's extraordinary description of the condition in a middle-aged subject fuelled intense research, it remained unknown to the general public. Novels depicting presenile dementia were practically non-existent. William De Morgan's *The Old Man's Youth* (1921) was perhaps the first narrative to bring into focus memory loss in old age as compared to middle age: the narrator Eustace John tells the story of his youth (while constantly doubting his memory) as well as that of his father's memory loss and fear of becoming a burden. At the same time, the narrative of old age as a social burden persisted, if we think about the short period of happiness and productivity in the life of F. Scott Fitzgerald's 'The Curious Case of Benjamin Button' (1922). All in all, understanding of accelerated memory loss as an illness of the elderly steadily increased. In 1922, G. Stanley Hall (1844–1924) mentioned 'senile dementia' in *Senescence: The Last Half of Life*, a key text on ageing during the first quarter of the twentieth century. He conceded that 'senile insanity or atrophy of the brain is certainly more common and appears earlier', but deliberations on the 'dawn of old age in women' took considerably more space.[93]

By the 1940s, a rising number of research publications sharpened awareness among scientists that, in cases of Alzheimer's disease, symptom onset often coincided with the individual's incipient menopause. These observations nourished assumptions that Alzheimer's disease might depend on hormonal alterations: its prevalence was higher in women.[94] Such deliberations were paralleled by a growing psychiatric interest in menopausal effects during the 1930s and 1940s, an interest to which the publication of Helene Deutsch's *The*

Psychology of Women: A Psychoanalytic Interpretation (1944–5) contributed.[95] Against this backdrop, fictional works soon began to centre on female rather than male individuals marked by memory loss. Among them, we find William Trevor's Mrs Maylam in *The Boarding House* (1965), Paul Bailey's Faith Gadny in *At The Jerusalem* (1967), Elizabeth Taylor's *Mrs Palfrey at the Claremont* (1971), Kingsley Amis's Marigold Pyke in *Ending Up* (1974) and Barbara Pym's Marcia Ivory in *Quartet in Autumn* (1977). Norah Hoult's *There Were No Windows* (1944) I consider to be one of the first works of its kind.[96] I take a closer look at this narrative here for two reasons. It suggests that, on a cultural level, memory loss in old age continued to be connected to nineteenth-century conceptualizations of insanity. In addition, hallucinatory symptoms are central to the protagonist's illness experience. This is important to our understanding of dementia in fiction for two reasons: such symptoms held a prominent place in fiction much earlier than in clinical practice, and they continue to be an established part of dementia fiction.

The novel's structure traces the 78-year-old writer Claire Temple's increasing isolation and growing confusion. We easily get lost in the narrative's progressively more convoluted structure, as an increasing number of chapters and subchapters fans out through the text unsystematically. We are drawn into 'the dark night of the imagination' (183): that is, Claire's confused inner world, and her remaining awareness in the narrative's overgrown third and final part, which covers half the text. Hoult explores and anticipates this void in part two of the novel. There, we see Claire's isolation through the eyes of the visitors to her house. These visitors are Edith Barlow, who 'was not looking forward to her fortnightly Sunday luncheon with her old friend … which had now become merely a trial to flesh and spirit' (91); Mrs Sara Berkeley, Claire Temple's former secretary, who hopes to get hold of Claire's 'oceans of waste-paper' (129) for her war-time charity work; and Mr Francis Maitland, who 'had been warned that old Claire had gone completely ga-ga, that it was too embarrassing and quite too dreary to go and visit her these days' (143). Joining Claire one by one, chapter by chapter, they give us the full view from 'outside' (89) onto Claire's condition; 'of her becoming such a bore, and such a dreary lachrymose bore' (91), 'who was now so much of a lunatic' (134) and 'half-mad' (162).

Each visitor has their diagnosis of Claire Temple's condition ready, ranging from her being mad, bad and wicked to simply sad and depressed. These diagnoses suggest how the nineteenth-century changes in the definition of insanity continued to linger in the popular discourse surrounding mental illnesses in the early twentieth century.[97] At the same time, in their awareness of Claire's former activities, those close to her find it difficult to diagnose senility in her condition, because 'senility sat back heavily in its arm-chair, saw little, and heard little, was lifted into bed and out of bed' (179). Literary critics have developed their own views on Claire Temple's illness. Julia Briggs, in her afterword to the novel, reads Hoult's account first and foremost as a narrative about the trials and tribulations of ageing, while deliberations on dementia feature only insofar as she mentions Iris Murdoch's 'comparably sad, non-fictional ending[ ]'.[98] Nick Hubble and Philip Tew, by comparison, believe that Hoult 'captures the repetition, circularity, dislocation and irritations of dementia', but they do not historically contextualize such a reading.[99]

I see Claire Temple's symptoms in the same lineage as those of Auguste D. and Johann F., though at a later age (note that I do not aim for a diagnosis in present-day terms, but strictly refer to Alzheimer's descriptions made in 1907 and 1911). Claire's 'screaming, scream after scream' (174) for fear of being abducted results from her fear of being 'robbed and poisoned' (281) and trying unsuccessfully to dress herself (307). In addition, just as Alzheimer stressed Johann F.'s impoverishing of vocabulary and eventual, total loss of spontaneous speech (361), Hoult uses Claire's loss of language and literary world to characterize her protagonist's condition. Claire's continued citing from literary works implies her loss of identity – both in its witnessing to Claire's living in the past and in its intertextual reference to Shakespeare's *King Lear*. Her name dropping and reference to long-deceased authors as her colleagues place her near the living dead.

Also, Claire's freedom of spirit brings her close to the feeble-minded women Patrick McDonagh identifies in Victorian narratives, who are also depicted as sexually available in inappropriate ways.[100] I find this reading of McDonagh's theories particularly helpful, because notions of feeble-mindedness give us access to Hoult's narrative exploration of Claire's illness experience. To illustrate this point, I want to take a brief look at William Faulkner's Benjy in *The Sound and the Fury* (1929). Benjy is one of four children of a Southern aristocratic family in decline after the American Civil War. I introduce him here, because it would seem, as Suzanne Nalbantian has observed, 'that Faulkner creates the mentally retarded character Benjy in order to depict an uncontrolled intrusion of memory which completely breaks down rational barriers between the present and the past'.[101] I propose that Benjy's memory is the kind of associative memory that Freud had in mind when putting forward his ideas about how experiences, and the associations they create, leave memory traces that underlie mental pathology. This type of memory is conceptually close to the hallucinatory associations Hoult creates in order to draw the reader into Claire Temple's condition.

Claire's final days are set against the background of 'the madding war' (153), the Blitz, which Briggs identifies as echoing Claire Temple's 'inner chaos and fear'.[102] In setting her story in the Second World War, Hoult develops a twofold way of opposing Claire's lack of present-day perception to her enhanced living in the past (her long-term memory remains intact). Claire associates the London bombing with Zeppelin attacks of the First World War. And she links blackout nights to 'darkest London', an image that was engendered, for Victorian society, by the industrialization-related growth and urban poverty in the wake of nineteenth-century demographic and population shifts.[103] Hoult can depict Claire Temple's less and less steady grasp of the present by letting her move between the present Second and the past First World War. The linking perceptions are those of her past experience: London's deprivation and destitution during the First World War. These oscillations across time, and her old age, make Claire 'a specimen of pre-war … pre-last war, of course – literary-cum-social hostess' (179). Hoult then further enhances Claire's living in the past, as Claire, in believing to live in the present of the First World War, constantly moves back to *her* past: the lost Victorian Age.

With the arrival of Miss Jones, a companion and caregiver for Claire, we appreciate the emotional need related to Claire's living in the past: 'Her chief way of escape previous to the advent of Miss Jones had been into the past … in which … she had been admired and entertained … [Miss Jones] barred the entrance to that house of the past … putting her in full mind of the unhappiness of her present lot' (250–2). Why is Claire Temple so concerned about living in the present? She tells herself: 'If you're not careful, they'll put you into a mental home. You have lost your memory, and that's what they do to people who have lost their memories' (252). Her imagination running wild, she hears 'the harsh noise of iron gates clanging behind her; she passed through a door, and turned just in time to see the bolt drawn behind her; she screamed; someone muffled her mouth with a huge hand and dragged her on; dragged her into a padded cell, where no one would hear her screams' (254). Claire's mental pictures of an asylum are fuelled by fears of involuntary detention and, linked to this, being 'certified as a raving lunatic' (288).

Just as definitions for mental conditions had shifted throughout the nineteenth century, the understanding of how individuals suffering from such illnesses should be cared for also changed.[104] Ending up in any institution was considered a disgrace, because it meant that a family had not the means to look after an individual. It was the custom of the upper classes to hide cases of feeble-minded family members, rather than admitting them to institutionalized care.[105] In addition, towards the end of the nineteenth century, fears had grown about wrongful detention. Wilkie Collins, for instance, explores memory loss in relation to wrongful detention in Laura Fairlie's character in *The Woman in White* (1859), and Mary E. Braddon broaches the role of detention in a private institution rather than prison in *Lady Audley's Secret* (1862).

Most problematic, however, was the involuntary aspect of detention. In England, the power of the state to incarcerate the 'furiously mad and dangerous' first appeared in the eighteenth-century Vagrancy Laws: individuals could be confined subject to the decision of a magistrate.[106] These fears were appeased by the 1890 Lunacy Act: two doctors' certificates were now additionally required to effect incarceration of the insane in private as well as public asylums. But medicalizing the process of psychiatric detention did not lift the stigma attached to being certified. Also, substituting the phrase 'mental hospital' for 'lunatic asylum', which the 1930 Mental Treatment Act achieved, did not remove the stigma attached to institution and the certified.[107] Read in this light, Claire Temple's supplication to her doctor, 'You won't certify me, will you?' (261), is more than understandable. It illustrates how preconceptions of the elderly needing institutionalized care became amalgamated with ideas of mental impairment.[108]

Claire Temple's fears continue to be fuelled by the degenerationist and social Darwinist notions that linked the mentally ill to the child, animal, savage and elderly and that determined their treatment by institutionalization: she feels treated 'as a wild beast' (308).[109] After the Great War, theories of degeneration, which since Morel had framed the classification of dementias, started to go out of favour.[110] This is partly explained by the fact that the fin-de-siècle period was conceptually coming to an end by 1940 – and with it societal concepts of degeneration as a characteristic feature of that period.[111] However, its most atrocious eugenicist consequences entered general awareness only at Hoult's time of writing, as her character Dr Fairfax reflects on in regard to the 'sacrifice of the individual to the claims of the state, the race, or the hero leader' (305). In addition, treatment practices did not necessarily have a good reputation. This was the case because attitudes towards the insane initially resulted from compensating for the absence of medical therapies.[112] For example, in the conviction that senile dementia was 'of its very nature incurable', Kraepelin covered therapeutic intervention in less than half a page in his textbook (632).[113] Such marginalization exemplifies the limited care the elderly received before geriatric medicine and gerontological research began to make headway in the 1940s.[114]

In summary, dementia as an illness was slow to become an issue within wider cultural awareness. For a long time what now appear to be hallmarks of dementia remained part of the characteristics of old age and ageing, and went unnoticed in popular culture for most of the nineteenth century.[115] Late Victorian society lacked the readiness to accommodate and appreciate the elderly and their need for care. In such a context one could argue that Galsworthy's Timothy had to be a secondary character.[116] In fact, Timothy's presence (in his absence), from beginning to end, his hovering over the family clan and the early-twentieth-century reader, epitomizes the issue. His passivity and apparent absence from text and life suggest the lack of awareness of dementia as a growing challenge for a population that was beginning to grow old at the end of the nineteenth century.

In the tradition of social realism, the Edwardian Galsworthy delineates dementia as one form of degeneration and links it to post-Darwinian myths of decline and degeneration. Throughout this study, his 'scrupulous depictions of external appearances', as I note in Timothy Forsyte, will remain with us.[117] We may think of Hoult's novel as revisiting Galsworthy's motifs. Yet Claire Temple's case is more firmly cemented in the context of contemporary scientific work and a growing general awareness of ageing as a societal challenge. Furthermore, the medico-scientifically depicted cellular disintegration becomes a theme ever more widely elaborated in fiction, a rhetorical strategy and image shared between brain science and creative fiction.[118] Galsworthy gives voice to different characters to map the dynasty's breaking apart; Hoult uses a similar strategy in order to bring out Claire's disintegration.

Only during the 1980s would fiction bring home the inner workings of the mind of subjects with Alzheimer's disease, exploring notions of associative memory to the full. Hoult's method of narration and her strictly limiting third-person perspective gives away information and understanding only bit by bit. It gives us, in the wake of early-twentieth-century stream-of-consciousness narration, a first glimpse into the mind of the patient. Thomas G. Pavel observes that

> modernist novels often take place in a pervasive atmosphere of emptiness and sorrow. The characters do not entirely know who they are or why they act the way they do. Focusing on their fragile psyches, the all-powerful writer leads readers through an intricate stylistic labyrinth whose gloomy incoherence,

biting irony, or uncanny intellectual sophistication emphasizes its distance from real life. The links between characters and their society grow weak or disappear, and the figure of the loner ... increasingly holds center stage.[119]

In this regard, modernist narratives like Hoult's do not only anticipate the 'partial view' from which both patients and caregivers must present their case.[120] They have created the terms in which we understand consciousness and mental illness today. They push us closer to the patient's lived experience – from which we remain removed in social realist narratives. As such, modernist narratives should also be taken as foreshadowing a more profound understanding of the patient experience towards mid-century. Modernist narration could be understood as having opened the patient's experience to the medical arena in a way that Freud perhaps could not achieve during a period preoccupied with purely materialist approaches to dementia; when the psychoanalytic case history remained 'an institutionalized other for clinical medicine ... at the margins of medical discourse'.[121] With a growing awareness of ageing as a socio-medical challenge, old-age psychiatrists would become increasingly interested in the patient's experience, as I explain in Chapter 3.

In this cultural history, we could consider John Galsworthy the prophet of the twentieth-century broader cultural dementia discourse. Galsworthy focuses on single characters, whose individual destinies echo historical and social processes.[122] He illustrates that during the period of the decline of the British Empire, societal changes occurred that included a growing number of elderly people, whose prevailing conditions encompassed memory loss. Norah Hoult subsequently depicts the first consequences of these economic and cultural changes and anticipates sociopolitical concerns about the rising number of older people and age-related diseases, which basic scientific research and applied medical practice were addressing. Later in the twentieth century, we see writing that explores how attention to the elderly as needing care turned them into an entity requiring clinical management, without any recognition of their individuality.

Part II

The ageing perspective

3

Culture shapes politics shapes science

Researching old age: From medical science to old-age psychiatry

> Old age in America is often a tragedy. Few of us like to consider it because it reminds us of our own mortality. It demands our energy and resources, it frightens us with illness and deformity, it is an affront to a culture with a passion for youth and productive capacity. We are so preoccupied with defending ourselves from the reality of death that we ignore the fact that human beings are alive until they are actually dead. At best, the living old are treated as if they were already half dead.
>
> Robert N. Butler, *Why Survive?*[1]

Robert N. Butler (1927–2010) was an American old-age psychiatrist whose work on the social needs of the elderly profoundly influenced healthcare politics in the 1960s and 1970s.[2] In 1974, this development culminated in the Research on Aging Act, Butler becoming the first director of the newly emerging National Institute on Aging (NIA). The NIA has shaped scientific and medical research on ageing and old-age diseases both nationally and internationally, and its actions have strengthened a medico-scientific dementia discourse and formed a parallel sociopolitical and healthcare dementia discourse, during the 1980s. In his Pulitzer Prize-winning *Why Survive? Being Old in America* (1975), Butler engaged with popular beliefs about ageing and old-age diseases, among them dementia and memory loss.

Butler's observation that the elderly were being treated as half-dead reminds us of early-twentieth-century cultural images of the dementia patient as living dead, such as Galsworthy's Timothy Forsyte and Hoult's Claire Temple. It suggests that popular concepts of ageing and dementia had become linked by the mid-1970s. This chapter tells the story of how this amalgamation happened.

I will trace the continued evolution of a cultural dementia discourse that centres on loss and decline, charting it against the background of ageing as a growing challenge to healthcare between the mid-1940s and the mid-1970s. During this period, age-related medical research – geriatrics and old-age psychiatry – began to become a distinct field. I will illustrate three mechanisms that supported how popular concepts of ageing and dementia became linked. First, gerontology as the science of ageing employed terminology that strongly relied on early-twentieth-century medico-scientific concepts in the images of decline, degeneration and loss it adopted. Second, old-age psychiatry relied on materialist descriptions in how it blended concepts of ageing as illness with those of senility or dementia as old-age illness and so perpetuated notions of loss and decline. Third, on a cultural level these discourse developments contributed to dementia becoming a prevalent attribute for older people.

Until the 1870s, the elderly were considered incurable, because old age and the decline of the body meant the same thing.[3] In the discourse of ageing, degeneration referred to tissue alteration and cell loss. Therefore, as late as 1904, a general theory of old age – considered to begin between 50 and 60 years of age – might be expressed in the following way: 'a general atrophy of the organism' explained by the loss of cells in kidney, liver and brain and cells showing 'signs of degeneration in the form of deposits of fat'. The immunologist Elie Metchnikoff (1845–1916), who used this terminology, believed that one's immune system attacked the body's own tissue weakened by age and described old age as a 'chronic disease which is manifested by a degeneration, or an enfeebling of [nervous cells]'.[4] Similar expressions were employed by Alois Alzheimer and his colleagues in relation to dementia. Around 1900, both dementia and ageing were considered pathological: conditions of degeneration and enfeeblement.

In the early twentieth century major advancements in the medico-scientific understanding of old age came from the United States. They evolved around the idea that debilities typical in later life were caused by decline in old age.[5] The Austrian-born American Ignatz Nascher (1863–1944), who in 1909 coined the term 'geriatrics', defined disease in old age as a pathological process that occurred in a normally degenerating body.[6] This degeneration became normal, because, as in Georges Canguilhem's maxim, 'statistical frequency expresses not only vital but also social normativity'; with an increasing lifespan in an ever-increasing number of individuals, biological processes related to ageing had to become normal.[7] This means that ageing remained linked to a process of degeneration, but began to be seen as only one period in life, wherein diseases were more likely, statistically and biologically. One of these illnesses was senile dementia.

In the United States in the 1940s attention to health and illness in old age intensified further. By 1941, leading figures of gerontology and geriatrics were convinced that research regarding the elderly should cover three distinct areas: 'the biology of senescence, geriatric medicine, and the sociology of an aging population'.[8] The related change of perspective – to consider the sick elderly as patients – spread across the United States and eventually Europe.[9] In Britain, the geriatrician Marjorie Warren (1897–1960) was the first figure publicly to lament the neglect of the elderly in hospitals.[10] Her writing perpetuated the conceptual link between illness in old age, and ageing and dementia as conditions of degeneration. In particular, Warren considered 'senile conditions of all types, including all the degenerative conditions, from simple uncomplicated senescence to the terminal stages of senile dementia'.[11] That said, her efforts, together with those of other geriatricians, laid the groundwork for what in 1947 became the British Geriatrics Society. A year later, the National Health Service came into being, enabling the growth in Britain of geriatric medicine and research related to old people's welfare and service provision as well as medical conditions.[12] This research was keenly interested in what was normal as compared to pathological in older people, and I will show how the ways in which this research relied on organic understandings of dementia further strengthened notions of ageing as loss.

As one of the first socio-medical studies conducted around the middle of the century, Joseph H. Sheldon's *The Social Medicine of Old Age: Report of an Inquiry in Wolverhampton* (1948) addressed this issue.[13] Sheldon (1893–1972) surveyed the health status of older people living in their own home. Based on his observations, he defined elderly people as 'normal' when they led 'for the most part an active life, with full retention of their faculties and without undue emotional disturbance' (117). He further distinguished individuals with 'slightly impaired' faculties (118), who showed 'a tendency to apathy ... depression of spirits; a slight failure of memory' (122). 'Cases with severe mental deterioration', in turn, showed themselves 'forgetful, childish, difficult to live with' or even 'demented, very difficult to live with' (123). Importantly, Sheldon concluded that 'the feeling of being still necessary to the world and of having something to do is essential to the mental health of old people' (117).

Building on Sheldon's insights, Kenneth Hazell aimed to outline 'medical and social problems and the link between the two' in *Social and Medical Problems of the Elderly* (1960).[14] Such research was mainly driven by two factors: fears about the ageing of the population, and the awareness of old-age poverty in the context of post-war prosperity and the welfare state.[15]

Considerations of 'productive capacity', in fact, played a major role in how the elderly were viewed negatively during the 1960s and 1970s. In one of his books on the societal challenges of ageing, Trevor H. Howell, for example, stated that 'old people consume more of our national wealth than they produce ... in fact from the economic point of view most old people are parasites'.[16] When considering Butler's definition of ageism as manifesting itself in 'discriminatory practices in housing, employment and services of all kinds', the growing geriatric healthcare initiatives by themselves risked contributing to this kind of stereotyping of the elderly.[17] In Barbara Pym's tragic-comic *Quartet in Autumn* (1977) the healthcare worker Janice thinks in such marginalizing terms about Marcia Ivory. She wonders 'what kind of job could somebody like Marcia do? ... keeping a house in order needed a certain attitude of mind', and later muses that 'sometimes you almost felt it wouldn't be a bad idea to shove [all these retired people] off all together somewhere'.[18]

Hazell's 1973 edition of *Social and Medical Problems of the Elderly* is the first major medical work that I have discovered to make substantial reference to dementia in clinical practice (as compared to medico-scientific research activities at the beginning of the century). Dementia as one form of 'organic mental syndrome' was described in terms which Alois Alzheimer had used in relation to Auguste D.'s behaviour:

> The patient is likely to show gross loss of memory, intellectual weakness, lack of concentration, grasp and attention, and to be hopelessly disorientated as to time and place ... Emotionally, he may be quite out of control, weeping one moment and laughing at another, or flying into petty tempers. Or the patient may show emotional loss with gross apathy and indifference to his surroundings, undue drowsiness and perhaps incontinence. (1973; 145)

Alzheimer's disease, as a condition of socio-medical concern, emerged only in the early 1970s, as a newly introduced chapter on 'Mental disorders in the elderly' in the 1973 edition of Hazell's book suggests. This slow change might be related to the fact that there was no agreement on who should be responsible for cases of dementia: welfare services, psychiatry or medical geriatrics.[19] Medical scientists focused on the specimen, that is, molecules, mechanisms and methodologies related to Alzheimer's disease diagnosis. Old-age psychiatrists, in turn, paid attention to behavioural aspects. However, the science-based emphasis on the condition's organic nature was imported into, and perpetuated by, old-age psychiatry during the 1960s. An analysis of the studies presented at the first international conference featuring Alzheimer's disease in Lausanne in 1967 reveals that the patient was considered a carrier of behavioural symptoms of loss that could be related to cerebral losses.

Work discussed at this conference focused on evidence for mental events grounded in brain tissue studies. It reveals a reciprocating relationship between geronto-psychiatric approaches and medico-scientific research. This relationship reinforced concepts of decline and dissolution. Researchers charted patients' 'progressive demential dissolution' and presented evidence of '[spectacular] similarities between the behaviour of the demented and of children'.[20] These similarities included, among other aspects, 'disintegration of body awareness', 'loss of recognition' and impoverished vocabulary.[21] As noted previously, the idea of mental illness as aligned with childhood was initially expressed during the 1850s and developed further by social Darwinists. By the mid-to-late 1960s, it had become scientifically entrenched: expressed as a 'hierarchy in disintegration which proceeds in an inverse order to that of infantile evolution', where disintegration moved 'from voluntary action towards automatism'.[22]

The images drawn by old-age psychiatrists at this conference consolidated the medico-scientifically cultivated narrative of loss and expanded this concept of loss to the entire individual (as McMenemey's work had anticipated). By the end of the conference, however, delegates had to acknowledge that they were unable clearly to delimit normal from pathological ageing processes; nor could they offer 'precise and common definitions for the various degrees of senile decline'.[23] Both gerontologists and geriatricians in Europe and America considered these two issues to be of central importance. Their attempts to delineate the boundaries between normal ageing and disease in older age, together with growing efforts to improve the situation of older people, ultimately drove the creation of the NIA in 1974.[24]

Coinciding with the NIA's creation, neuroscientists like Robert Katzman (1925–2008) advocated the elimination of age of onset as the criterion separating senile dementia from Alzheimer's disease, writing that

> the argument that Alzheimer disease is a major killer rests on the assumption that Alzheimer disease and senile dementia are a single process and should, therefore, be considered a single disease. Both Alzheimer disease and senile dementia are progressive dementias with similar changes in mental and neurological status that are indistinguishable by careful clinical analyses ... We believe that it is time to drop the arbitrary age distinction [younger or older than 65] and adopt the single designation, Alzheimer disease.[25]

In other words, Alzheimer's disease should become the term used for all cases of dementia – both senile and presenile. The move prepared by Katzman galvanized public awareness of, and led to increased funding for, Alzheimer's disease. This process went hand in hand with scientists increasingly seeking contact with the popular press.[26] Katzman's 'killer' metaphor would remain prominent, eventually serving as a central figure of speech in a growing scientific and cultural dementia marketplace. By 1985, the NIA had established ten federally funded Alzheimer's disease research centres to coordinate clinical, behavioural and laboratory research. This fast rise in biomedical research suggests that Alzheimer's disease advocates could convincingly pitch finding a cure as a necessary and good thing. However, they articulated less successfully the need for caregiver relief.[27] As I will discuss later, this strategy drove the neuroscientific, non-psychiatric description of the patient, while the caregiver's struggle for support increasingly marginalized the patient.

By the late 1960s, cultural representations linked ageing and dementia in their images of decline and loss. I will read Paul Bailey's novel *At The Jerusalem* (1967) against contemporary socio-medical work to show how research-driven attention to the elderly as needing care turned them into an entity requiring clinical management, without any recognition of their individuality. Bailey's narrative is both product and reflection of this development.[28] But *At The Jerusalem* does more. It is a fictional counter-narrative to the patient being objectified by the 1960s' geronto-psychiatric discourse that equated ageing with senile dementia. It is a riposte to what Sally A. Gadow called the 'failure of gerontology to attend to the subjectivity of aging'.[29] I would even describe it as the beginning of a counter-discourse which patients would develop in the 1990s: a counter-discourse to the objectification conditioned by the NIA's healthcare campaigns of the 1980s.

At The Jerusalem: Dementia defines the elderly in 1960s new realist fiction

> Loneliness is the cause of much misery and also the root cause of much ill health and disease in the elderly. It leads to depression, a disinterest in what is going on around and, coupled with a poor memory, presents many of the features of mild dementia.
>
> Kenneth Hazell, *Social and Medical Problems of the Elderly*[30]

In 1973, Kenneth Hazell described the socio-medical causes of dementia in terms that summarize much of what Bailey's Faith Gadny endures in a home for the elderly.[31] Paul Bailey returns to earlier images of the forgetful senior, destined to die in a confined space, and broaches the burden of elderly caregiving. But he sharpens these images in the light of an increasingly ageing population, especially since, by the 1950s, the majority in the United Kingdom would live well into their sixties.[32] In fact, Bailey had not thought of either Alzheimer's disease or dementia when imparting memory loss to Faith Gadny: he 'simply wanted to make Faith look old'.[33] Bailey narrates dementia as dispossession on a societal scale: as the loss of independence and dignity, freedom and will. Dementia as the loss of one's mind, and the loss of a life worth living, is presented as the result of the devaluation of the elderly in institutionalized contexts.

In a flashback, the narrative's second part shows the reader the reason for Faith's predicament: her perceived loneliness in her stepson Henry and his wife Thelma's house before she is sent to live at The Jerusalem. Set outside the time and space of the chronological sequence of events in the narrative, we can take this section as Faith's reminiscing about the past. This reminiscence is triggered by her being 'sentence[d] to solitary confinement' within the home, which in turn is covered in the narrative's final part.[34] In reliance on Robert N. Butler's definition, I would describe Faith's life review as

> a naturally-occurring universal mental process characterized by the progressive return to consciousness of past experiences, and particularly, the resurgence of unresolved conflicts; simultaneously, and normally, these revived experiences and conflicts are surveyed and reintegrated. It is assumed that this process is prompted by the realization of approaching dissolution and death, and by the inability to maintain one's sense of personal invulnerability.[35]

Faith's unresolved conflicts lie at the root of her having been sent to a home for the elderly in the first place. They are intimately related to the time in Henry and Thelma's house, where she moves after her husband's and her daughter Celia's deaths.

Since the mid-1940s, it has been commonly accepted among geriatricians and social researchers that enabling older people to live in their own home for as long as possible would help them stay fit and active, because they retained their independence for longer.[36] The resulting need for more support in the community for the elderly became clear in the 1950s, and socio-medical concerns about the elderly at home rose in the 1960s. With or without assistance from their families, elderly people, like Faith, had to adapt to new forms of dependence. Given that 'in seventy years she had never sunk so low [that] she had to accept charity' (98), Faith fears being seen as a burden. She refuses to disengage from social life 'out of a desire to remain useful', consequently 'dust[ing] and polish[ing]' (100), and offering to mind her grandchildren Edna and Michael.[37]

In this context, Faith's wish to pass on some of her books to her grandchildren powerfully symbolizes her increasing feelings and fears of uselessness and dispossession, not least as these books end up 'in a dustbin' (152). The thrown-away books therefore embody Faith's increasing loneliness and become an image of her losses, both material and intellectual. Helen Small notes that 'old age and decline are not simply biological facts … but highly responsive to a person's perception of their ability to go on possessing certain goods crucial to their sense of themselves as themselves – including favourable historical, political, and intellectual environments'.[38] Just as Timothy's dummy books had anticipated the image of the dementia patient as an empty shell, Faith's thrown-away books could be read as her mental deterioration and emptying, also because she wonders herself: 'If I tried to read would I reach the end of a line? Would my brain desert me halfway?' (152). As an image of lived experiences, acquired knowledge and wisdom, discarded books ultimately stand in for Faith's losing herself: she considers herself treated by Henry and Thelma 'like trash, waste' (41).

Faith's feelings of low self-esteem follow her sense of loneliness – her perception that she was just 'a guest in [Thelma's] house' (119) – battling to get 'through the day, talking, eating, dressing, trying to sleep' (137). These indicators of depression are accompanied by the first signs of memory loss.[39] Eventually, Faith's decline becomes physical. She wets the bed. In this state, she is sent to The Jerusalem, feeling that 'she had dragged her body about [Thelma's] house. The only suitable description. You could hardly say it was living' (140). She feels, is led to feel, like the living dead.

Read against what we come to understand as Faith's past, the narrative's opening broaches what social medics like Hazell were concerned about. Faith will decline into senile dementia in The Jerusalem:

> Then there was a dazzle of green and white, white and green. Then the colours separated, became clear: the white was above, the green below. Tile, she saw, followed tile … there were graves … Beyond the graves, women were seated on benches, enjoying the sun. One woman's skirt was drawn up above her knees … The sun shone on the graves, the women, the pair of knees. Like the tiled walls, they became confused before her. She had to blink to make certain that a gravestone wasn't grinning. (9)

Faith enters 'a home for the elderly and dying' (14). Its maze-like corridors and tiring staircases match Hazell's description of old-type institutions that were originally workhouses and, as such, not suitable for very old and frail individuals (1965; 243).[40] It contributes to Faith's unsuccessful restoration in her 'new life' (10).

Faith has the impression that 'the painted brick walls would have suited a prison more' (19). These descriptions also included what researcher Erving Goffman described as the loss of autonomy and self-control, ideas of 'permanent dispossession' and 'civil death', and the fear of falling into a category beyond the 'normal'.[41] Faith loses autonomy and independence, is constantly being taken by the hand, only to 'ma[k]e her hand slide free' (18) for it to be 'gripped' again (19). She feels treated 'as if I was a thing' (26); and interactions with caregivers remain superficial and non-empathetic. Especially the matron's smile 'was more like a tic' (59), as she, again and again, 'smiled, shut the smile off quickly' (e.g. 11, 12, 13, 57, 167). Faith is fully aware of the impact on her state of her surroundings, 'the dinners, the graves, the tiles, the doctor ... the humiliation' (36), and will hold their impact at bay as best she can. Setting herself apart from the women, whose ridden-up skirts reveal their lack of self-awareness, Faith keeps an eye on her appearance (48).[42] She refuses to comply 'with the official role expectation of the "happy" and adjusted inmate' and destroys the few photographs that remind her of her past, rather than sharing them during the 'photo session' (55) held by others.[43] Such behaviour contributes to her not being considered normal any more, rather 'a child' (82), 'simple' (83), 'mad' and 'dotty' (92). Eventually, she is isolated from the home's community.

With reference to Hazell's initial observation, Faith is unquestionably located within a social network – first in the form of a family and then the community of women in the home. But these networks do not work for her. She sees herself in 'a world of enemies. Henry. Thelma. Edna.' (139) and feels repelled by some of the elderly, whom she refers to as 'creature[s]' (66) and characterizes as 'crazy' (68). This feeling of loneliness leads her to disengage and withdraw from the women's chatter. She becomes isolated first emotionally, then physically. Faith's relocation to a single room in the home's attic matches concern about the power relations that were played out in the life of elderly. Kathleen Woodward argues that 'the aged have been forgotten and hidden from sight in nursing homes and hospitals by the narcissistic younger social body, by those in power'.[44] Faith's sense of self is threatened by her social isolation. Her initial feelings of loneliness lead to what Hazell termed disinterest and feelings of depression. Her physical isolation from the home's network, then, reduces her sense of self-worth, which derives from relations with others. Consequently, once Faith lodges in isolation, the narrative slows down to observe her memory loss, mental absence and physical neglect. Physical isolation leads to dementia, that is, the exacerbation of what were initially loneliness-related symptoms.

As previously observed in the context of Timothy Forsyte's final appearance, the small room's sparse fixtures anticipate (together with Bailey's staccato style) Faith's dissolution: 'A small window, she saw. A bed. A table. A chair. A bowl of chrysanthemums.' (91). When living at Thelma's, she had simply sat 'for long stretches, staring' (118), appearing to be 'living with the dead' (116). Now, she feels 'helpless; too weary to look' (156) but is aware of being 'old … [with] her poor body' (169). Eventually, during a party she is allowed to attend, Faith feels drawn into a 'black hole' (181) and begins to scream and 'use her hands' (181). The inmates call her 'mad', 'an animal' (181), and demand she be sent away 'to a bin' (182). Faith's dementia, by now manifest, eventually leads to her further, final, isolation. She is sent away in an ambulance. Reading this scene against Butler's remarks on 'hospital tactics in handling the elderly', we must assume that she is taken to a 'hospital for the mentally ill' and might 'conveniently die in the ambulance'.[45]

Bailey claimed that 'the ending of the novel is both tragedy and victory' – a statement I would like to consider in relation to R. D. Laing's assertion that 'madness need not be all breakdown. It may also be break-through. It is potentially liberation and renewal as well as enslavement and existential death'.[46] Laing endeavoured to change the treatment of psychiatric illnesses between the 1960s and 1980s. He proposed that a patient's behaviour and experience be considered within his or her specific social context.[47] In Faith's case, her reaction helps her to escape the social norms and rules she experiences as oppressive and limiting: nursing home–imposed structures that don't match individual patients' needs. Her behaviour liberates Faith from the objectifying patient images Bailey explores in the women's chatter. At the same time, this ending brings out the losses incurred by the patient, once such objectification has influenced healthcare considerations. It brings out what a contemporary reviewer described as 'society's ways of dumping its human garbage'.[48]

Bailey's new realism is powerful in bringing home this message. The women's and nurses' conversations are reported in excruciating detail, almost cruel detachment. Imparting no feelings, the invisible narrator treats Faith with exactly the same lack of empathy she experiences within the institution. In addition, the textual arrangement suggests that neither matron nor nurses ever become aware of her past and reasons for her situation and behaviour. The reader, by comparison, is given access to Faith's past as conditioning her behaviour. This means, we take in much more fully the deleterious effects institutionalization has on Faith's mind. We could consider this constellation to represent the period's continued disregard for the patient's narrative and personal illness experience, a view that is amplified when noting the exhortation of a contemporary physician, namely: 'Listen to the patient's story – he is telling you the diagnosis.'[49]

This brings me to a further dimension to the dispossession pictured by Bailey, appreciated all the more when considering Helen Small's thoughts on the historical figure Cephalus in Plato's *Republic*. Cephalus's authority 'rests on his age and his place in society', explains Small; it rests on 'his ability to speak from the perspective of a long and prosperous life'.[50] Compared to this view of age, dispossession in dementia is further aggravated: a dementia patient's decreasing ability and – in Faith's situation – limited opportunities to speak further lessen her authority. It is left to the reader to identify the necessary actions that would prevent individuals like Faith Gadny from ending their life in an institution – mindless, dispossessed, demented. That said, as early as 1967, Villa and Ciompi suggested a 'dynamic approach leading to real human communication ... also with intellectually deteriorated patients'.[51] But the fate of fictional characters like Bailey's Faith and, a few years later, Taylor's Mrs Palfrey (1971) and Pym's Marcia Ivory (1977) suggests that such measures were rarely implemented. In conjunction with Thelma's claims, their fate also anticipates patient loneliness and isolation in the face of caregiver-centred approaches in the 1980s, which I illustrate in Chapter 4.

Kingsley Amis used comparable narrative tactics in *Ending Up* (1974), a short study of shared living for older people, preoccupied 'with mortality and the disintegration of the body, if not the personality, with age'.[52] George, for example, 'had nowhere else to go, except into a hospital ward' (11), and, for Adela, the cottage is 'far from a bad place to end up' (6). But Marigold, convinced that her memory loss 'is senility' (43) and that she is 'losing [her] poor little mind' (48), feels differently. Most telling in the context of Faith's fate, she senses 'aversion from the prospect of becoming hopelessly senile in the company of people who knew her' (60). Such feelings are perhaps not unexpected, considering that Shorty plays tricks on her memory and Bernard cherishes hopes 'of suggesting to Marigold that her insides had started to decompose!' (79). Reading the death of Amis's five characters in Tuppenny-hapenny Cottage against Faith Gadny's death by institutionalization further highlights how culturally dominant sociomedical and healthcare concerns for the elderly had become.

I have not only placed Bailey's narrative in the context of Hazell's studies, but read it alongside Butler's pioneering work. The similarities between the insights of a British and an American writer bring home that, by the mid-1970s, ageing, and dementia as a condition in old age, had reached a comparable level of preoccupation on both sides of the Atlantic. In fact, Bailey had written *At The Jerusalem* after having read John Updike's first novel, *The Poorhouse Fair* (1958).[53] Like Bailey's creation, Updike's fictional home offers lodging to those whom 'the world … had in the end discarded' and the poorhouse fair featuring in his fiction wants to give the elderly purpose.[54] By the early 1970s, socio-medical concerns regarding the elderly (whose core condition was dementia) had reached cultural awareness on an international scale. Their significance would be reflected in how the NIA, at its inception, sought to address two issues: large-scale funding for medico-scientific research on dementia and the support of long-term care for the ill.

In the next chapter, I pay closer attention to how the image of the elderly person as demented featured in the NIA-instigated social movement that transformed Alzheimer's disease from a rarely applied clinical diagnosis to a major medical condition in the Western world.[55] I ask how the power relations established between the young (and healthy) and the elderly (and demented) played out in the conceptualization of the patient in the healthcare context at a time when Alzheimer's disease began to replace senility in the public conception of ageing during the early 1980s.

4

The loss of self in healthcare and cultural discourse

Caregiver guides: Helpers in the face of loss and decline

> A chronic dementing illness places a heavy burden on families: it may mean a lot of work or financial sacrifices; it may mean accepting the reality that someone you love will never be the same again; it continues on and on; it may mean that responsibilities and relationships within the family will change; it may mean disagreements within the family; it may mean that you feel overwhelmed, discouraged, isolated, angry, or depressed.
>
> Nancy L. Mace and Peter V. Rabins, *The 36-Hour Day*[1]

In this section, I focus on how depictions of caregivers in the late 1970s and early 1980s perceived and dealt with dementia in their family, and I study the effects of the NIA's influence on the public understanding of Alzheimer's disease. I explore how caregiver manuals such as *The 36-Hour Day: A Family Guide to Caring for Persons with Alzheimer Disease, Related Dementing Illnesses, and Memory Loss in Later Life* (1981) contributed to the further development of the cultural dementia narrative. I show how these manuals perpetuated into the last decades of the twentieth century the unequal power relations between care provider and elderly individual established in the 1960s and 1970s, as they explored dementia as 'a heavy burden on families'. At the same time, this section explains how the first caregiver accounts contributed to the popular Alzheimer's disease narrative that emerged during the early 1980s. In particular, it illustrates how the language adopted in first memoirs overlapped with that of caregiver manuals, and how such narratives contributed to the cultural persistence of ideas of loss as they told of the involution and slow death of a loved parent or grandparent.

Rosalie Walsh Honel's *Journey with Grandpa: Our Family's Struggle with Alzheimer's Disease* (1988) was the first book-form memoir by a caregiver accessible to the English-speaking market. It was brought out by the same publisher as the first edition of *The 36-Hour Day*, and endorsed by the manual's co-author, Peter V. Rabins, who wrote the foreword to the caregiver's story.[2] This constellation underlines the strong educational value old-age psychiatrists saw in this narrative. Beginning in 1978, several years before any healthcare guide became available, Honel recounts her caregiving experience, in which social expectations had thrust the main responsibility for the patient on the family. In the 1970s, a steady rise of hospital costs and the consequential effort to limit costs in service delivery resulted in a growing demand on informal family care.[3] At the same time, many families felt an aversion to the 'nursing home', which occupied, as William F. May pointed out in 1986, 'the same place in the psyche of the elderly today that the poorhouse and the orphanage held in the imagination of Victorian children. Even those who never set foot in these facilities fear them as fate.'[4] This deep-seated aversion is also reflected in contemporary fiction. In Margaret Forster's *Have the Men Had Enough?* (1989), for example, the daughter Briget, who is a nurse, is adamant about the siblings caring for their mother at home, and the layout of the short-term care home Birchholme is reminiscent of Kenneth Hazell's descriptions of houses unsuitable for the elderly in the 1960s.[5]

Rosalie Honel was the primary caregiver to her father-in-law, while still having her own children to look after. She was, as Elaine M. Brody has described such carers, one of the 'women in the middle'.[6] In their manual, Mace and Rabins acknowledged this position, writing that 'daughters (and daughters-in-law) are "supposed" to take care of the sick. But the daughter or daughter-in-law may already be heavily burdened' (145). Honel's account was both the product and the reflection of this development. It was 'a story of tragedy', but also 'the story of heroism'.[7] It exposed the link between how the illness affected the patient and how caregiving burdened the adult child, as 'the balance [shifted] between what he could do for himself and what we needed to do for him' (149). In doing so, it set up 'the stereotype of "old people" as objects of care'.[8]

Honel's experiences were representative of contemporary caregiver issues more widely. Mario Monicelli's comedy *Speriamo che sia femmina* (Let's hope it's a girl), for example, broached the female family members' duty to look after an ageing parent with memory loss in 1986.[9] Elsewhere, the Japanese novel *The Twilight Years* (1972; trans. 1984) by Sawako Ariyoshi, to which Honel refers, portrayed a Japanese daughter-in-law's caregiver burden.[10] And Honel herself discovered during a school exchange event that 'it was obvious that the people in France are as concerned about the problem of Alzheimer's disease care as we are in the United States' (214). From its emergence as a theme of healthcare politics,

Alzheimer's disease posed similar challenges globally. In fact, what the present-day cultural and popular preoccupation with Alzheimer's disease refers to as the phase of Alzheimerization in the history of dementia took place in the early 1980s, and Honel's work is its earliest witness in multiple contexts: in healthcare, in the popular imagination and in medico-scientific discourse.[11] Honel's book contributed to the successful advocacy efforts that roused public awareness that change was needed in long-term care.[12] At the same time, it shows how the organic paradigm made its way into an emerging popular discourse.

The history of Frank Honel's diagnosis exemplifies how this advocacy effort turned concepts of ageing-related memory loss, so far termed 'senility' and 'dementia', into Alzheimer's disease in clinical practice. The initial diagnosis describes him as '"senile" and suffering from "hardening of the arteries"' (30). A few years later, a neurologist concluded 'that Grandpa did have senile dementia of the Alzheimer type' (164). To this Honel adds, in quoting from *The 36-Hour Day*, that the 'diagnosis confirmed what we had long believed. Parts of Grandpa's brain were being destroyed by some strange accumulation of "plaques" and "tangles"' (164). Mace and Rabins placed strong emphasis on the organic nature of the condition, when they contended that the 'changes that occur are not the result of an unpleasant personality grown old; they are the result of damage to the brain' (9). Such an assertion supported Honel in her conviction 'that Grandpa was not simply old. There was something wrong with his brain that made him forget so much and act so strangely' (90).[13] This explanation was in line with Butler's anti-ageist interest to counter public misconceptions about senility. With Katzman's political move in 1976, senility became related to a specific disease rather than ageing: to Alzheimer's disease as an organic condition of the brain.[14] In particular, Mace and Rabins wrote: 'The brain is a vast, complex, mysterious organ. It is the source of our thoughts, our emotions, and our personality. Injury to the brain can cause changes in emotions, personality, and the ability to reason' (21). They were using the medical model to preclude judgmental stigma about 'the unpleasant personality grown old'. Their emphasis on the brain as a 'vast, complex, mysterious organ' brought home to the reader how obviously the behaviour of patients had to change as a consequence of organic damage and how this change, therefore, had to be empathized with.

However, the 'empathetic technique' of *The 36-Hour Day* has been described as 'a two-part scheme': 'Only when family members ... have denied meaningful agency and subjectivity to the person with dementia, can these be reintroduced through the superimposition of their own subjectivity onto the experience of the other.' In addition, so the anthropologist Lawrence Cohen argues, the temporal language of the title of *The 36-Hour Day* conveys that the suffering is not that of the patient, who is here 'the agent but not the subject' of the disease.[15] These observations point us to the perspective from which Rosalie Honel was writing. She was closely watching the changes in personality of a loved and loving parent and grandparent, while having to bear the strain of caring for him. Honel draws us ever more into her burden, when, two-thirds into the text, she begins chapters in the present tense – a narrative strategy we also encounter in later life-writing such as John Bayley's *Iris: A Memoir of Iris Murdoch* (1998) and Andrea Gillies's *Keeper: Living with Nancy, a Journey into Alzheimer's* (2009). In addition, she places great emphasis on the chronology of events. Initially describing Frank as extraordinarily self-reliant and as focused on keeping active and useful, she ultimately depicts Alzheimer's disease as leading to Frank's progressive decline. After Frank's death, Honel ponders the times 'when he lost full human awareness and withdrew into a shell of pure physical existence' (240).

I emphasize Honel's use of the shell metaphor here not only because we already beheld it in the context of Galsworthy's writing. Its full implications were becoming apparent once carried into discussions on advance directives and living wills, which had to consider 'the relationship between the predemented self and the "shell" of that self left since the progression of dementia'.[16] Moreover, social scientists described 'an "unbecoming" process, due to the mental deterioration caused by the disease' in Alzheimer's disease, resulting in the patient being 'increasingly devoid of content'.[17] Notions of the patient's lack of self-recognition and failure to recognize others were reinforced by images of the living dead. These images, albeit in existence since the fin de siècle, obtained further meaning in a wider atmosphere of growing popular fascination with this concept seen in cultural texts such as George Romero's film *Night of the Living Dead* (1968).[18] The image of the patient as victim struck by a killer disease played into the same understanding of the patient as passive.

The idea of the patient as living dead dominated contemporary information campaigns. Such campaigns represented a 'health politics of anguish'.[19] They portrayed poster patients like Rita Hayworth, and in the early 1990s Ronald Reagan himself, as subjected to 'a devastating horror', in this way creating 'a new cultural demon'.[20] Celebrity faces successfully drew attention to the cause of Alzheimer's patients. But they were instrumentalized in a way that further stigmatized the condition in the mainstream.[21] Patients were not allowed to speak for themselves: they were presented as silent objects of pity without agency.[22] This agency-denying language rose to prominence once Alzheimer's disease had entered popular print media. An article in the *Washington Post*, for example, negotiated with Katzman's comparison of Alzheimer's disease to a killer, asserting that Alzheimer's 'changes our very soul, our very spirit. It lessens our humanity ... a person with a serious dementia is no longer human. He's a vegetable.'[23] A weekly news magazine reported on 'a slow death of the mind'.[24] In addition, a further caregiver guide propagated 'the loss of self', while claiming that patients 'have a great deal to teach us about living with the anguish of what one man called "a living death"'.[25] This contradiction in terms is embodied in the guide itself: it privileges, like *The 36-Hour Day*, the caregiver's perspective and experience.

This observation is noteworthy for the fact that a contemporaneous project paper set out principles of good practice for the care of older people with dementia.[26] It pointed out that 'failing mental mechanisms', the failure of coping strategies and the restricted power to communicate were stressful experiences for the patient (10). The paper asserted that, to 'understand the "experience of dementia", it needs to be seen from the elderly person's point of view – how it disrupts and complicates the process of living a full life' (9). Consequently, the authors matched action points to principles of care for the patient as an individual with human needs and rights (7–8). These action points were formulated in the first-person singular as claims and demands of the patient (14–23). Yet in 1985, family members still described 'care provided [in nursing facilities as] ignorant, indifferent, callous or even destructive'.[27] My argument here is not that scientific ways of thinking have influenced specific care habits. I merely suggest that certain habits and attitudes are symptomatic of a specific period, which is marked by particular scientific or medical perspectives on the condition. Take the following example: Sawako Ariyoshi's narrative was originally entitled *Kōkotsu no hito* (ecstasy man). But allusion to the patient's lived experience (and need to be active) was entirely overwritten in the American-published translation *The Twilight Years* from 1984; instead, the translation emphasized the caregiver's view of the patient's dissolution.

Caregiver burden and pain are intensely real at all times. But during the 1980s, they were highlighted at the expense of direct attention to the patient's personal experience. Both contents and layout of *The 36-Hour Day* confirmed the caregiver in their dominance within the carer–patient relationship.[28] Mace and Rabins certainly included information relevant to patients, recognizing that 'people, including those suffering from these conditions, may read this book' (xiv). But patients like Larry Rose, who published his illness experience in the mid-1990s, did not see their perspective reflected and needs considered in this manual, writing:

> *The 36-hour Day* was about the best one I read. There was a lot of good information there, but it was mostly for caregivers. I could not see where any of it applied to me. I am just not like most of the Alzheimer's patients I read about. I can still think, talk, walk, write and do almost anything I want.[29]

We do not gain the impression of such an independent and capable patient from Honel's account. She sees a child in her father-in-law. Among other things, she remarks that 'having Grandpa in our home at this time was like having another baby' (39). In a similar way, advice provided in *The 36-Hour Day* relies on notions of the patient as regressing into a second childhood: 'When we offer to help with something a person has always done for himself ... it is a strong statement that this person is not able to do for himself any longer, that he has, in fact, become like a child who must be told when to dress and must have help' (64). Notions of the patient as child may well originate from earlier medico-scientific as well as fin-de-siècle degenerationist concepts. But when we read from the perspective of the family caregiver we become aware of the full implications and added dimensions of this discourse. Subject to a 'change in roles', as Mace and Rabins wrote in 1981 (141), and Honel agrees (99), the adult-child caregiver loses a parent and, in turn, becomes, in Honel's words, parent to 'a two-year-old' (210). Engulfed by such feelings, Honel might find it difficult to give space to her father-in-law's own articulations.

Additionally, through a handful of Rosalie Honel's children's diary entries, we behold the pain that comes with Frank Honel's illness. Like her mother, Therese Honel believes that 'Grandpa is ... like a child' (73). But when she affirms that his 'happiness is the best thing about him' (76), she draws our attention more to her grandfather's remaining personality, rather than what the adult child – Rosalie Honel – sees as being lost. It is from the voice of a child – in Honel's account as much as in this cultural history – that we first learn of a positive aspect to the condition. Children share 'marginal positions within the family and society' with the elderly: they also both connect to cultural traditions of the idiot figure.[30] Ariyoshi's Shigezō, for example, turns to appreciating his daughter-in-law after having reprimanded her all his life. In a similar way, Pupi Avati's *Una sconfinata giovinezza* (A boundless youthfulness; 2010), though written significantly later, explores positive connotations of dementia within a parent–child relationship. Forster's *Have the Men Had Enough?*, by comparison, gives the narrative voice alternately to the caring daughter-in-law and her daughter. This strategy convincingly captures how the adult child's caregiver burden can lead to negative, identity-denying perspectives on the condition, because this burden obliterates appreciation of possible rewarding aspects of the condition reflected in the grandchild's thoughts – a perspective Honel (and many a caregiver after her) is able to take only after her father-in-law's death.[31]

The grandchildren's perspective emphasizes the patient's continued abilities rather than losses; its limited availability then reflects the origin of (and reason for) the popular beliefs about dementia in the early 1980s. Just as Rosalie Honel's voice prevails throughout the narrative, so too the caregiver's account about a lost former self took the lead in shaping a popular dementia discourse – which the media supplemented with images of patients as empty shell or living dead. This account was necessarily grounded in the caregiver's experience. But it also strongly related to the language employed in the limited support literature available. That said, caregiver advocacy in the form of printed narratives outside Honel's text and some paragraphs in the two manuals mentioned here, were rare in the 1980s. This means that caregivers during the 1990s not only relied on *The 36-Hour Day* (which by itself suggests the influence on the popular dementia discourse of such manuals). They themselves provided hands-on advice to fellow caregivers in their own writing, among them Carol Wolfe Konek's *Daddyboy: A Family's Struggle with Alzheimer's* (1991), Betty Baker Spohr's *Catch a Falling Star: Living with Alzheimer's* (1995), Frank Wall's *Where Did Mary Go?* (1996) and Ann Davidson's *Alzheimer's, a Love Story: One Year in My Husband's Journey* (1997). In this regard, Arthur Kleinman's *The Illness Narratives: Suffering, Healing, and the Human Condition* (1988) is exceptional in containing a brief account of how a spouse experiences Alzheimer's as 'a disease of the whole family'.[32] That the medical anthropologist included this condition in a work that would become canonical to the rising health humanities movement foreshadows the lead this discipline took in helping to reformulate the patient's and caregiver's position during the 1990s and early 2000s.[33]

The position of the patient eventually could change because those writing about them came to appreciate their family member's continued identity within a condition of the mind. This shift in emphasis also emerges from *The 36-Hour Day* as notions of the 'damaged brain' became, on several occasions, interchangeable with those of a 'damaged mind' (e.g. 46, 104). These developments suggest that, although practiced for over seventy years, the medico-scientific discourse of dementia as an organic condition of the brain, once imported into healthcare thinking, could not be maintained by those who experienced patient contact on a daily basis. As Cindy Honel wrote of her grandfather, the patient needed to be comprehended as a 'prisoner of the mind' (195), his changed behaviour explained by an illness afflicting the mind. This perspective is similar to Pinel's or Alzheimer's considerations, much in line with psychiatric thinking, and explored in even more detail in prominent fictional accounts of the 1980s.

Out of Mind: The postmodern novel delves into the mind of the patient

The ever-increasing presence of Alzheimer's disease in print media both suggested and stimulated the general public's growing interest in the condition. In 1979, Max Frisch forced the reader into the perspective of a first-person narrator with increasing memory loss in *Der Mensch erscheint im Holozän*, which appeared as *Man in the Holocene* in the *New Yorker* in 1980. In the same year, Samuel Beckett gave an account of an individual's end-of-life experience in the face of his 'mind never active at any time … now even less than ever so' in *Company*.[34] Here, I take a brief look at Beckett's text for two reasons: later Alzheimer's disease life-writing would refer to experiences Beckett described in his work; and Beckett's explorations of consciousness have implications for representations of dementia.[35]

Experiences expressed in these publications can be said to relate to a transition, during the post-war period, from modernity to postmodernity in the United States and Western Europe. This transition happened, so Jesse F. Ballenger observes, as modern concepts of coherence and ideas of selfhood, space and time became undermined by feelings of disorientation. Ballenger locates dementia patients between modern and postmodern selves, exploring aspects of fragmentation in Cary Smith Henderson's *Partial View: An Alzheimer's Journal* (1998).[36] The idea of fragmentation is central to how Beckett explores the protagonist's identity in *Company*. Also, while I do not suggest that Beckett's late prose represents a nosological account of dementia (i.e. one that is exactly true to diagnostic characteristics), its vision of human life is that of a postmodern condition. Dominated by incoherence, confusion and lack of control, this experience convincingly resonated with cultural images of dementia of the 1980s which I shall discuss shortly in this section.[37]

Beckett sets the 'old man['s]' (8) long-term memories against present-day 'repetitiousness' (9), and describes him 'as huddled with his legs drawn up within the semicircle of his arms and his head on his knees' (17). This portrayal reminds us of Alois Alzheimer's early description of the condition. In Beckett's work, shorter and shorter, less and less complete sentences depict 'mental activity of a low order. Rare flickers of reasoning of no avail. Hope and despair and suchlike barely felt' (29). Concurrently, we feel drawn into the protagonist's return to childhood. Short paragraphs of earliest childhood memories of 'a mother's stooping over cradle from behind' (31) are fused together with those suggesting the narrator's current lack of mobility, as he is 'crawling again and falling again. If this finally no improvement on nothing he can always fall for good' (31).

Again, I do not suggest that Beckett's *Company* describes an Alzheimer's patient. But I consider it possible that contemporary readers might have mulled over the protagonist's condition in this terminology – especially given how popular this term was becoming. Also, critics wrote of the protagonist as 'victim' within 'the physical confines of this prison'.[38] That said, if an association with Alzheimer's disease had been made, the imminent death of Beckett's narrator might have left too little room for the 1980s reader to relate to questions that began to be raised, namely: 'What do demented people experience? What does their condition mean to them? What is their reaction to it? What are their gratifications? What are their frustrations?'[39] The neurologist Joseph M. Foley asked these questions as late as 1992 and saw some of them answered in J. Bernlef's bestselling book *Out of Mind*. This fact and the narrative's broad international success invite a close reading of this text.[40]

Originally published in Dutch as *Hersenschimmen* in 1984, *Out of Mind* quickly gained international prominence. It featured in the 'Notable books of the year' section of the *New York Times* on 3 December 1989 and was translated into German as *Hirngespinste* (1986) and French as *Chimères* (1994). I am particularly interested in the narrative's initial Anglo-American reception following its early translation into English in 1988. Bernlef's novel adopted an unprecedented point of view. It narrated the experience of dementia exclusively from within the patient's mind. As will become clear shortly, the feelings Bernlef created obviously resemble the predicament of Beckett's protagonist. But what I particularly wish to explore is what this account reveals about the cultural image of dementia in the mid-1980s, at a time when a growing body of fictional texts centred around memory, post-war trauma and historical amnesia. I argue that the fictional deployment of dementia for the exploration of such memory imbued the condition with notions of extermination and post-war trauma and determined the use of Alzheimer's disease as a trope for memory loss and forgetting.

In *Out of Mind*, Maarten Klein is in his seventies when he realizes that

> year by year things happen to your body. Your feet lose their springiness. You go up and down the stairs once and you have to sit down to catch your breath again. Your eyes start to water when you look at one spot for a long time. The shopping bag moves more and more often from one hand to the other and you

meet fewer and fewer people's eyes. But this is different. More a general feeling of unease than a specific symptom. But no, it would be nonsense to think there is something really wrong. 'I'm still going strong!' (20)

At first sight, Bernlef tells yet another narrative of dispossession, decline and loss in old age. He pushes to centre stage the question of how demented people think and feel, by making it a felt experience for the reader. This felt experience goes further than the empathy solicited by Paul Bailey's or Norah Hoult's narrative strategies, because Bernlef exclusively confines the reader to Maarten's mind. Bernlef experiments with 'how much can be taken away from a human mind and still leave a narrator who can tell a tale comprehensible and attractive to the reader'.[41] He explores, as Steven Connor would say, 'the loose, the contingent, the unformed and the incomplete in language and experience'.[42] During a period concerned with questions of selfhood at the end of life, this experiment pushed the reader to ponder psycholinguistic implications of denying patients' continued identity once they cannot verbally articulate themselves.

In the absence of chapter headings or distinct line spacing the reader is exposed to the patient's uninterrupted stream of thoughts. The only hiatuses are nine section breaks, marked by their italicized respective first sentences. These sections, in turn, consist of an overwhelming number of longer and shorter paragraphs, whose variable lengths depict Maarten's fluctuating attention span. Across the entire text, paragraph length also mirrors Maarten's decreasing ability to narrate coherently. While the initial paragraphs stretch to up to four pages, the final sections compress the protagonist's narrative (and actual) dying in paragraphs as short as four or fewer lines – a strategy taken up later in life-writing like Thomas DeBaggio's *Losing My Mind: An Intimate Look at Life with Alzheimer's* (2002). The only landmarks of orientation (for the reader just as much as the patient) are Maarten's conversations with his wife Vera, as he 'is begging her to understand something I do not understand myself' (38). These conversations with Vera are the anchor through which we understand what actually happens, one whose representation is grounded in the caregiver-centred approach of the 1980s. But as Maarten becomes less and less able to communicate, Vera's words turn into disconnected echoes in Maarten's mind, and eventually disappear with Maarten's fading awareness, identity and personhood.

Primarily, Bernlef uses the protagonist's loss of language to characterize the patient's dissolution. Maarten Klein gradually loses his fluency in the English language, which he had learned to master when the Dutch couple moved to Boston after the Second World War. At first pleased that he can speak 'in fluent

English' (109–110), he later dissociates from his English-speaking environment, becoming aware that 'only English is spoken' (119). Once moved to a hospital environment he muses that 'the thought of an interpreter doesn't occur to them...I am the only survivor of my own language' (120).[43] The loss, also mirrored in the growing number of incomplete sentences as well as suspension points (i.e. '...') in the text, goes further as Maarten has to make a real effort 'to say "Vera", say it, VeraVeraVeraVeraVeraVera until I hear it...hear how my voice drifts away...gone is gone' (121). Such suspension points appeal to the visual imagination; they show 'the pauses, hesitations and endings of dementia', and they reflect how language becomes 'destabilised, fragmented and fractured for dementia sufferers'.[44]

Bernlef wrote his narrative during a period marked by a growing body of psycholinguistic work. In the early 1970s, the literary theorist Luce Irigaray studied how cortical lesions manifested themselves in the patient's production and handling of language, and asked whether specific alterations in the use of language could help distinguish the demented from the psychotic or aphasic.[45] That a major literary theorist decided to bring dementia into her theoretical project indicates the broadening of disciplines interested in dementia during the 1970s, and further foreshadows the growing attention to narrative in dementia in the 1990s. The worsening of language skills also featured in the 'global deterioration scale (GDS) for age-associated cognitive [decline] and Alzheimer's disease', which Barry Reisberg developed in the early 1980s. On this scale, stages ranged from 'word and name finding deficit[s]' to the point when 'all verbal abilities are lost'.[46] Bernlef's narrative suggests that with the ability to articulate himself in spoken word severely curtailed, Maarten's self-awareness is quickly lost. The first-person pronoun disappears from the text, as he becomes 'a thin, transparent point in space' (125). The second-person pronoun then strips Maarten of awareness of his identity as a gendered person, completed once 'guards remove this body ... [and] take it to a space where there are beds' (129).

Maarten's perceptions resonate with the feelings of W. G. Sebald's Jacques Austerlitz, whose 'entire life sometimes appeared like a blind spot without continuity', and who 'could not imagine who or what [he] was'.[47] The purpose of Sebald's *Austerlitz* (2001) is bent towards the protagonist's recovering of childhood memories in search of his parents' fate in Nazi-German occupied Czechoslovakia. But his identity search remains as tentative as the reader's contact with him, which is mediated through the monologue of an unknown narrator talking about him in a text even less structured than Bernlef's. *Out of Mind* and *Austerlitz* both emerged during a period marked by an 'explosion of memory discourses'.[48] Sebald's narrative, in particular, has been considered 'a literary meditation on the workings, and failure, of memory – a literary reflection on possible modes of Holocaust memory'.[49]

Looking at Bernlef's novel from this angle illustrates how post-war memory discourses added further meaning to dementia. Their language directly related to the imagery of the already existing sociocultural dementia discourse. For Bernlef, dementia becomes the killer that medico-scientific writing had coined and which the popular press perpetuated. More precisely, Bernlef links the long-standing medico-scientific idea that dementia enhances the vividness of long-term memories at the expense of short-term recall to the understanding of dementia-related hallucinations. In this way, he makes Maarten victim to the most troubling and unsettling experiences of his youth, which had been lived in the face of the atrocities committed by Nazi Germany during the Dutch occupation. In hallucinatory associations, Maarten connects present events to past experiences. He mistakes a tranquillizer injection for a liquid food infusion during war time (73) and later for a Nazi's attempt to kill him (110). Such memories visit the patient with increasing frequency and intensity. This process matches the uncontrolled remembering experienced by Saul Bellow's Artur Sammler, a Holocaust survivor, who moved to New York City to forget this 'dead life'.[50] It also compares to Anton Steenwijk's attempts to forget the assassination of his family by Nazi Germans in Harry Mulisch's *The Assault* (1985).[51] Steenwijk organises his life around '[burying] all that, the way one buries the dead' (61). Yet, his subconscious and chance encounters with others who had been involved in the crime keep confronting him with his past. This very combination of memory triggers matches both the neuropsychological structure and the narrative purpose of Maarten's hallucinations.

Bernlef's war metaphor turned the 1980s dementia patient into the image the NIA had propagated. But it did so by comparing the experience of dementia to that of the Nazi-German occupation and, by extension, the Holocaust, at a time when the fortieth anniversary of the end of war in Europe was being commemorated. Reliving memories of having 'been liberated' (109) towards the end of the novel, Maarten realizes that he has been 'occupied from within…my liberators have occupied me' (124). The condition which has besieged him will lead to his death, and only in his death will he find freedom – like Steenwijk, who reckons that 'life on this planet was a failure … Not until it ended, and with it every single memory of all those death throes, would the world return to order' (146). Dementia has become the equivalent of both torture of the mind and post-war trauma. Patients like Thomas DeBaggio would later compare their condition to a 'death sentence' and the 'holocaust of my brain' to evoke the most atrocious objectification of human life and characterize themselves as victims of a terminal disease.[52]

Bernlef's novel ends as 'spring … is about to begin' (130), the season symbolizing the renewal of life. This ending provokes the question as to whether the narrative professes a positive view of dementia, given that Maarten ponders 'can this be called progress while in fact it is regression' (125). This question seems pertinent also because the unidirectional nature of the disease pathology would make it difficult for patients to tell triumphalist narratives in the apparent absence of a possibility for satisfying closure.[53] Maarten's statement mirrors the kind of 'crabwalk' Günter Grass explored in much of his later writing, namely his narrators' 'steal[ing] ponderously back and forth across characters, plot, and time'.[54] Maarten Klein's dementia is Bernlef's appropriation of the 'crabwalk' motif, that is, Klein's – as much as Sammler's, Steenwijk's and Austerlitz's – negotiation with the past and his integrating this past into the present.

Bernlef's use of dementia to explore historical amnesia fills the cultural image of the condition with traumatic notions of war, murder, resistance and extermination. At the same time, Bernlef could call on an entire generation's shared understanding of memories to engender empathy for a patient's illness experience. And this experience relates to both the pain and loss remembered and the fragility of identity in the present. As early as 1983, American pastor A. Ralph Barlow likened the effect on family relationships of Alzheimer's disease to a sense of disconnectedness in modern society – a perspective that illustrates how Alzheimer's disease became a metaphor for social ills during the 1980s.[55] In fact, the major societal anxiety about the ageing population of the 1960s and 1970s had, by the 1980s, turned into a concern for dementia and Alzheimer's disease. For both Bernlef and Galsworthy, dementia reflected large-scale societal concerns. For Galsworthy, dementia was an image. But at the time of Bernlef's writing, dementia had turned into a reality in need of images. In then using the condition as an image in the context of an entire generation's post-war trauma, fiction like Bernlef's prepared its use as a global trope for memory loss and forgetting.

In our efforts to trace discourse developments we also need to be clear about where Bernlef locates dementia in relation to ageing, and how Maarten Klein's dementia differs from Faith Gadny's condition. Bailey's Faith continues to be aware of herself and rejects prejudices about the ageing. Her initial positive self-image, as opposed to how she is perceived by others, resonates with Woodward's observation that there exist 'two contradictory representations of old age – one which is aggressively negative ... and is projected onto the old by the middle-aged, and one which is fundamentally positive and is *embodied* in the figure of an old person'.[56] Faith's is the lived awareness of what Butler defined as ageism, and her loss of memory is Bailey's way of depicting her surrender to this ageist discourse. By contrast, Maarten is only initially aware of ageist behaviour, noting the 'tone they usually adopt here when they address someone over sixty. Amiable condescension mingled with distaste' (52). Likewise, he loses awareness of his bodily self, becomes a stranger to himself, as he takes his reflection in a window pane for an 'old man in pyjamas look[ing] at me, imitat[ing] a live man with his hollow black eyes' (112). Bernlef's Maarten appears to suffer from a condition of the mind only. Bailey's Faith, by comparison, remains aware of her condition as an embodied person throughout. This gendered representation aligns with one of the main differences between the male and female bildungsroman, namely the 'heroine's manipulation of the gaze of others as well as her control over her self-image'.[57] And it anticipates how especially female patients will rise to the challenge of writing about a condition that is perceived as afflicting the male associated mind.[58]

We have come across the idea of the patient as stranger and 'other' already in the fin-de-siècle social Darwinist discourse. The abjection of the patient typical of that period was less prominent in the 1980s version of the image of the patient as stranger – both in fiction and the popular press.[59] By the time the patient was cared for in the family, the personal relationship with that patient came to structure the image: the struggle of family members with 'their commitment to the person they knew and their increasing detachment from the stranger that person has become'.[60]

Two decades after Bailey, Bernlef is only marginally concerned with dementia as a condition of ageing, the protagonist concluding, 'You become forgetful. It's part of old age' (20). Also, Bernlef does not aim to bring out dementia as a nosological entity, as authors would aim to do two decades later. Although the narrative's contemporary reception was quick to describe it as 'a novelistic study of Alzheimer's', it really only prepared the ground for dementia to become a narrative device several years later.[61] But Bernlef succeeded in making the experience of dementia convincing by building on two things. First, he could count on the reader's familiarity with the dementia storyline, which after nearly a decade of information campaigns had reached general awareness. With Bernlef's text, the reader, for the first time, could be forced into the patient's point of view without losing the plot in Vera's disappearance. And second, Bernlef built on the narrative possibilities of modernist and postmodernist approaches to memory loss. In locating the narrative in the patient's mind, the condition became, on a cultural level perhaps earlier than on a medico-scientific level, one of the failing mind.

Literature creates the terms in which we think about phenomena like decreasing self-awareness and consciousness, and I would not even want to start arguing for the dominance of fiction or linguistics in establishing that, as Paul John Eakin has argued, serious impairment in the ability to tell or understand stories severely impacts on our sense of self.[62] Suffice to say that Bernlef's depiction of Maarten's continued awareness of his state follows and gains authority from the emerging (though debated) 'evidence of awareness and sensitivity' in advanced Alzheimer's patients.[63] Yet, it remains a nagging thought that Bernlef's composition might be flawed due to 'some obvious inconsistencies in the narrative techniques'. Alexander Zweers finds that the reader 'accepts that even within the heavily damaged brain of the protagonist[ ] an observing instance can "somewhere" remain active'.[64] And for Anita Desai, the reason why 'we willingly suspend our disbelief in Maarten's ability to narrate so coherently his growing incoherence' lies in Bernlef's prose: it makes us 'participants in [Maarten's] tragedy'.[65]

Be that as it may, with reference to Bernlef's text, the neurologist Foley acknowledged that 'clinical experience is never as deep as fiction'. His contribution to *Dementia and Aging* (1992) virtually invited patients to make their own claims and come out with their narratives, in order to contribute to policy changes:

> It is important to identify functions that are lost, but even more so to identify functions that are preserved. Dementia per se does not always deny patients the right to participation in decisions about their own care, their own life, or their own health ... we must recognize that individual demented persons ... each remain a person, with their own gratifications and frustrations, their own unique background, and their own unique destiny.[66]

Such an exhortation is remarkable, especially since, during the period from 1991 to 2001, not a single popular magazine article focused on the perspective of the individual who experienced the disease.[67]

At the same time, the growing need to hear and see the condition more from the patient's perspective is perhaps attested to most clearly by the 1993 television film by Tony Harrison, *Black Daisies for the Bride*. Combining drama, music and poetry, this documentary traced the personalities of three women patients in the Alzheimer's ward of High Royds Hospital in West Yorkshire. The film propagated the image of the patient as 'baby', 'victim', 'shell' and 'lost self', and the condition as a 'blizzard'.[68] In frequent above-eye-level takes, it subjected patients to the viewer's scrutiny without definite consent. In addition, a voiceover conveyed the caregiver's perspective, further removing agency from the patient. Considered as a whole, Harrison's work was probing, as Lucy Burke puts it, the boundaries between 'viewer and viewed, subject and object ... carers and patients'.[69] It was also the first visual narrative about the patient – the caregiver not featuring centre stage in the film recordings themselves, perhaps in response to first-patient narratives appearing in book form from 1989 onwards.

Robert Davis's *My Journey into Alzheimer's Disease: Helpful Insights for Family and Friends, a True Story* (1989), Diana Friel McGowin's *Living in the Labyrinth: A Personal Journey through the Maze of Alzheimer's* (1993) and Larry Rose's *Show Me the Way to Go Home* (1996) pioneered the life-writing of Alzheimer's patients, at a time when the public considered them devoid of a social self and victims of a killer disease. Reflecting the social isolation of patients, the early reception of these texts almost exclusively focused on how their narrators strove to assert their identity as individuals.[70] These patient-authored publications appeared not only in response to an ageist and caregiver-centred healthcare dementia discourse. They also developed a counter-narrative in response to fast-paced medico-scientific research which had begun to revolve around Alzheimer's disease from the early 1980s onwards. Butler's original desire to move the care for, and curing of illnesses of, the elderly centre stage turned into a development that stigmatized older people even more. It has, indeed, been argued that the response of the healthcare system itself to the increasing older population was partly responsible for catastrophic views of ageing.[71] The early 1990s chastised the curative medicine of the 1980s for not prioritizing enough quality over quantity of lived years at a time when the number of elderly kept increasing.[72] In the tradition of the grey lobby of the 1970s, which advocated attention to the needs of older people as such, patients would assert themselves increasingly during the 1990s.[73]

Since the 1960s, medico-scientific research into diseases of old age had steadily intensified, unnoticed by the general public. This changed in the mid-1970s, once medical scientists propagated the single designation of Alzheimer's disease for dementia regardless of the patient's age. In addition, NIA funding strategies, inherently focused on diseases of old age, popularized Alzheimer's disease because it was a disease that is common in old age.[74] Eventually, the NIA became the most important federal agency of Alzheimer's disease research.[75] These strategies and events contributed to medical sciences forming an independent entity. In their growing specialization, they cultivated, as I show in the next chapter, a dementia discourse removed from the patient as individual and separate from geronto-psychiatric initiatives which were beginning to sharpen their understanding of the ageing patient as needing empathetic care.

Part III

The cognitive picture

5

The narrative of loss in a growing biomedical and literary marketplace of Alzheimer's disease

Medico-scientific progress during the 1980s and 1990s continued to advance the idea that Alzheimer's disease was an organic condition of the brain. The neurosciences consolidated the turn-of-the-century concept of degeneration in the biomedical dementia discourse, dressing it up anew as neurodegeneration. What is termed the period of biomedicalization, in fact, relies on the same principles of investigation and focuses on the same neuropathological features as had been relevant to Alois Alzheimer and his immediate colleagues. The efforts of three medical scientists to translate, during the 1980s, the research papers pertaining to 'the early story of Alzheimer's disease' suggest to me that the research community itself appreciated this continuity.[1] Following the huge investment by the NIA in Alzheimer's disease, much of what had been research on dementia until the late 1970s turned into research on Alzheimer's disease from the 1980s onwards.[2] *The Alzheimer's Disease Standard Reference*, edited by the old-age psychiatrist Barry Reisberg in 1983, convincingly documents this continuity.

Emerging lines of biochemical and genetics research now influenced the characterization of the neuropathological features (especially cellular degeneration, plaques and tangles) that Kraepelin had established as central to the disease pattern, and this characterization further cemented the narrative of degeneration, decline and loss as it involved images of the broken down architecture of brain cells. Throughout the 1950s and 1960s, electron microscopic magnification enabled the identification of senile plaques and neurofibrillary tangles at a structural level much finer and deeper than light microscopes had allowed for at the end of the nineteenth century. During the 1980s, these discoveries spurred the biochemical and, slightly later, genetic characterization of these features. Jesse F. Ballenger has placed the leading studies into their historical context, but I am inclined to disagree with his view that electron microscopy 'transformed the familiar and unproblematic pathology of dementia into something new and strange'.[3] Rather, I see this line of research as directly continuing the work of the 1930s and 1940s that described an organic condition of tissue loss and cellular degeneration based on microscopic observations. For example, Arne Brun's contribution to the 1983 reference compendium on Alzheimer's disease reads like a summary of this early-twentieth-century neuropathological work. Reviewing evidence on 'widespread atrophy', Brun underscored that special techniques bring out 'numerous senile ... plaques (SPs) and neurofibrillary tangles ... neuronal loss ... and blurring of cortical cytoarchitecture ... Affected neurons often appear to be in stages of terminal degeneration ... The number of SPs is also reported to parallel the severity of dementia'.[4] Medical scientists now set out to break these morphological features down into smaller and smaller units, pursuing a 'neuromolecular style of thought' based on the principle that mental states expressed themselves in neural processes that could be 'anatomized at a molecular level'.[5]

I will thus explore how the 'aggressive program' on Alzheimer's disease pursued by the NIA impacted on patient and illness perception throughout the 1980s and 1990s, as it accelerated research on the cause or causes of the disease and the identification of risk factors and their early detection.[6] My concern here is neither to judge the validity of various disease hypotheses nor to give a comprehensive overview of the various lines of investigation that medical sciences opened during the 1980s. Rather, I will focus on developments in relation to initial biochemical investigations and genetics research. In salient research papers of the 1980s and their popular scientific echoes, I will trace, in the first two sections of this chapter, how concepts of Alzheimer's disease as a neurochemical communication disorder and as an inevitable, because inherited condition deepened ideas of loss in relation to dementia. These developments, together with what David B. Morris called 'genuine warfare against the biology of old age', significantly influenced how Alzheimer's disease increasingly became understood.[7]

As I chart these developments, I specifically address where patients found themselves on a sociocultural level in the 1990s. At that time, images of Alzheimer's disease as a condition in old age prevailed in a growing marketplace – a marketplace of dementia accounts as much as for service users (by which I mean patients and their carers) – but emerging genetics research became engaged in identifying the cause of early-onset forms of dementia. This question is crucial because the 1990s saw a surge in popular science writing on Alzheimer's disease, which made scientific research all the more accessible to the general public. And, as I will address in the final section of this chapter, this public included a steadily growing number of caregivers, who, in their own writing, were negotiating their parents' forgetting at a time when memory became – as we have seen in the context of Bernlef's narrative – a key cultural and political concern.

Neurodegeneration: The biochemical narrative of lost molecules, pathways and communication

With biochemical investigations of the 1980s, Alzheimer's disease became linked to notions of dying neuronal networks. The growing silence and isolation noted in socio-medical and cultural descriptions of the patient in the 1960s found a biochemical echo in illustrations of dying neurons in the 1980s. These biochemical studies described the loss of specific molecules in the brain, termed 'neurotransmitters'. The emerging idea of a breakdown of cellular communication was paralleled by ever-finer descriptions of collapsing cellular architecture and networks. The language employed in these studies can be traced back to the nineteenth century.[8] But communication between cells as such came under scientific scrutiny only during the twentieth century.[9]

During the 1800s, lesion theories had fuelled enquiries into possible links between bodily functions and specific brain areas. With the rise of the neurochemical sciences in the 1960s, this thinking expanded to the molecular level. Neuroscientists identified distinct brain areas as being dominated by specific molecules – neurotransmitters typical for that brain area and produced in a set of cells prevalent in that area. A decrease in the amount or activity of the neurotransmitter was taken as indicating cell death in that brain area. It also served as molecular explanation for the loss of function of that brain area and, in consequence, the symptomatology of a disease. Alzheimer's disease research focused on cholinergic cells which are so called because they use acetylcholine as their neurotransmitter. In their choice of language, the first research papers that identified cholinergic cell death as central to the pathology established Alzheimer's disease as specifically a neurochemical communication disorder.

The first biochemical studies related to this field of enquiry were published in the mid-1970s. They detected changes in molecular activity especially in two brain regions: the cerebral cortex (which harbours our ability to think and reason) and the hippocampus (which is fundamentally involved in memory and learning). The authors reported that especially cholinergic neurons were lost in

Alzheimer's disease and concluded that the condition was 'a cholinergic system failure'.[10] In 1982, the American neurologist and psychiatrist Peter J. Whitehouse and colleagues gave further weight to this concept.[11] They located the loss of cholinergic neurons to a specific area at the bottom of the brain known as the nucleus basalis of Meynert (nbM). This is important, because the 'large neurons in the nbM project directly to the cerebral cortex [and] similar neurons … project to the hippocampus' (1237), that is, they connect to higher brain areas via elongated nerve cells. These observations led the authors to formulate the 'cholinergic hypothesis': that cholinergic cell death could explain impaired 'memory processes' and an altered 'cognitive activity' (1238).

These observations can be read alongside Laura Otis's research on nineteenth-century networking metaphors. This can illuminate how scientists came to use the projection metaphor and how it advanced their grasp of the condition. Otis's notion that nineteenth-century scientists' 'images of social bonds affected the way they saw those in the body' is directly applicable to the projection and connectivity concept Whitehouse and colleagues established.[12] Their metaphor had been prepared in the socio-medical and – if we think, for example, of Bailey's Faith Gadny – literary realm during the 1960s. The concept of the lost network conveyed knowledge on two levels: the patient's loss of social integration and her cerebral dissolution. I would go further and suggest that these concepts provided the basis for a neuroscientific understanding of the patient's loss of identity, joining Otis in her conclusion that 'then, as now, the [networking] metaphor invited all who encountered it to rethink their identities'.[13] Based on these new insights, cognitive disturbances were beginning to be linked to communication issues, as, for instance, in the language of neurologist David A. Drachman: 'Interruption of this neuron–neuron interaction may result in decreased connectivity, resulting in *irretrievable "atrophic" loss of information.*'[14] In other words, the loss of neurons became equated with the loss of contents on a cognitive scale, a move quickly reflected in cultural images of dementia, for example, Michael Ignatieff's novel *Scar Tissue* (1993).[15]

Ignatieff's narrator links 'the circuits … going down' to the gradual loss of his mother's articulation and her ability meaningfully to communicate with him. He observes: 'Every day I come in here and one more has gone. She could walk three months ago. She could eat her food three months ago. She could talk. Now the words are going' (160–1). From there, it is a short step for him to explain to himself that she 'has left her self behind' (161). For Bailey, dementia was a condition of lost identity because of lost social connectedness. For Ignatieff, it is a condition of lost identity because of the lost ability to convey intelligence. This understanding emerges from the narrator's meeting with Moe, a patient with amyotrophic lateral sclerosis (ALS), a form of motor neuron disease. Usually a fast progressing terminal neurological disorder, it exclusively affects voluntary muscle control, including the ability to speak, but leaves mental acuity intact.[16] Linked to a computerized system, Moe tells the narrator of wanting to remain conscious.[17] The narrator appreciates that Moe wants 'to see everything, to feel everything, to hold on to his awareness of life until the very last moment; and to him this consciousness – whatever its desolation – is worth fighting for to the last instant' (138). Moe is fully aware, integrated, connected. His continued awareness is the cognitive triumph over the silence of the narrator's mother – her Alzheimer's disease.

I want to linger on this difference between Alzheimer's disease and ALS, because it may help us better appreciate the persistence of communication breakdown metaphors in the popular dementia discourse. The sharp contrast fashioned in Ignatieff's novel between ALS as an illness of the body and Alzheimer's disease as a condition of the mind has become commonplace in cultural notions of each condition. As late as 2004, Jonathan Weiner's popular scientific portrayal of ALS, *His Brother's Keeper: One Family's Journey to the Edge of Medicine*, builds around the sharp contrast between the mind-defined ALS patient Stephen Heywood and the author-narrator's mother with Alzheimer's disease. Similarly, novels such as Michelle Wildgen's *You're Not You* (2006) depict ALS patients as autonomous. Kate reminds her caregiver Bec of her role as amplifier rather than interpreter, averring, 'This is me ... You're not you right now ... Remember it's me talking.' Her ultimate intellectual autonomy and control are portrayed in her requesting Bec to omit calling for help as she suffocates and dies.[18] Also in Hazel McHaffie's *Right to Die* (2008) a highly intellect-driven ALS patient ultimately devises his own assisted suicide, making close friends and family accomplices without their knowledge.

The neurochemical view of lost networks and communication alone cannot account for the medico-scientific import of loss into contemporary ideas of Alzheimer's disease: the question of intellectual control is essential. This will become evident from looking at narrative representations of Parkinson's disease. The neurochemical model of disrupted networks, pathways and projections directly applies to Parkinson's disease, which progresses at a rate, and afflicts a population, similar to that afflicted by Alzheimer's disease. Diagnosed with Parkinson's in 1989, Joel Havemann tells his story in *A Life Shaken: My Encounter with Parkinson's Disease* (2002).[19] The senior editor of the *Los Angeles Times* describes how Parkinson's 'disrupts the brain's circuitry' (35) and expands from brain to body. He acknowledges the 'relentless march' (113) of his illness, pictures it as a 'robber' (108, 158) and alludes to the prospect of 'slowly slid[ing] away' (113), as 'the disease's victories' force him gradually to give up his independence (104). These are the notions of decline and dispossession we know from the cultural Alzheimer's narrative. But most revealing is Havemann's awareness of the 'threat of dementia' (114), given that some forms of Parkinson's co-present with dementia. 'I count myself lucky', he writes, 'that my brand of Parkinson's disease is the least destructive kind. And I regard myself as luckier still that insufficient dopamine production in the substantia nigra affects mostly movement rather than mental faculties. The symptoms of many other chemical imbalances in the brain are mental' (44).[20] Alzheimer's disease is Havemann's first example, and two consecutive chapters starkly contrast his excelling in work with how his mother-in-law's 'mind had permanently deserted her' (69). In this he agrees with other Parkinson's patients, among them Sandi Gordon, who feels as if 'my body were moving in slow motion. My mind was racing ahead full speed'.[21]

It is a widespread feature of cultural representations of Parkinson's disease that it is explored as a condition of control and balance easily lost. Havemann is not the only instance of a writer reminding us of his continued effort to 'control' the disease (122, 150), to ensure he remains 'self-sufficient' (157). In Jonathan Franzen's *The Corrections* (2001), the tyrannical patriarch Alfred Lambert glides into the 'relief of irresponsibility'.[22] Or take Yaron Zilberman's movie *A Late Quartet* (2012). When the string quartet's eldest member, Christopher Walken's Peter, has to retire from his role as grounding cellist due to Parkinson's disease, the ensemble comes close to dissolution; it is Peter's wisdom that saves the group's coherence. What emerges from these reflections is that identity and self were considered intact when enacted by an autonomous and meaningfully communicating person within a social network.

In summary of these comparative readings, as the twentieth century came to a close, Alzheimer's disease was considered a condition whose cognitive component meant the patient's lost control and living a social death. Images we have come to know from socio-medical and cultural descriptions of the 1960s were bolstered by neurochemical insights during the 1980s. Soon they would be underpinned also by popular scientific descriptions, as I will demonstrate in one of the next paragraphs.

Thinking in terms of the trajectory of the disease discourse, it is crucial how firmly these concepts were grounded in the neuropathology of the condition. Whitehouse and colleagues reported findings in brains selected based on the coincident detection of the very behavioural aspects and the pathological features defined by Alzheimer and propagated by Kraepelin: neurofibrillary tangles, senile plaques and degeneration of cells. I stress the continued interest in these features here, because the language that began to develop around them at this point further consolidated the concept of a neuronal communication disorder; it fed on, and further bolstered, concepts of system failure and network breakdown. This partly emerged from an understanding that the death of cholinergic cells was associated with neurofibrillary tangle formation.[23]

By the mid-1960s, electron microscopy studies had described neurofibrillary tangles as bundles of filaments.[24] Scientists now wanted to understand what these filaments were made of and, in 1986, identified 'a normal brain cytoskeletal protein' (termed 'tau' or τ) as an important component.[25] The involvement and impairment of a cytoskeletal protein (a molecule stabilizing the architecture of a cell) in Alzheimer's disease appeared to explain, in structural as well as functional terms, Drachman's earlier idea of atrophic loss of information. According to the American neurologist Dennis Selkoe and his colleagues, abnormal tau in nerve cells could 'result in instability ... loss of effective transport of molecules and organelles, and, ultimately, neuronal death', an understanding formalized in the 'tau hypothesis' in 1991.[26] The representation of this hypothesis in popular science further solidified notions of Alzheimer's as a communication disorder. In *Decoding Darkness: The Search for the Genetic Causes of Alzheimer's Disease* (2000), for example, the neurogeneticist Rudolph E. Tanzi described the role of tau in terms that remind us of the language Otis identified in nineteenth-century scientific and social networking concepts:

> In its normal soluble state, tau is essential to a cell's infrastructure. In essence, it forms the cross-beams of a cell-wide span of railway tracks – filaments called microtubules – which are constantly laid down and taken up for the purpose of shuttling types of protein cargo around a cell ... Corrupted tau might mean the kiss of death for a cell.[27]

The epidemiologist David A. Snowdon went even further. He explicitly related these neuropathological features to social death, writing that the spread of tangles through different brain areas 'parallels, in part, the general pattern of the progressive loss of mental, physical, and social functioning that occurs in Alzheimer's patients'.[28] The further investigations of senile plaques, by comparison, eventually led to the evolution of the medico-scientific dementia discourse on a quite different level; but like neurochemical studies they further sharpened perceptions of the patient's loss and involution.

On genes and genealogy: The patient as specimen, carrier and type in research and popular science

By the mid-1970s, both light and electron microscopy-guided studies of plaques had revealed that they contained amyloid.[29] This insight led several laboratories to apply themselves to finding the molecular source (i.e. the precursor) of the amyloid-forming peptide. They did so in the conviction that amyloid beta was causing brain cell death in Alzheimer's disease.[30] This search soon centred on genetic investigations.[31] Throughout the 1990s, these studies identified gene mutations that could either inevitably cause Alzheimer's disease or substantially increase the susceptibility to the disease. Amyloid beta played a central role in these studies. In the language of contemporary medical and popular science, it became the molecular culprit of the killer disease, which the NIA had placed under investigation fifteen years earlier. I believe that this depiction of amyloid largely explains what Margaret Lock described as the 'staying power' of the amyloid hypothesis in discourse developments; and it perpetuated, as I argue in this section, the narrative of Alzheimer's as a killer disease and the patient its victim.[32] This narrative was further enhanced as genetic testing potentially transformed every individual into a classified entity.

During the late 1980s, the genetic information for amyloid beta was matched to the chromosome, which earlier work had identified as holding a defect causing familial Alzheimer's disease.[33] A chromosome is a packed structure in the nucleus of the cell that holds the genetic instructions of an organism, the DNA. A defect or mutation can alter or delete part of the information held on a chromosome and lead to pathological manifestations. For example, the amyloid precursor might suddenly be processed differently, yielding a form of amyloid (amyloid beta) that, instead of remaining soluble, deposits as plaques. In 1991, researchers around the British geneticist and molecular biologist John A. Hardy identified such a mutation in the DNA of subjects in whom early-onset Alzheimer's disease had been confirmed after autopsy.[34] These findings contributed to the formulation of the 'amyloid cascade hypothesis', which claimed that amyloid beta was 'the causative agent' of Alzheimer's disease.[35]

The mass media response to Hardy's findings had already picked up on this: a headline in the *New York Times* in February 1991 promised that the causes of Alzheimer's disease would soon be found. The article told a narrative of genes doing 'their deadly work' and of 'sticky balls [of amyloid beta fragments] outside dead and dying nerve cells in the brain', closing on Selkoe's belief that 'blocking of amyloid will be a therapy'.[36] The popular scientific perspective on these research endeavours amplified the idea of amyloid beta as the molecular version of Alzheimer's, the killer disease. For Tanzi, amyloid fragments were 'the toxic killer of neurons' (202). Most importantly, researchers themselves now adopted the figures of speech for this very molecule that the media previously had used for the disease as a whole: the amyloid precursor protein turned into a 'robber' (108).[37] The metabolic product resulting from a specific genetic alteration became the molecular surrogate of the condition itself. This development can be explained by the fact that, as José Van Dijck put it, the gene 'became reified as the central locus of vital activity, and molecules other than DNA were ... ranked as secondary'.[38] Genetics research thus appeared to be able to hunt down amyloid to the innermost core of the patient's body.

Geneticists themselves framed their investigations in this way. In the early 1990s, two further genes were identified – presenilin 1 and presenilin 2 – as they presented in a form of Alzheimer's disease with onset in the forties.[39] Selkoe introduced presenilin as a further 'guilty gene', which highlights how geneticists pitched their work in terms of 'a classic detective story' or an 'adventure plot' for the benefit of a general audience, on whose faith further public funding relied.[40] Rudolph E. Tanzi expanded on the researcher's 'detective work into Alzheimer's' (xvii) and the 'hunt' for genes and proteins (106, 130), and the neurologist Daniel A. Pollen described, in 1993, the ongoing work on presenilin 1 as 'the quest for the genetic origins of Alzheimer's disease' in the subtitle of *Hannah's Heirs*.[41] Centrally involved in studies on inherited forms of Alzheimer's disease during the 1990s, Pollen described the family tree and the autopsy information of Hannah's descendants as 'a map for deciphering one of the genetic defects of familial Alzheimer's disease' (133). But with respect to the meaning and desired effect of the map metaphor, Van Dijck observed that it suggests 'it is only a small step from ... locating genetic aberrations to correcting those diseases'.[42] Indeed, in an engaging and generally accessible register, Pollen drew parallels between research on Huntington's disease (a familial, incurable neurological disorder that affects both motor function and cognition) and Alzheimer's disease, writing that 'in Huntington's disease, just as in familial Alzheimer's disease, many of the primary steps in seeking a cause and then a cure came from the efforts of the families actually involved' (120). With Bruno Latour, we should consider Pollen's popular science publication a functional part in the networking strategies with which investigators reach outside the laboratory to garner support for their research activities; a book like Pollen's would particularly target those who help in the 'belief and spread of the facts'.[43] As will become clear in one of the next paragraphs, Pollen's writing contributed to the public learning about genetics research on Alzheimer's disease, and it fostered belief in the potential of this new research strand to identify the cause of the condition. Focused on these goals, it was perhaps unavoidable that such presentations also triggered expectations for a cure, which pharmaceutical research has, to the present day, not been able to meet.

Pollen distinguished between individuals 'who escaped the defective gene' (34) and those 'carrying the abnormal gene' (65). Such terminology reinforced the idea that Alzheimer's disease engulfed the entire body of the person who, in Tanzi's words, was therefore 'genetically doomed' (38). This notion is also explained by the effect an ever-deeper molecular gaze would have, if we follow Foucault's claim:

> The gaze plunges into the space that it has given itself the task of traversing ... In anatomo-clinical experience, the medical eye must see the illness spread before it ... as it penetrates into the body, as it advances into its bulk, as it circumvents or lifts its masses, as it descends into its depth. Disease is no longer a bundle of characters disseminated here and there over the surface of the body ... it is the body itself that has become ill.[44]

Further findings supported the idea that Alzheimer's disease pervaded the entire body. By the mid-1980s, amyloid peptide was found in blood vessels and cerebrospinal fluid, and researchers began to identify a relationship between pathological changes in the brain and metabolic events in the periphery that might mirror these changes. Serum, blood constituents and even skin cells appeared to grant access to the understanding of central nervous processes, in the form of biomarkers of and diagnostic tests for signs of its culprit causes. By 1994, predictive genetic testing for Alzheimer's disease became a possibility. But given that no treatment was (nor yet has become) available, such tests only produce uncertainties.[45]

Charles P. Pierce's *Hard to Forget: An Alzheimer's Story* (2000) explores these feelings of uncertainty.[46] It reveals the consequences for identity and personhood of the fact that genetic knowledge about Alzheimer's disease 'outstrips' the availability of pharmacological tools.[47] The American writer gives an account of his 'family's story' – his father was diagnosed with Alzheimer's disease in 1985 and the father's four siblings developed it, too. Pierce's narrative is the search for medico-scientific evidence for inherited forms of Alzheimer's disease that would be of potential relevance to his own prognosis. The writer sees his own future as determined by paternal nature and nurture as well as the medico-scientific response to such determination:

> I saw in my father my walking future. It stepped into every corner of my life, like the subtle spread of a cold fog. We're all a little wary of the ways in which we find ourselves becoming our parents ... I may move through the stages myself – a specimen, until I am a lower primate, until I am a brain down the hall. My father was a specimen. Then he was a lower primate. Then he was a brain down the hall, not far from Dan Pollen's lab. (97)

Pierce draws heavily on what he calls Pollen's 'definitive account of the search for the Alzheimer's gene on chromosome 14' (203). This search relied, as Pollen detailed, on brain tissue analysis and the securing of 'specimen[s]' (169) from descendants of the same family tree. Pierce's reference to a lower primate might remind us of post-Darwinian ideas of degeneration. But its connection to medico-scientific specimens essentially suggests concern with objectifying practices.[48] Subjected to procedures of measurement, the patient turns into what Hacking termed 'an object of knowledge', knowable in genetic terms.[49] Pierce eventually decides against being tested, concluding: 'I am a specimen ... For all practical purposes, for all the good it will ever do me, I already know all I need to know' (173). Having Alzheimer's disease cannot be reduced to a gene locus nor can it make an identity.[50]

The genetic model drastically impinged on the patient's understanding of identity and individuality. This comes into sharp relief when we read Pierce's decision against Jean Baréma's in *The Test: Living in the Shadow of Huntington's Disease* (2005).[51] Aware that Huntington's disease ran in his family, the journalist Baréma saw himself confronted with a 'time bomb ticking away that will inevitably explode' (14). He faced 'a fifty-fifty chance of becoming a vegetable within the next few years' (32) and of being treated like a 'regressing adult' (45), while his brother and sister 'are slowly dying' of the same condition (75). Baréma's rendering of Huntington's typifies many of the cultural images we relate to Alzheimer's disease. Both Baréma and Pierce write memoirs that belong to a new generation of illness narratives which, according to Anne Hunsaker Hawkins, move away from 'medical science as invincible in its march to eradicate disease'.[52] Pierce foregrounds the conceptual impact the disease label has on his father's as well as his own social situation. Understanding this label in terms of 'three of the genes ... [that] are death sentences' (94) turns being tested for Alzheimer's disease into what Baréma describes as 'the life test. The death test' (96). As cures for Huntington's (as well as Alzheimer's) disease continue to be elusive, similar accounts persist to the present day, patients describing their diagnosis as a 'curse'.[53]

Ethical concerns regarding genetic testing came to be considered even more pressing once researchers identified mutations other than those predicting inevitable early-onset forms of familial Alzheimer's disease. The risk of having inherited an increased susceptibility to developing the condition at a later age became a matter of concern, channelling attention again towards Alzheimer's as a disease of older people.[54] This development was facilitated by epidemiological research. Epidemiologists maintained that, next to genetic factors, high levels of mental and emotional stress as well as environmental depletion could present additional risk factors. The healthy ageing discourse that developed in the wake of this work pitched brain health as a life project, but it also sustained the view that Alzheimer's disease was failed ageing.

David A. Snowdon, in 1990, attracted significant funding from the NIA for a large-scale epidemiological study with nuns.[55] The study added to the body of work that reflected a growing preoccupation with cognitive performance towards the end of the century. Its central claim was that a high level of linguistic ability (in this study reflected in idea density and complexity in writing) in early life can buffer cognitive decline; that, in reverse, lower levels of linguistic ability early in life could predict 'poor cognitive function' later in life (528). Patients indirectly were made responsible for their own later disease. And this led to the false hope that prediction could be tied to prevention.

The Nun Study dominated epidemiological Alzheimer's disease research throughout the 1990s. But it was Snowdon's move, with *Aging with Grace: The Nun Study and the Science of Old Age: How We Can All Live Longer, Healthier and More Vital Lives* (2001), to offer a popular account that turned 'cognitive reserve' into what the general public had to appreciate as a powerful asset against developing Alzheimer's disease. In popular terms, having or not having 'brain reserve' determined the separating out of future patients from the future healthy population, because 'a stronger brain has more of a reserve' and is 'more efficient, with better processing capacity' (45). Inevitably, the reader would have to picture the brain of an Alzheimer's patient in 'lesser', 'weaker' and 'poorer' terms. In another project related to the Nun Study, Snowdon carried this polarization further. He extrapolated from observations related to Alzheimer's disease to the healthy ageing of those with 'positive emotions' (194), and his eventual emphasis on 'longevity and successful aging' (216) seemed to insinuate that Alzheimer's disease was a condition of lacking success earlier in life and of failed ageing.

Cognitive reserve dominated public health initiatives that promoted healthy cognitive ageing since the mid-2000s. These initiatives went hand in hand with the promotion of brain health by the Alzheimer's Association, which in turn contributed to Alzheimer's disease becoming a major public health concern in the new century.[56] Concurrently, more and more research and popular scientific outputs concerned themselves with questions of lifestyle and diet as determining cognitive health in ageing. *The Myth of Alzheimer's: What You Aren't Being Told about Today's Most Dreaded Diagnosis* (2008) by Peter J. Whitehouse is perhaps the most prominent example. It encouraged readers to 'tak[e] control of their own brain aging' by building up a cognitive reserve. This take on cognitive fitness resonated with the goals and interests of a growing cognitive health industry, which advertised its tools as helping people to 'think faster, focus better, and remember more'.[57]

This popular scientific perspective on cognitive reserve encouraged lifestyle considerations. But it did not fundamentally change the situation of those diagnosed and labelled with the disease. Once positively tested for specific risk factors, the prospective carriers of 'amyloid burden' were left with the fear of falling victim to an incurable disease, of becoming a burden to their family and of being seen as losing their self and identity.[58] Guidebooks like Whitehouse's lifted some of the stigma attached to memory loss in old age, because it challenged readers to think about brain ageing 'not as a disease, but as a life-long process fraught with challenges'.[59] But this change in perspective has been slow in coming and, to the present day, keeps shifting in incremental steps only. As recently as 2017, the director of the New York Memory and Healthy Aging Services, Gayatri Devi, made the case for dementia as continuous variations in cognitive ableness. Embracing the concept of cognitive reserve, Devi argues that Alzheimer's disease is a 'spectrum disorder that presents with different symptoms, progresses differently, and responds differently to treatment, with different prognoses, for each person'. But confirming the slow change in wider cultural understanding, she laments that we have 'not yet learned to associate Alzheimer's disease with functioning individuals, although this is, in fact, the majority of patients'.[60] Concurrently, concern about the worldwide marketization of cognitive fitness is mounting, with Peter J. Whitehouse, for example, calling for 'a counter-narrative that reinforces broader multifactorial notions of brain health'.[61] Such concerns are particularly valid, given that brain fitness technologies have further enhanced the 'ambiguous image' of the ageing brain as 'improvable and "plastic" but also inevitably in decline'.[62]

The idea of the brain as plastic and the patient as functioning individual gradually seeped into the wider cultural discourse, with literary writing beginning to envision Alzheimer's disease as a spectrum disorder from around the mid-to-late 1990s. I will attend to these developments more fully in Chapter 6 and show how fiction that featured dementia as a condition of continuous capabilities returned agency to the patient.

Death in Slow Motion: Past identities, lost plots and old age in caregiver life-writing

During a period marked by a significant rise in popular science writing, caregivers began to be faced with scientifically sanctioned, desolate descriptions of their family member's condition – as encountered in popular writings such as Tanzi's: 'Dying nerve cells. Snuffed synapses. Severed pathways. Fallen brain regions. Impaired mental and corresponding physical abilities' (75). In this section, I want to think more fully about the position of the patient in the stories of caregivers at the time of these medico-scientific developments. What alternative concepts could caregivers hold onto as neurochemistry and genetics sanctioned socio-medical and popular beliefs about the patient's social death? To do these questions justice, we must include another factor that increasingly influenced the cultural dementia narrative: the marketplace. I already mentioned that scientific research could not remain free of external influences on its practices. The competition for public money furthered scientific outreach, which, in turn, held out hopes to patients and caregivers that the condition's cause or causes would soon be identified. Concurrently, popular scientific speculations about cognitive reserve created a market for preventive interventions. The production of literature could not remain exempt from such external deformations. During the period of biomedicalization the number of caregiver memoirs rose quickly, serving a readership increasingly aware of the condition but also in need of readable plots and redemptive endings.

During the 1990s, memory turned into 'a cultural obsession of monumental proportions across the globe'.[63] Caregivers became 'archivists' like the French writer Annie Ernaux, or turned 'hunter' and 'detective', and set off on an 'inquiry [back to the past]' like the American writer Mary Gordon and her conational, sociologist Lillian B. Rubin.[64] That such a wealth of adult-child narratives concerned with the personal past as part of post-war history emerged, in both Europe and America, especially during the 1990s, is perhaps best put into words by Lisa Appignanesi. Born to Jewish parents in Poland and brought up in France and Canada, she deliberates on the 'politics of memory' as follows:

As we approach the millennium, it seems we are all preoccupied by memory. To construct a future, we need to unearth new narratives of the past. Pasts which have been buried by repressive regimes and left to fester or pasts transformed by Cold War politics; or simply pasts relegated to the limbo of latency because they were too painful to think about.[65]

Appignanesi's *Losing the Dead: A Family Memoir* (1999) shares similarities with Linda Grant's *Remind Me Who I Am, Again* (1998). As the descendant of Jewish immigrants, the British writer searches for identities for herself and her mother, who is losing her memory to vascular dementia: 'in her brain resided the very last links with her generation'.[66] In keeping with the contemporary understanding of identity as intricately linked to the ability to narrate, Grant explicitly ties notions of an intact identity to an undamaged awareness of one's own past. She describes her mother as 'continuously inventing for *herself* (and the rest of the world) a coherent identity and daily history' (156; emphasis original). With reference to Grant's narrative, Nicola King observes that it is 'not only the *content* of memories, experiences and stories which construct a sense of identity'; rather, the self in these narratives also depends 'upon assumptions about the function and process of memory and the kind of access it gives us to the past'. In the absence of her mother's memory, Grant – together with other adult children whose narratives King analyses – tells 'two life stories, her own and her mother's, which modify and reinterpret each other'.[67] Grant brings this meaning of memory work for the identity of both the writer and the written-about to the fore, and she places it in relation to the post-war trauma that fiction writers had begun to work through over a decade earlier: 'the whole thing about memory is that it's not just one member of a family losing their memory. And for the Jewish community it's even more complex because while all cultures are to do with memory, none more so than the Jewish community in which everything is about what was' (269).

But if we were to 'collapse memory into trauma', Andreas Huyssen cautions, we would mark memory 'too exclusively in terms of pain, suffering, and loss'.[68] We have seen this understanding of memory, for example, reflected in Bernlef's account: the trauma of the past is so much stronger than dementia-induced forgetting that would or could conceal this very trauma. Memory is comprehended as an archive in these narratives, and I here address how archival notions of memory in caregiver accounts influenced the dementia discourse. I show how stories of the patient's emptying memory played on two ideas: the narrative of what Drachman termed 'irretrievable "atrophic" loss of information', and the patient as living dead.

Like other adult children, Eleanor Cooney carries out an 'archaeological job' in the search for the cause of her mother's Alzheimer's disease.[69] *Death in Slow Motion: A Memoir of a Daughter, Her Mother, and the Beast Called Alzheimer's* (2003) reminisces on the American novelist's 75-year-old mother, the writer Mary Durant. It relies heavily on the metaphorical language perpetuated since the NIA's first information campaigns. Reference to the illness as beast turns the patient into its victim exactly like the killer disease had done in the 1980s, and the book's reception and marketing plays on the same imagery, promising an 'account of a hopeless and heroic battle'. At the same time, Cooney's excruciating honesty about her burnout and stress-related alcohol and drug abuse anticipates adult-child narratives of the later 2000s. These adopted an activist tone in the caregiver's struggle to receive appropriate healthcare support, among them Andrea Gillies's *Keeper* (2009) and Sally Magnusson's *Where Memories Go: Why Dementia Changes Everything* (2014). Much of Cooney's account illustrates the socio-economic and healthcare situation, as it presented itself around the turn of the century. *Death in Slow Motion* is the outlet of Cooney's travails caring for her mother-turned-child (35). She knits together her struggles to find a suitable nursing home and her search for the cause of her mother's illness. But to ascertain the cause of Durant's illness, for Cooney, really means to identify the root cause of her own caregiver depression. Discovering the reasons for her burnout, in fact, would, in Cooney's eyes, justify why she betrays her mother in placing her in a nursing home (111).

From the outset, Cooney is convinced that the death of Durant's husband Mike 'was the cause of her mental deterioration' (2). Cooney stresses that Mike's 'mind was going in every direction' (166) in the process of deciding to undergo (eventually unsuccessful) life-saving surgery. She particularly underlines: 'His decision to go ahead with the surgery was based, finally, on a sensible, practical, hard look at the statistical chances … that his condition would, untreated, lead to stroke, early senility, ignominious decline. In other words, living death' (166). Reading these deliberations in the context of Cooney's negotiation with her mother's dementia brings out just how reduced in free choice she believes Durant to be, how limited in mind in contrast to Mike's limitations in body. But as Cooney probes her theory of Mike's death as the cause of her mother's illness, she has to look farther into the past. Her narrative reads like a celebration of the powerhouse relationship between 'Mike and Mary', to whom she dedicates her memoir. The loss of this relationship again typifies Alzheimer's as a condition of lost integration and lost communication.

Durant wrote her 'masterpiece, [which] got rave reviews, won some prizes' (121), together with Mike, namely *On the Road with John James Audubon* (1980). Cooney's choice of language, when she reviews Durant's writing 'about Audubon's decline' (123), establishes continuity between the mother's past (as well as her meaningful writing in the past) and her present state. This need to create life-story continuity emerges from the experience of illness as biographical disruption.[70] For example, the title of Cooney's final chapter, 'Sunset, Children ...', corresponds to the final lines in her mother's book (124), on whose taste and sensitivity Cooney comments: 'She knew what fine writing that was, knew it was sheer inspiration to end their amazing saga with those two words' (124). Picking this quotation from Durant's text suggests that Cooney acknowledges her mother's own life as an 'amazing saga'; it also implies Cooney's awareness that she learned her trade from her nurturer (81). Durant's 'powerful gift for character development' (7) now falls to Cooney herself. She has to write for her mother, because Alzheimer's develops her mother's (as well as her own) character during the eighteen months they share together.

Against this perceived loss, the narrative's early chapters centre on Cooney's vivid recollections of Durant as a loving mother. But most of all, Cooney's book contains an appendix, a space separated entirely from the daughter's writing, for a short story she had discovered in 'a rich stash of [Durant's] lost writings' (214). In posthumously publishing for the mother, Cooney's writing becomes a tribute to her mother 'as the whole person and ferociously good writer that she was' (215). But as Cooney's book turns into a device for maintaining Durant's identity as a writer and former self, this act underscores Cooney's perception of the patient as a 'living dead' (140, 166), unable to narrate and communicate. Cooney herself confirms this reading, when explaining that, in this appended narrative, her mother 'comes back from the dead' (214), continuing: 'This is my mother. And if you don't believe that Alzheimer's is the equivalent of death, try living with it for a year and a half. It's worse than death. In the long interval before actual physical death, the afflicted person is insidiously replaced by an imposter' (215). The narrative's final chapter then depicts Cooney's perception of her mother as a living dead most acutely, when Cooney returns to Connecticut 'to empty [her] mother's house' (200). The chapter's content and metaphorical language come very close to Galsworthy's use of the house as image. Constance Rooke comments that 'the house represents the self in society and contains (through memory and memorabilia) the "furniture" of identity'.[71] Cooney is determined not to 'throw away a single book' (211), because these books, together with 'thousands of notes, typed or scribbled – research, ideas, references, inspirations, ideas, ideas, ideas' (212), embody Durant's former fullness of life.

Cooney honours her mother's legacy. But the narrative arrangement matches what Barbara Ehrenreich several years later criticized as 'retreat from the real drama and tragedy of human events'.[72] Cooney articulates her story, as Kathlyn Conway would put it, from a 'position of authority outside the actual experience'.[73] Looking back on her caregiving experience, she can write about her learning, while avoiding talk about her mother's complete bodily and mental disintegration. She can offer a 'culturally validated narrative of triumph over adversity' – with a storyline that is, given the condition's trajectory, not available to the patient.[74] This culture of survivor literature would make it harder and harder for patients to find their own voice and reposition themselves in this mainstream narrative. It is, in fact, truly significant that Cooney places the account of Durant's actual final days in an epilogue. This textual separation symbolizes the narrative place and time where Durant, in Nancy Reagan's words, has 'gone where I couldn't follow' (232). There, Durant is seen to lead an existence without a past, as 'she doesn't mention Mike anymore … Terribly sad, but merciful. For her, and for me' (225). Forgetting has turned into a disease-inflicted kindness. Also Appignanesi's observation that her mother increasingly 'denied what she didn't want to know or chose to forget' hints at more redemptive notions of forgetting towards the end of the century, a perspective increasingly found in much later life-writing, including, for example, Dana Walrath's *Aliceheimer's: Alzheimer's through the Looking Glass* (2016).[75] In the overwhelming awareness of societal ageing (and increased risk for everyone to fall ill with dementia), forgetting becomes redemptive. The kindness of forgetting would become a version of the narrative of loss that could appeal to a mass readership for whom Alzheimer's disease had come to embody the death of the self.

Presenile dementia, while fiercely pushed in genetics research, failed to receive a broader literary reflection in its own right at this point in discourse development (this changes towards the end of the first decade of the new century, as I will show in Chapter 7). In *Death in Slow Motion*, presenile dementia only features in passing when Cooney becomes 'intensely aware of [two women] who are not a hell of a lot older than I am. They're in their fifties, and have early-onset Alzheimer's' (230). In spousal memoirs, however, presenile forms of dementia find fuller examination. These spousal accounts of early-onset patients published in the late 1990s draw on very similar forms and categories of representation as the ones just highlighted for late-onset dementia in Cooney's book. This suggests how dominant the cultural image of Alzheimer's disease as a condition of old age continued to be at the end of the century.

Margarita Retuerto Buades's husband, José Antonio Gómez-Bárcena, was 56 years old when diagnosed with Alzheimer's disease in 1996. The Spanish lawyer's account, *Mi vida junto a un enfermo de Alzheimer* (My life next to an Alzheimer's patient; 2003), was one of the earliest caregiver narratives written in Spanish and published in Spain.[76] It echoes much of the need to share the actual experience of caregiving found in early adult-child accounts like Rosalie Honel's *Journey with Grandpa*. But unlike these early adult-child accounts, Buades's narrative is less concerned with a chronological exposition of events. Instead, it focuses on specific aspects of the caregiving experience which all emphasize that Buades has 'lost [her] husband' (89). 'Visits and friends' (95), for example, while emphasizing again the patient's shrinking social network, laments how 'at the same rate as my husband's condition worsened, these visits became more and more and more spread out' (105). And in the chapter 'From wife to caregiver' (79), Buades confesses: 'I miss my husband' (81). These images are not dissimilar to the dimensions of loss that Wayne Booth identified as 'universally pertinent' in ageing, which I referred to in the introduction: the loss of friends and partners in life.[77] In addition, Buades compares her husband's behaviour to that of a child (e.g. 31, 66, 72), an image also taken up in a national newspaper feature.[78] These similarities to adult-child narratives bring out the impact of the caregiving experience in Buades's account: the spouse with dementia, much like the parent with dementia, is considered a child. This is significant for the fact that Ann Burack-Weiss has described a 'history of reciprocity, along with a sexual history' as separating 'the situation of partners from other family care relationships' in contexts other than dementia.[79]

Most importantly, though, Buades's reliance on medico-scientific and popular accounts links José's condition to one of ageing. She references, among others, David A. Snowdon's *Aging with Grace*, seeks authority through mentioning José's clinical documents and specifically refers to evidence for the organic, neurochemical basis of cognitive deficits. Influenced by Reisberg's work on age-associated cognitive decline, Buades places memory loss alongside the 'inevitable and normal consequence of ageing' (54). In addition, she brings out how the condition 'progresses much more aggressively in younger people' (65). All things considered, I read Buades's rendering of her husband's condition as a cultural confirmation of, and commentary on, Bradley T. Hyman's observation in relation to the genetic dementia research of the 1990s. Hyman states 'that genetic factors lead to quantitatively worse disease but not to a qualitatively different pattern of brain involvement'.[80] Even though Buades writes about a middle-aged patient, the images she uses, and the concerns she has, overlap with those of adult-child caregivers, whose patients are elderly. Her strategy for surviving 'all those years at war with Alzheimer's is to think that my husband is dead' (17).

Incidentally, Buades's reference to John Bayley's *Iris* (1998), in which Bayley writes about his wife's illness when she was in her seventies, is also significant for her perception of José as old.[81] For while I do not suggest that Bayley's account gave her ideas about how to write about José, the sociocultural reach of Bayley's account made it a welcome source of comparison for caregivers across various cultural contexts and traditions, and this dominance helped strengthen the narrative of the patient as old, dependent and childlike.[82] Rudolph E. Tanzi's reference to Bayley's account as giving the reader 'a sense of an Alzheimer patient's terribly limited world' (74), in turn, indicates a growing oscillation of language and images between the literary and popular scientific realm, and points to the central role literary writing assumes in amplifying a particular version of a narrative.

Bayley's text and the dementia narrative it portrays rose to extraordinary prominence in part because of its fast adaptation for the big screen in 2001.[83] Bayley starkly contrasts the couple's life together in the narrative's 166-page-long first part, 'Then', to a diary-like, forty-page rendering of the 'Now'. This textual arrangement underlines the Oxford don's perception of loss and provides the basis for the film's key narrative device. Richard Eyre's movie uses extended flashbacks to make most tangible emotional and physical disruption as well as discontinuity; it does not offer a 'life-course perspective'.[84] Rather, in constantly switching between the 'young' (Kate Winslet) and 'old' Iris (Judi Dench), the film contrasts Murdoch's fulfilled and youthful life as an intellectual and Bayley's struggle for her attention, with Murdoch's inability to continue to write in the present day of her old age and illness.[85]

Bringing Alzheimer's disease to the big screen enhanced clinical (if not medico-scientific) appreciation of the patient's and caregiver's experience, also because the Alzheimer's Association influentially endorsed the film. We might read the positive reactions of geriatricians and psychiatrists to the film in the tradition of the geronto-psychiatric work on senility in the 1960s and 1970s.[86] But they were also the product of whole-person approaches to dementia care that began to emerge during the 1990s – developments that became instrumental in giving a voice to patients themselves.

As Alzheimer's disease became a household word during the 1980s, and wider societal attention grew for neurochemical and genetics research, the patient became increasingly marginalized. Pictured as suffering communication breakdown and objectified in caregiver accounts that zeroed in on the burdens of care, those who experienced the illness directly had been pushed ever farther away from the centre of attention. As mentioned previously, the writings of patients began to be published in 1989. However, it took another decade of growing patient organization, linked to a rise in psychosocial work on Alzheimer's disease, before patients would begin to be perceived (and perceive of themselves) as political agents. Agents who were able and authoritative enough to establish their own version of the illness – removed from the socio-medical, neurochemical and healthcare narratives of loss as well as caregivers' pessimistic accounts, amid a growing marketplace.

Patients were additionally pushed to develop a counter-discourse by a further evolution of the medico-scientific view of Alzheimer's disease and its related portrayal in the mass media. With the advent of in vivo imaging methodologies, illness progression could be 'seen' in the living brain. As Tanzi put it, for the community of medical scientists and clinical neurologists 'now to actually *see* direct evidence, to be able to actually watch the disease progress in a living person, is quite amazing. It gives us a way of tracking the rate and severity of the disease's progression; of knowing, for the sake of applying drugs, who will get the disease sooner than later' (245–6; emphasis original). Only there were no drugs available.

6

Neurotechnologies and narrative examine the failing mind

The visual exploration of the brain and fascination with the mind

In the 1990s, Alzheimer and his disease have become the flavour of the decade ... Clinicians increasingly face demands from basic scientists to provide subjects for research studies who meet 'strict DSM III-R diagnostic criteria' of the disorder. The race between pharmaceutical companies to find the 'cure' for Alzheimer's disease is fierce, and established scientists have been recruited to the tasks.
G. E. Berrios and H. L. Freeman, *Alzheimer and the Dementias*[1]

Since the early 1980s, medico-scientific research had been working at an international level towards elucidating the condition's pathogenesis. With new imaging technologies becoming available for medical practice, a patient's biochemical and genetic make-up could now be assessed in conjunction with real-time observation of functional and structural changes in the brain. In the wake of these developments, as summarized by G. E. Berrios and H. L. Freeman, the medico-scientific narrative of decline and loss attained a new level of perceived truth. Visually observable brain shrinking became the correlative, if not surrogate, of a patient's decreasing performance, because cognitive functioning scores could be correlated to data collected from the brain at work.

Imaging methods led to an ever more detailed study of the mind, while conditioning and enhancing a growing preoccupation with cognitive prowess. A US government initiative additionally heightened public awareness of brain research during what George W. Bush had proclaimed the 'Decade of the Brain', the period spanning 1990 and 1999.[2] The emergence of the first drugs used for the treatment of Alzheimer's disease also furthered this cognitive focus. (These drugs were intended to counter neurochemical deficits while enhancing cognitive performance, but they would reveal themselves as largely ineffective: they did not halt disease progression and alleviated cognitive symptoms only to a limited extent. Yet, essentially they continue to be the only drugs available for Alzheimer's disease.) These developments, I argue, reformulated dementia as a cognition- and performance-limiting process, based on organic changes now observable in vivo.

The imaging techniques of particular relevance are: computer tomography (CT), a method building on X-ray technology; magnetic resonance imaging (MRI), which assesses the environment of hydrogen atoms in the body; positron emission tomography (PET), which traces the uptake of specifically labelled glucose (abbreviated FDG), and hence cellular activity; and single photon emission computed tomography (SPECT), which measures blood flow. When these methods started to gain prominence in the 1980s, CT and MR images were considered to represent tissue anatomy and biochemical make-up, while the PET-FDG image was taken as a 'map' that reflected neuronal activity.[3] The implications that Van Dijck has identified regarding the meaning and effect of the map metaphor where it applies to genetics are relevant here too: its equating to a map underlines how the research community increasingly considered the visual image as the most reliable form of evidence. MRI technology in particular represented, as sociologist Kelly A. Joyce elaborates, a *cultural icon – a sacred object* on which questions about personal health and identity revolved.[4] Early MRI studies provided evidence of the patient's shrunken brain.[5] And, in 1990, a group of US scientists around William Bondareff advocated the use of MRI for 'following the course of dementia' over time, because they saw differences in disease severity.[6] Such work was seminal in discourse developments, because medical images were valued 'as Evidence in its quintessence'.[7]

In keeping with Foucault's insight that the clinical gaze detaches the body from a person, this research brought home how patients could be turned into abstract entities. Presenting assessment scores and computed lesion-brain ratios, three tables and six graphs adumbrate, in the paper by Bondareff et al., the complex procedures of weighing, measuring, dividing and multiplying that yielded the numbers standing in for the individuals subjected to these procedures. The imaging specialists' focus on visually collected and mathematically processed ratios of brain measurements further distanced the observer from both specimen and individual. I emphasize that, despite the perceived reductionism of these procedures, I do not attribute to them any particular role per se in discourse developments. Bruno Latour and Steve Woolgar convincingly present the data processing as part of a creative process and explain that, for the scientist, the outputs obtained in such a process remain in 'direct relationship to "the original substance."'[8] Rather, I am interested in how these procedures contributed to what Ian Hacking would term new ways of classifying and knowing patients, and how such classification played into concepts of loss related to archival notions of memory.[9]

The resulting diagrams suggested potential correlations of each individual's lesion-brain ratio to various cognitive test scores. Specifically, the 'magnitude of these lesions ... correlated closely with the severity of dementia indicated by scores on the Blessed Dementia Scale and the Folstein Mini-Mental State Examination' (47). These cognitive tests were intended to quantify the 'degree of intellectual and personality deterioration' in elderly individuals, as they categorized the cognitive performance of patients along a numerical scale.[10] Originally developed in 1968 and 1975, they came to be prominently used once pharmaceutical compounds required assessment. Tacrine and donepezil were the first drugs (termed pro-cholinergic because they were designed to support cholinergic activity) approved in the United States and other countries, in 1993 and 1997, respectively. A pharmacological study of one of these compounds illustrates how the language evolving around these drugs (and the testing of their efficacy) heightened concepts of dementia as a cognitive condition and the patient as under-performing.

In a US report from 1998, S. L. Rogers and colleagues evaluated the efficacy and safety of donepezil.[11] Their description of inclusion criteria and methods for assessing drug efficacy brings home two points: the language regarding Alzheimer's disease had become strongly performance focused, and the description of the patient followed principles of systematic clinical categorization and classification. The diagnosis of 'probable AD' was made

> according to criteria outlined by the National Institute of Neurological and Communicative Disorders and Alzheimer's Disease and Related Disorders Association (NINCDS-ADRDA), with patients also fitting DSM-III-R illness categories ... Patients had scores on the Mini-Mental State Examination (MMSE) of 10 to 26, and a Clinical Dementia Rating (CDR) score of 1 (mild dementia) or 2 (moderate dementia) at both screening and baseline. (137)[12]

Also clinical outcomes were expressed in strictly standardized terms based on a battery of five different tests: 'the cognitive portion of the Alzheimer's Disease Assessment Scale (ADAS-cog) and a Clinician's Interview-Based Impression of Change scale that included caregiver supplied information (CIBIC plus) ... the MMSE, patient-rated quality of life (QoL) scale, and the Sum of the Boxes of the

CDR scale (CDR-SB)' (137). The authors dedicated almost a quarter of the eight-page-long paper to the description of these research methods, and I have quoted from them at length to convey their style and purview.

For the purpose of Bondareff's research, patients were tested similarly (47). Based on mathematical and statistical operations, the authors hypothesized that the visible changes they observed 'might affect cognitive functioning; perhaps ... by interrupting networks [that] are the subject of contemporary discussions of the neurophysiologic basis of learning and memory' (51). With such speculations, the authors took a significant step towards the claim that visual data that reflected a loss in brain cells could stand in for cognitive performance and memory capacity. Five years later, in a Japanese study, Kazunari Ishii and colleagues explicitly linked visually collected information to impairment in a brain area involved in memory operation (the hippocampus).[13] This step deserves particular mention because they published their evidence in colour.

The use of colour rather than grey scales to display differences in recorded intensities in the brain had been in use since the early 1980s. But only when scientific journals supported publication in colour, and such publication became affordable for research labs, did coloured brain sets become the norm. Anthropologist Joseph Dumit points out that choosing a specific set of colours to represent linear activity values in specific brain regions is entirely arbitrary. For him, this very arbitrariness reinforces the sense that 'these regions are internally coherent, separate from their neighbors, and therefore able to adequately represent the "functioning of the task" in question'.[14] This means that colouring per se added to the perception that a specific area of the brain was structurally damaged and its function impaired in Alzheimer's patients. In reverse, the image could be seen as the correspondence of patients' cognitive impairment. Ishii and colleagues opted for a scale ranging from blue to red, with red representing the highest degree of intensity. The absence of the signal colour red in those areas where 'decreased' cerebral blood flow was observed (1163) visually determined the loss in the research subject's memory.

The images were selected with the aim of enhancing textual arguments in both scientific and popular articles.[15] Colour made research results more easily accessible to mass media exploitation, but this process was reciprocal, as researchers soon felt the pressure to make their scientific publications more engaging. Lisa Cartwright has superbly illustrated how, across the twentieth century, knowledge and subjectivity in science are 'entangled with a broad range and mix of cultural and representational practices'.[16] Her insights inform my understanding that images of the living brain opened up medico-scientific descriptions of Alzheimer's disease to largely unfiltered mass media use of the condition as meaning cognitive deficit. Articles frequently incorporated colourful brain images or, at least, made reference to the usefulness of imaging technologies to 'detect progressive brain cell degeneration' in the brains of patients in very early stages of Alzheimer's disease.[17]

Like the general public, researchers were convinced of the 'certainty' conveyed in these visible findings, which was 'heightened by the 'status attributed to large, expensive, complex technologies'.[18] Lennard J. Davis has made the important observation that science and scientific medicine are themselves not objective disciplines of knowledge acquisition but part of a 'new problematics of obsession'.[19] By the turn of the century, the evolving technology could generate dynamic blood volume images to assess brain activity during stimulation, such as cognitive testing. In keeping with Dumit's statement that brain areas circumscribed by colour were taken as functional units, the image itself now answered for the subject's cognitive performance.[20] But such brain imaging also revealed the sobering fact that long-term treatment with drugs like donepezil did not improve 'regional cerebral blood flow' in patients.[21]

This brings me back to the clinical trial conducted by Rogers and colleagues in 1998. Their study concluded by stating that future trials were 'necessary to determine if donepezil has significant effects in delaying deterioration or actually improving functional outcomes for AD patients' (144). Cognitive abilities had become the all-encompassing parameter by which to define patient well-being. In fact, for Hacking, one of the reasons that memory loss in Alzheimer's disease is so emphasized in the final decade of the twentieth century is that objective quantitative tests for memory loss can be so easily defined.[22] This is especially important because once so tested, therapeutic approaches proved largely unsuccessful – counter to expectations nourished by the myth of the transparent and knowable body, resting on the assumption 'that seeing is curing'.[23] Based on similar assessments, a British long-term clinical trial found donepezil not to be cost effective in 2004.[24] This study eventually led the National Institute for Health and Care Excellence (NICE) to no longer recommend the use of pro-cholinergic drugs in mild to moderate cases of dementia in Britain.[25] During a period marked by a prominent fascination with the brain, whose visual scrutiny seemed to give access to the 'neural correlates of the activities of mind itself in real time', cognitive performance had become a question of healthcare budgets.[26]

Such decisions were influential for the further development of the status of patients. They were made at a time when, as we have seen, cognitive health and performance in older people was increasingly promoted in public health strategies such as the US Healthy Brain Initiative in 2005.[27] Caregivers witnessed their parent's or partner's loss of memory and mourned their social death. These losses appeared sanctioned by imaging data and testing outcomes, and they were enhanced by a popular fascination with cognitive performance and the visual representation thereof. As Stephen G. Post observed, 'Nothing is as fearful as AD because it violates the spirit (*geist*) of self-control, independence, economic productivity, and cognitive enhancement that defines our dominant image of human fulfillment. Deep forgetfulness represents such a violation of this spirit that all those with dementia are imperiled.'[28]

Post's deliberations on the situation of Alzheimer's patients in a 'hypercognitive society' are particularly poignant when considering that imaging technologies grappled with the same question: the separation of normal from pathological processes of ageing. In the search for diagnostic tools with which to distinguish between 'normal age-related cognitive decline' and levels of cognitive performance 'that precede the onset of dementia', a series of constructs had been devised since the mid-to-late 1960s. The most durable term, 'mild cognitive impairment', entered the lexicon of literature on ageing and dementia in 1988.[29] The move to introduce a category of 'mild' impairment enabled practitioners to lift the ensuing stigma of lacking cognitive performance from the ever-growing number of older people, who feared the diagnosis of Alzheimer's disease, because the demented individual was – as popular writing had brought home – less able, less independent, less productive and less conforming with society's expectations of performance.

The growing cultural interest in these neurotechnologies both paralleled and conditioned a marked increase in literary explorations of a broad range of neurological conditions from the late 1990s onwards. For American literary critic Marco Roth, this rise of the 'neurological novel' was closely linked to the fact that the press, like the *New York Times* in its sciences pages, eagerly took up the reductionism of mind to brain.[30] Ian McEwan's *Saturday* (2005) is a good example.[31] It centres around the psycho-neurologically challenged Huntington's disease patient Baxter, whose disease-typical behaviour becomes indistinguishable from that of a socially disadvantaged criminal aggressor (91). Baxter falls short in terms of intellectual abilities. He becomes a mirror in which the values of a cognition-driven society are reflected – those of his opponent, the high-performing and successful brain surgeon Henry Perowne. For Perowne, Baxter merely has 'a diminishing slice of life worth living, before his descent into nightmare hallucination begins' (278).

This fictional representation brings home the deleterious impact on personhood of what Raymond Tallis termed the 1990s 'culture of neuromania'.[32] With the biomedicalization of Alzheimer's disease, Joseph M. Foley much earlier had seen that the dementia patient's chances to be considered an individual with personhood and identity were at stake. He wrote that 'we must know not just about cognitive capacity but also about awareness, feelings, and emotional reactions to the personal and social consequences of dementia'.[33] He had expressed these concerns as early as 1992, which signals how anxieties about identity in dementia foreshadowed the exponential rise in medico-scientific research on Alzheimer's disease and its coverage by popular science and mass media throughout the 1990s.[34]

The patient's feelings and experience, eventually, attracted increasing attention in the final decade of the twentieth century. Partly, this change was conditioned by the said cultural fascination with 'neuronovels' and a rising number of genre novels, among them 'thrillers[ ] of amnesia'.[35] Countless popular novels (including detective fiction) and an endless number of televised serials continue to bring out the condition as a mirror, in which values of the cognitively, romantically or economically excelling chief protagonist are reflected. Still, I contend that the narrative exploration of Alzheimer's disease as one of the failing mind, especially during the 1990s, eventually opened the way for patients to articulate themselves and reclaim their authority and identity. I cover such patient life-writing more fully in the final section of this chapter. First, I will examine three early pieces of dementia detective fiction. Their unreliable narrators keep these narratives anchored in the tradition of the novel of consciousness. Yet, it is through their renewed focus on plot as tracing mindful performance that these narratives would be of notable service to patients and their care.

The Dying of the Light: Detective fiction claims back patient authority

In 1992, one of the first narratives carrying 'Alzheimer's' in their title appeared in print – Hans Dieter Mummendey's *Claudia, Alzheimer und ich: Kriminalroman* (Claudia, Alzheimer's and I: A detective novel).[36] Judging from the core image explored in the narrative's opening chapter, the story did not promise a radically different approach to dementia: an allusion to the patient's living death as the 64-year-old narrator, undergoing a scanning procedure, wonders, 'Is this already the scenario of interment, of cremation?' (10). But the story's subtitle opened the condition to an entirely new genre. Mummendey develops his mystery consistently from the perspective of an ageing person, whose reliability is questionable from the outset. The narrator himself explicitly ponders, 'Does my brain keep up with reality?' (10), and later comments on his forgetfulness (15) and age-related chattiness (15). In this narrative mode, the patient turns to telling of his much younger wife Claudia's death. He expands on how Claudia perceived increasing signs of 'imperfection, lack, weakness, illness' (61, 87) in him and how she created situations which suggested his inability to remember (69). Based on such information, the reader begins to wonder how Claudia might have died, especially since the narrator keeps emphasizing her sportiness and robust health.

Bernlef's narrative is also told from the patient's perspective, but Maarten's reliability is never an issue for the reader – both for Vera's presence and because we are led to read *Out of Mind* as a convincingly charted documentation of the narrator's Alzheimer's disease.[37] The narrative's free indirect style reinforces this perception. In comparison to Bernlef's Maarten, Mummendey's narrator is both actor and witness, and therefore has control over the truth or not that is revealed. With a detective novel in hand, the reader relies on the patient-turned-narrator to solve the mystery. But anxious to see its solution before Alzheimer's has gone too far for the narrator to impart his knowledge, the reader experiences a completely new feeling with a dementia novel in hand. We read for the plot, against the disease and its trajectory.

Between the 1940s and the 1980s, the tools and strategies of modernist and postmodernist fiction enabled the narrative exploration of the workings of the patient's mind and emotional experiences. These approaches were, as Peter Brooks has argued, permeated by a suspicion that 'plot falsifies more subtle kinds of interconnectedness'.[38] This means that modernist writing narrated the condition from within the patient's mind, relying on the condition's own plot, which medical and social scientists as well as psycholinguists had characterized in terms of decline and loss. Paralleled by the growing emphasis – by both medico-scientific research and the adult child's memory work – on chronology, genealogy and temporality during the 1990s, plot re-emerged as a structuring principle in narrative. In this context, no narrative relied more on plot than the detective story, given that 'everything in the story's structure ... depends on the resolution of enigma'.[39]

Lisa Zunshine points out that 'we open a detective novel with an avid anticipation that our expectations will be systematically frustrated, that we will be repeatedly made fools of, and that ... we will be fed deliberate lies in lieu of being given a direct answer to one single simple question that we really care about (i.e., who done it?)'.[40] Put differently, the questionable reliability of dementia patients makes them characters highly suitable for detective fiction. At the same time, detective fiction, in driving towards dénouement, is perhaps the genre most suited for challenging this supposed lack of reliability. I here advance the argument that it was exactly such plot-related expectations that encouraged the reader to question societal preconceptions against patients as unreliable and lacking in authority. The genre's requirements for red herrings as well as dependable clues led authors to appreciate dementia patients' narrative possibilities. By the same token, it opened readers to the narrative capabilities of patients and heightened societal responsibilities for listening and granting authority to their accounts. In short, beginning with Mummendey's novel, the plot-reliant detective story of the 1990s returned agency and control to the patient.

Plotting is the narrative's core device and the patient's central pursuit in Michael Dibdin's *The Dying of the Light: A Mystery* (1993).[41] Rosemary Travis dictates the plot, and she misleads detective and reader. In so doing she awakens fellow patients 'from their catatonic stupor and paranoid delusions' (132), instigating their improved care and awareness of continued agency. Dibdin's mystery is both hilariously funny and playfully critical. It reveals itself as a parody on the continued lack of quality care, a satire on society's views on the elderly and memory loss as their main condition, and a (Travis's) travesty of the contemporary form of the detective novel. First and foremost however, *The Dying of the Light* can be read as an allegory for memory loss in the elderly – a cognitive condition where fact and fiction merge. To identify the murderer of Rosemary Travis's friend Dorothy Davenport, the reader's interpretive ordering must achieve more than sifting reliable from unreliable figures. It must probe the relevance of different narrative layers dominated by these figures: Rosemary Travis's fanciful storytelling; detective Stanley Jarvis's unenthusiastic investigating; both their impressions conveyed in free indirect style; and, perhaps the only layer in which to find trustworthy clues, the views of a hardly perceptible narrator.

From the outset, Dibdin plays with the reader's expectations. He portrays Rosemary and her friend as the elderly storytellers, who are perceived as tedious by those who are young. Yet, the narrative's first part is set in an environment in which storytelling becomes a necessary pastime. Befitting its name, Eventide Lodge is a place where time stands still. The clock on the mantelpiece has stopped at ten past four, preventing the plot from moving forward, like the darkness which engulfs the place when lights are turned off at 9.30 every night. Rosemary and Dorothy 'make a fictional virtue of the factual necessity' (49), taking to plotting their own 'comforting narrative' (45). For their 'murderous web of intrigue' (56), they turn their fellow inhabitants 'into cardboard characters whom they manipulat[e] to suit their whims and the twists and turns of the story' (59). In this diversion between the friends, 'it was always Rosemary … who kept all the strands of the plot in play while still managing to accommodate – and thus to some extent control – the real horrors which surrounded them' (44). However, once Rosemary is removed from conversational interaction, she feels her control – her identity as plotting individual – slip away. With the news of Dorothy having to go to hospital for cancer treatment her sense of audience drops, because Dorothy's part was 'to fill in the gaps … to spot the errors … to approve and criticize, suggest and reject' (45). But Rosemary does not give up at the loss of her social network. She does 'rage against the dying of the light', as the book title's allusion to Dylan Thomas's 1937 poem for his dying father anticipates.

On the day Dorothy is found dead, Rosemary's ability to plot reaches a new level. She becomes the narrative interface in Dibdin's network of fact and fiction, relaying between the different fictional layers. With Dorothy's death, a real mystery needs to be solved in part two of the novel, requiring an actual detective, who would separate clues from red herrings – in an environment whose closed society makes all its members potential suspects and unreliable.[42] An avid reader of detective fiction (65), Rosemary is aware of the rules of the genre and sees the need to assert her selfhood in a plot apparently no longer of her own design. Put differently, the reader replaces Dorothy as Rosemary's audience, observing the altercations ensuing between Rosemary and detective-inspector Stanley Jarvis.

But if the pleasure in reading detective fiction derives from 'observing the mastermind's work', Dibdin dashes his readers' hopes for an authoritative inspector entirely: Jarvis has no interest in the case.[43] Uninspired, 'he consulted his thoughts. They were empty' (78), 'executed a doodle in his notebook' (79) and later 'added an elegant curlicue' (81). As an invisible narrator observes Jarvis, the reader, like Rosemary, begins to wonder: 'You *are* in charge, aren't you?' (93; emphasis original). And Jarvis's authority is fully undermined when he feels the need to 'tower[ ] over the elderly woman, swaying back and forth in the manner cultivated by the constabulary for the purposes of impressing the populace' (96), and silently prays 'Don't let me hit an old lady' (98). Thrown back onto Rosemary's initiative, the reader begins to appreciate the patient's resourcefulness. Rosemary coordinates her fellow patients' interrogation by the detective, setting the scene for a 'classic English detective story, with its unique blend of logic and fair play' (98), in which all are 'gathered together in the lounge of this isolated country house to face the detective's probing questions. One of us is guilty, but which? Can the sleuth succeed in unmasking the murderer before he – or she – strikes again?' (112–13).

Rosemary positions herself as the centre of a network of information. In this network, she regains dominance through language, because, as Otis would express it, she 'tries to know and control others by controlling their words'.[44] Her threat – 'if you leave now, I will be the next to die' (120) – forces Jarvis to take the matter in hand, even though convinced of Dorothy's suicide (87). As the narrative accelerates towards the mystery's dénouement, Jarvis returns to Eventide Lodge in the third and final part of the narrative, to identify the testimony of one of Rosemary's fellow residents as 'fiction from beginning to end' (135) and Rosemary as 'the author' (135) of it. As it turns out, Rosemary has been plotting the case all along, putting Dorothy's suggestions in her suicide note into action, namely 'to make it look as though I were the victim in one of our whodunnits' (142).

Rosemary is the real mastermind of the piece of fiction in hand. She plotted the case and laid out the clues for Jarvis to pick up and the reader retrospectively to arrange. She performs theory of mind. She uses – I am adapting Alan Palmer's definition here – her awareness of the existence of Jarvis's mind, the knowledge of how to interpret her own as well as the detective's thought processes.[45] But Rosemary's design is not only the key to the reader's pastime. We honour her plotting, because we have to go over the story again to revise our interpretations; we have to use our own theory of mind, aiming to follow Rosemary's as well as Jarvis's mind to avoid losing the plot.[46] In this regard, our readjustment goes beyond the separating out of witnesses and culprits: like Jarvis, we must come to terms with the fact that Rosemary does not fit our preconceptions of an unreliable, because elderly and perhaps forgetful, storyteller.

Confirming this reading, Dibdin presents Jarvis and Travis as plotting equals in their animated conversation covered in the novel's final part. For Rosemary, Jarvis turns out to 'be someone of a rather higher calibre' (147); and Jarvis explicitly tells Rosemary of the moment when he 'first started to take [her] seriously' (137). Once the inspector – as the detective fiction's key integrative figure – values the patient, Rosemary is lifted onto a higher level of authority. This ingenious change in dynamics invokes notions of whole-person approaches to dementia care, which advocate that 'much of personal identity belongs ... to the mutual recognition between two people'.[47] It also forms the centre of Dibdin's social criticism. With her newly gained authority, Rosemary can finally live her social self, which goes far beyond that of a cardboard character in a piece of detective fiction: she can lecture Jarvis for not having approached the 'case in the open-minded and impartial manner befitting a detective' (136).

But it is the joint action of Travis and Jarvis – the phonetic similarities between their names anticipating their cooperation – that solves the narrative's real concern, after we have followed the red herring of Rosemary's plotting: the patients' well-being. It is through Jarvis's actions that Rosemary becomes aware of Eventide Lodge's true impact on the patients' mind and spirit (131). In keeping with W. H. Auden's view that 'the job of detective is to restore the state of grace in which the aesthetic and the ethical are as one', Jarvis, at last, turns out to be the perfect inspector.[48] Initially sorry for the home's director Anderson 'stuck out here in the middle of bloody nowhere with a bunch of oldies well past their sell-by dates' (87), he eventually charges him and his sister for 'wilful cruelty and gross neglect' (132). He brings home the narrative's core message that care is, to some extent, a cure. However, it is only through Rosemary's plotting that Jarvis is led to expose himself and the reader to the patients' environment in the first place. Through his focus we appreciate the home's smell of mould (77). We suddenly see everywhere 'the same miscellaneous assortment of third-hand furniture, the same oppressive volume of chilly grey light, the same sense of desolation and decay' (107). Together with Jarvis, we hear Anderson and his sister Letitia Davis considering the elderly 'dribblers and bed-bespatterers' (80), with 'brains in their bums' (91) and without any future, let alone credibility:

> All the residents of this establishment are shortly destined to become the victims of a ruthless and anonymous killer against whom the combined forces of civilization have so far proved powerless. What more natural than that they should seek to contain their terror by recasting themselves as characters in a nice cosy whodunnit, threatened not by impersonal oblivion but a fallible human murderer, acting in a recognizable manner and for comprehensible motives, whose identity will be revealed in the final chapter? (81)

Dibdin also satirically overstates Anderson's and Davis's alcohol abuse and their physically and verbally abusing the patients. In this way, he satirizes the continued lack of quality patient care in the 1990s. Dorothy's choice to end her life could also be read as a response to the bleak outlook drawn, and caused by, caregivers like Anderson.

In Dorothy we encounter Dibdin's other powerful patient advocate. Her right to self-determination could be taken as a fictional response to contemporary evidence suggesting an increasing trend in elder suicide.[49] As such, Dibdin's choice with Dorothy and Rosemary to explore both end-of-life possibilities is most effective. In conjunction with the novel's ending – Rosemary asks explicitly, 'How's that for a happy ending?' (151) – Dorothy's choice may be viewed in relation to David C. Thomasma's argument that while 'a good death is a benefit to all of us, the duty to help bring it about can be fulfilled through loving care of individuals, proper control of our medical technology, and community support through hospice programs'.[50] The ending makes for a comedy of manners, given that it 'contains an expulsion of the socially undesirable which insures the continued happiness of those remaining'.[51]

We could read *The Dying of the Light* as a response to the National Health Service and Community Care Act from 1990, which regulated the assessment of patients' needs and abilities as a precondition for them to receive care and services. It suggests that many institutionalized individuals could live within the community for longer, if adequate healthcare were provided in that community and if healthcare were more focused on communicating with the patient. This perspective is in line with agendas, developed since the 1990s, that centred on care and campaigned for the celebration of neurodiversity. According to Dana Lee Baker, both agendas connect to the question of personal identity and promote patient-centred care. Dibdin's narrative appears exceptionally well placed to bring home the potential of celebrating neurodiversity: it develops the individual with the condition into a three-dimensional character with continued capabilities and identity.[52] The play-like narrative relies on the continued interactional abilities of older people. Almost entirely set in the home's lounge, the narrative's confined space of action encourages the reader's perception of a drama unfolding in the continued present tense of a play on stage with a central focus on conversation. Dibdin's emphasis on verbal exchange appears particularly pointed at a time when a growing body of work developed around differences between psycholinguistic concepts of breakdown and sociolinguistic appreciation of continued abilities in cases of dementia.[53] As explored in Bernlef's *Out of Mind*, psycholinguistic studies had particularly concentrated on language breakdown, explained by worsening cognitive skills. Sociolinguistic studies, from around the mid-1990s, rather emphasised remaining linguistic capabilities and assigned an important role to the patient's environment in sustaining these abilities.

Still, the question arises as to how realistically patients can be portrayed in their progressive decline, if they are meant to represent proactive characters.

Baker, in fact, cautions that celebratory agendas frequently emphasize the higher-functioning individuals, while excluding those who appear less capable.[54] Accordingly, the argument could now run that the condition of Mummendey's narrator or Dibdin's protagonist simply had not gone far enough so that these patients could still be presented as active. In line with sociolinguistic approaches to dementia, psychologist Steven R. Sabat would, I believe, counter this objection, arguing that patient understanding comes down to appreciating remaining abilities and to figuring out how those abilities 'can be recognized and supported by others'.[55] Martin Suter appears to take this issue to heart in *Small World* (1997), while similarly broaching the topic of improved patient-centred care.[56]

The Swiss writer credits a vulnerable patient with continued abilities, even though Konrad Lang is much less able than Rosemary Travis to perform theory of mind. Without dedicated support, the secret Konrad holds would not be revealed. At the same time, the disease trajectory drives the plot: the mystery can only be uncovered once Konrad accesses his earliest childhood memories (51). Miriam Seidler considers Konrad to be a flat character, because his condition represents the continuation of his lifelong subjection to the stepmother.[57] But he is at the centre of all other characters' attention. The matriarch Elvira Senn, 'alarmed by his detailed memories' (45), wants to murder him before he can reveal her crime; his stepbrother Thomas tries to escape him because Konrad's detailed memories make him doubt his own cognitive abilities; and Simone, the young wife to unfaithful Urs, Elvira's heir, focuses her caring instincts on Konrad once she notices the growing tension in the Senn clan.

Related to Simone's efforts, a large part of the narrative explores Konrad's recovery of memories, particularly in response to the study of photographs from his own past. This therapeutic approach had featured in the clinical literature since the mid-1990s, and was part of what developed as patient-centred care towards the turn of the century.[58] Together with Simone, the reader sides with Konrad, who needs 'to be treated like a human being' (23), rather than sent to a nursing home. Simone arranges for highly qualified day- and night-care personnel, a dietician, physiotherapist, occupational therapist and a neurologist in the Senn villa's guesthouse. Konrad 'blossomed' (170) from the first day in this caring environment, 'showing longer phases of presence, improved ability to concentrate and communicate' (171). As care becomes key to Konrad's memory recall, the reader becomes privy to the crucial need for patients' dignified care.

Accomplishing the 'fusion of the popular in terms of subject, audience, and form', a piece of detective fiction could impart this message on a much larger scale than caregiver manuals like the revised 1999 edition of *The 36-Hour Day* could do with its 'Guidelines for Selecting a Nursing Home' (464–75).[59] Dementia detective fiction was, in its reach as much as in its cognitive approach, the cultural reflection of two trends: significantly intensifying dementia research and an expanding body of interdisciplinary work centred on cognitive sciences.

In the final section of this chapter, I show how patients themselves attempted to influence the dementia narrative at the interface of science, medicine and the arts towards the turn of the century; how they located themselves in a slowly changing atmosphere that credited them with continued capabilities; how they fashioned their 'return from the living dead'.[60]

Who Will I Be When I Die? Patient life-writing around the year 2000

For Christine Boden, writing *Who Will I Be When I Die?* (1998) is her reaction to the fact that 'sufferers seem to be ignored – on the assumption, perhaps, that we are "too far gone" to care'.[61] She is mindful that there are not many books written by people with Alzheimer's disease, 'as generally they tend not to be aware of their gradual deterioration, nor to be able to document the changes happening to them. There are several written by carers', she emphasizes, but these 'have a very different perspective' (ix–x).

As part of a slowly growing body of patient life-writing, Boden's narrative strongly emphasises the patient's continued agency and voice in a discourse that, for twenty years, had been dominated by the caregiver's view of the patient as passive. Eleanor Cooney and Margarita Retuerto Buades perhaps told narratives of the patient as living dead, because they experienced the loss of a parent or partner. At the same time, accounts about a patient can perpetuate ideas of the patient's social death, because the existence of these accounts per se embodies the unspoken assumption that the patient is unable to communicate. Only once patients began to occupy centre stage in fictional texts did their autonomy and independence become more visible. The emphasis on self-determination in these texts is shared with that in patient accounts which appear towards the end of the century. Written with the purpose of self-affirmation in a cultural climate that hails cognitive prowess, the tone and intent of such life-writing significantly differ from that of genre fiction. But like the examples of detective fiction I discussed, these texts imply that Alzheimer's disease is not necessarily a condition of loss alone and restore patients' narrative and social identity.[62] They emphasize 'that the self persists in dementia'.[63]

Who Will I Be When I Die? affirms the patient's continued personhood, just like the three early patient accounts introduced in Chapter 4 – Robert Davis's *My Journey into Alzheimer's Disease* (1989), Diana Friel McGowin's *Living in the Labyrinth* (1993) and Larry Rose's *Show Me the Way to Go Home* (1996). Like Davis, McGowin and Rose, Boden challenges the perception of patients as losing the plot and needing to be written for. But she goes further in her direct negotiation with objectifying healthcare practices and in her claim on a say in healthcare policymaking. In particular, Boden's demand for the patient to be seen as a person with continued identity and voice strongly resonates with contemporary biopsychosocial approaches to Alzheimer's disease, approaches that attend to the biological, psychological and social dimensions of the condition and inform interaction with the patient. At the same time, Boden's narrative illustrates how the patient's voice continued to be squeezed between popular concepts of the condition as performance-limiting and societal images of Alzheimer's disease as pertaining to old age: diagnosed with Alzheimer's disease in 1995, at the age of 46, the Australian civil servant felt 'far too young to get an old people's disease' (3).

Reading Boden's account in the context of the growing interest in narrative in medicine and healthcare in the 1990s shows how the growing attention to discourse analysis by healthcare professionals and linguists began to change the view of the patient, as patients themselves asserted their continued identity in reliance on an organic understanding of their condition. Boden's wish for patients to be treated 'very much as people, not just as medical cases' (31) matches concerns, raised by medical humanities and bioethics scholar Kathryn Montgomery Hunter in 1991, about 'the professional shortsightedness that sees maladies rather than people as the objects of medical attention'.[64] Boden voices the value to the patient of an empathetic doctor, when she illustrates her encounter with the practitioner, who provided a second opinion on her diagnosis: 'He sat behind a desk in a small but comfortable office, listening, asking, rather than expressing immediate opinions and statements' (8). This characterization of a clinician comes close to Rita Charon's later depiction of narratively trained medical doctors, who should 'donate [them]selves as meaning-making vessels to the patient'.[65] It also underlines Boden's continued ability meaningfully to articulate herself – a perspective taken up by Malcolm Goldsmith, who asserts that 'communication with people with dementia *is* possible' and advocated by healthcare professionals such as the Australian nurse Julie Goyder, who seeks time for conversation with her patients.[66]

As early as 1987, social psychologist and gerontologist Tom Kitwood (1937–1998) had propagated ideas of a 'malignant social psychology' that amplifies the ill-being of individuals with dementia.[67] In *Dementia Reconsidered* (1997), Kitwood contended: 'Alzheimer victims, dements, elderly mentally infirm – these and similar descriptions devalue the person, and make a unique and sensitive human being into an instance of some category devised for convenience or control.'[68] These concerns invited the patient to take control of developing a counter-discourse. And patients were, indeed, beginning to achieve a real shift in perspective. Goldsmith cited earlier patient narratives – Davis's *My Journey into Alzheimer's Disease* (1989) and McGowin's *Living in the Labyrinth* (1993) – and Boden, in turn, praised Goldsmith's study for his attention to patients' accounts in her second narrative (she was by then remarried and called Bryden).[69] In this awareness, Boden becomes an instructor herself, directly addressing the reader and caregiver with the claim that proper care and support can turn patients into proactive people: 'Don't hide us away – involve us, let us still experience the joy of living, with the help of your memory, your abilities, and your patience' (53).

Boden strongly relies on an organic definition of the disease as explaining behavioural changes. Her account builds on the understanding that 'some cells of the brain become diseased and tangled, confused and no longer able to function. The cells affected are those making up our personality, our behaviour, our thinking, our memory' (51). She uses such descriptions to emphasize that 'we are not mad, but sick' (169). In this way, she attempts to claim back (as in many pathographies of the 1990s) the grounds for the patient's individuality within the process of biomedical labelling.[70] Boden's claim is especially noteworthy because it highlights her struggle to position herself between a materialist explanation for her symptoms and an appreciation of her identity through a whole-person approach. She fears the reductive nature of the former, wondering whether her remaining personality would still be recognized as an entity separate from the condition. At the same time, she is afraid of potential stigma entailed in whole-person approaches, because she could be reduced to an unpleasant person grown old.

Boden clearly acknowledges: 'I have lost that vibrancy, the buzz of interconnectedness, the excitement and focus I once had. I have lost the passion, the drive that once characterised me. I'm like a slow motion version of my old self – not physically, but mentally' (49). That a patient in 1998 should adopt the century-old language of loss and decline brings out how difficult it was for individuals with dementia to assert themselves authoritatively without having developed a language of their own. More so, it brings home the core of the patient's experience. Alzheimer's disease *is* a condition that imparts feelings of losing one's self – as Auguste D. had told Alois Alzheimer in one of her many interviews: 'I have, so to say, lost myself.'[71] Boden fears 'that slowly there might be less and less of "the old me"' (47). In the face of this imminent loss, Boden shares the understanding that a patient's lived past structures their illness experience in the present. This perspective formed the core of the psychosocial efforts made by Kitwood and the Bradford dementia group – and informs patient-centred care to the present day.[72] Asserting her unbroken identity and personhood despite Alzheimer's disease, she explains:

> These memories form the kaleidoscopic perspectives of all the many expressions of my being over my lifetime: as a child, daughter, grand-daughter and sister, as a student and young adult, as a wife and mother, as a friend, as a researcher, an editor, an information officer, policy manager and senior public servant, as a member of St George's church and a Cursillo team member, and as a writer of this book. (49–50)

This statement imparts the central claims made by geriatricians and psychologists, who increasingly advocated whole-person approaches to dementia, namely that 'people should be treated as wholes, with attention not only to their biology, but also to their psychology, their social and ethical concerns, and the cultural and spiritual aspects of their lives'.[73]

It is exactly Boden's spirituality around which her understanding of a continued identity and biography centres. This understanding helps her to give an account that entirely offsets Arthur Frank's view of Alzheimer's disease narratives as representing 'chaos' narratives.[74] Boden's telling about her 'emotional, physical and spiritual journey with Alzheimer's disease' (ix) underscores how little she is '"losing it"' (58, 60, 65), how much she is aware of a plot in her life. The journey myth Hawkins identified in much illness life-writing does not easily hold for people with dementia, since the illness experience is one of no return, into darkness.[75] Except that Boden explicitly pictures her illness as 'an adventure into the unknown' (85). The narrative's final part, 'Thank God God's in charge' (125–46), is Boden's life review as much as it is her search for continuity in a life history disrupted by illness.

Like the Miami pastor Robert Davis in *My Journey into Alzheimer's Disease* (1989), Boden sees continuity in her growth as a Christian within and through her illness. She considers her purpose in life to be 'to explain Alzheimer's in a way that lots of people might be able to understand this physical disease' (137). By integrating her condition into a belief system, she deflects, as Claude Couturier would do in *Puzzle: Journal d'une Alzheimer* (Jigsaw: Diary of an Alzheimer's patient) six years later, any responsibility for her condition. Couturier lifts the cause of her memory loss onto a literally unreachable level, when she expresses the hope that 'God at least knows what becomes of my lost neurons and that, above all, he makes good use of them'.[76] This is an important step if we consider that notions of self-infliction emerged in epidemiological research at this time and how genetic and visual evidence was seen as equating patients as specimens to their very illness.

The dualistic purposes of Boden's narrative position her text at a turning point in dementia life-writing: the move from assertions of biographical continuity to increasingly sophisticated purposes in the new century.[77] More and more, patients presented themselves as authority on and within their condition. Christine Bryden's most recent *Will I Still Be Me?* (2018), for example, reads like a manifesto for the patient's continued sense of self in the lived experience of dementia.[78] As such, people living with dementia first relocated themselves at the centre of a cultural dementia discourse, supported by a growing interest in psychosocial approaches to the condition; and thereafter they repositioned themselves in a healthcare discourse. But although vocal writings like Bryden's are on the increase, this shift continues to be slow in coming. Wendy Mitchell's very recent experiences suggest exactly this. Diagnosed with dementia at the age of 58, Mitchell had been working with the British National Health Service for twenty years, but still cannot get the support she needs. 'What chance do others with dementia have?', she wonders; 'I still have a valuable contribution to make and I'm not ready to write myself off.'[79]

In an environment of multiplying patient assertions, from around the turn of the century, more and more caregivers incorporated accounts by the told-about patient within their narratives – a phenomenon mirrored in fiction where caregiver and patient become narrative equals. In Wolfgang Held's *Uns hat Gott vergessen* (God has forgotten us; 2000), for example, fictional diary entries alternate between the husband Markus and his wife Monika with dementia. Or take one of Ulrike Draesner's short stories. The reader plunges into a patient's ruminations, piecing their meaning together only once the narrative voice shifts to the caregiver in the second half of the account.[80] One of the earliest caregiver memoirs of this kind was Kim Howes Zabbia's *Painted Diaries: A Mother and Daughter's Experience through Alzheimer's* (1996).[81] Zabbia includes entries from Lou Howes's notebook, which she kept from the time of her diagnosis to 'record her feelings and her constantly changing symptoms' (54). In essence, these entries trace, as Vaidehi Ramanathan notes, 'a mental disappearance [which] is an analogy of death but far worse, for while the body lives some sense of an "essential self" has gone'.[82] Zabbia herself offers an alternative view on the condition not only by letting her mother speak for herself. She develops the story of caregiving in parallel to her struggling to excel as an artist, in which she gains support from her mother, 'soak[ing] up some of her positive attitude' (139). Similarly, Reeve Lindbergh offers extracts from the pilot's former writing in *No More Words: A Journal of My Mother, Anne Morrow Lindbergh* (2001); Elinor Fuchs includes snippets of conversations with her mother at the beginning of each chapter in *Making an Exit: A Mother-Daughter Drama with Alzheimer's, Machine Tools, and Laughter* (2005); and Arno Geiger reports page-long conversations with, and aphorismic wisdom imparted by, his father in *Der alte König in seinem Exil* (The old king in his exile; 2011).

The cultural fascination with the visual helped these developments. Zabbia wanted to paint what her mother felt (124), supplementing her book with drawings about the stages of her mother's condition, drawings which she herself described as 'art with an emotional impact' (73). Jeannette Montgomery Barron, in turn, published a kind of fashion diary of her mother. Each double page shows an item of clothing or accessory of her late mother, accompanied by Barron's personal reflections on the opposite page. These reflections tell of Barron having 'missed her [mother] for years before her death'.[83] At the same time, this memoir celebrates the mother's life: printed on glossy paper throughout, Eleanor Morgan Montgomery Atuk comes to life on every page as Barron tells of her mother's love for a particular piece of clothing or a special occasion when a sparkling accessory was used. Similarly, Phillip Toledano records *Days with My Father* (2010), the photographs focusing on interaction with, rather than the passive portrayal of, the parent.

In the wake of a growing body of patient life-writing, patient portrayals would become more identity-affirming, both in textual and visual representations, with photographs shifting (though not exclusively) the emphasis from chronicling to moment-in-time recording. An increasing body of artistic work concerned with Alzheimer's disease, inter alia, in the framework of exhibitions, as well as attention to late-life creativity further enhanced a dualistic perspective at the turn of the new century.[84] By contrast, I am not aware of contemporaneous visual productions by caregivers that centre on brain recordings that imaging studies produced. Considering their increasing presence in popular science writing and mass media reporting, I take the carer's choice not to rely on raw medical scans as further evidence of the identity-affirming dimensions increasingly found in visually enhanced caregiver writing.

But in making these arguments, I would caution that an ambivalent or more positive interpretation of visual representations is perhaps only possible once the beholder of these images is aware that a shift in cultural discourse is taking place – or, as John Berger puts it, 'the way we see things is affected by what we know or what we believe'.[85] Already in 1991, Kathleen Woodward had claimed that 'socially, politically, physically – the meanings of old age are changing. On the whole our literary tradition has lagged behind, continuing to produce predominantly dark portraits of aging'.[86] Both Dibdin's portrayal and Boden's narrative illustrate the dichotomous terms in which dementia increasingly came to be understood – at a time when scientific research correlated visual brain recordings to the patient's cognitive loss, as caused by guilty genes, articulated in failed network concepts and recounted in a growing body of caregiver life-writing. We first beheld such ambivalent views in relation to redemptive takes on forgetting in some caregiver narratives of the 1990s, and I consider this growing sociocultural ambivalence towards the condition in more depth in the final part of this study.

Part IV

The whole-person prospects

7

The dichotomy of Alzheimer's disease

On Christmas Day 2014, the latest film by actor, director and producer Til Schweiger, *Honig im Kopf* (Honey in the head), was released in German cinemas.[1] The movie is about a character called Amandus Rosenbach (Dieter Hallervorden) and his granddaughter Tilda (Emma Schweiger) who are travelling from Hamburg to Venice, where Amandus and his late wife had spent their honeymoon. In picturesque nature shots the sun is seen shining, the beautifully arranged mise-en-scène focuses on the characters' smiles and the rhythmic score carries the audience away. Building on thirty years of public learning about Alzheimer's disease, the film works by merely hinting at the patient's disorientation in stock scenes: the hand fumbling with a door knob, the shaving foam being squirted onto the tooth brush, the aimless searching for the correct word. By the same token, the movie spares us the later stages of the condition: incontinence, aggression, disintegration. More precisely, from the moment Amandus does 'lose it', meaning that he does not recognize his granddaughter any longer (1:58:07), Tilda becomes the narrator (2:01:21). She has told the story all along, but we become aware of the retrospective nature of her account only when Tilda speaks about her grandfather's final happy years in the closing minutes of the film. Emotionally charged slow motion artificially extends these joyful times: Tilda's mother takes time off work so that Amandus can stay with them, and Tilda's father spends more time with the family.

Schweiger's film avoids a deeper consideration of caregiver strain, both physical and emotional. Instead, loss and decline are experienced by others: the patient shooting with a stolen police gun (12:03), the patient pushing the ignition button to set off a firework display at a garden party (56:21), the patient driving the son's car (01:10:05) – all these scenes create mayhem. They produce 'fireworks' in a comedy that additionally relies on a number of sexist and misogynistic clichés: the laid-back son with a high-powered job, the stressed-out daughter-in-law with no sense of humour and her disgracefully ageing mother with inappropriate sexual drive. Amandus cracking jokes about doing sums in a cognitive testing interview (50:51), or insinuating the sexual symbolism of a cucumber in conversation with a nun (1:47:40) show Hallervorden as the slapstick comedian well-known to the German audience – merely grown old himself.[2] Given the cinematic memory of the audience, as Raquel Medina would put it, viewers feel all the more emotionally involved in the fate of Hallervorden's Amandus.[3] Contrary to what Sally Chivers importantly observes for mass media, literature and film, Schweiger does not use Alzheimer's disease as a magnifier of what people fear most about old age, namely that it 'could manifest itself ... in an apparent loss of sense and self'.[4] Hallervoden's Amandus has freedom of spirit, lives in the moment and enjoys life. Memory loss – any kind of loss – does not centrally define him.

Schweiger's film was a blockbuster success in Germany. It continued to be on screen eight months after its release, with television talk shows discussing the condition in binary but lively terms.[5] The film can be criticized as commercially exploiting the socio-economically pressing topic of Alzheimer's disease in a summer's fairy tale for the big screen during the touchy-feely Christmas period. As a comedy, the film lacks any sense of social reality, as Joachim Kurz, who lost his own father to Alzheimer's disease, asserted in a column for the internet presence of German arthouse cinemas.[6] In response to such criticism, Schweiger explained that he wished to complement the perspective taken by David Sieveking's *Vergiss mein nicht: Wie meine Mutter ihr Gedächtnis verlor und meine Eltern die Liebe neu entdeckten* (Forget me not: How my mother lost her memory and my parents rediscovered their love) two years earlier, a big-screen documentary which had drawn attention to the condition and different ways of approaching care and end-of-life choices.[7]

This cinematic context cannot sufficiently explain how a dementia patient could feature centre stage in a big-screen comedy. *Honig im Kopf* sits with a large body of feel-good films about spirited ageing – including Jon Turteltaub's *Last Vegas* (2013), Dan Fogelman's *Danny Collins* (2015), Nancy Meyers's *The Intern* (2015), Zach Braff's *Going in Style* (2017), Roger Goldby's *The Time of Their Lives* (2017) and Bill Holderman's *Book Club* (2018) – many of which reveal the favouritism towards aged men that Chivers observes in mainstream cinema.[8] But given how strongly concepts of loss and decline have dominated the cultural dementia narrative across the twentieth century, Schweiger's move must be further contextualized. The final part of this study is concerned with the question of where the disease and its sufferers are located in the first two decades of the new century.

Writing about the immediate present precludes truly historical considerations. However, key concepts can nevertheless be seen developing in medical science, healthcare and cultural representations, even though some of these may not be as durable as what I have referred to as the organic paradigm or the cognitive picture. Schweiger's film is both a product of and a reflection on some changes that have taken place since the start of the new millennium. These changes are partly rooted in postmodern views of the condition as I have illustrated them in Chapters 4 and 6. They are also carried by a rapidly growing marketplace that continues to develop around the condition, especially in rich societies. This marketplace both creates and responds to the priorities of the ultimate consumer – patient, caregiver, the health services, society at large and whoever feels threatened by ageing-related memory loss. It is a space inhabited by various domains that are concerned, each in their different ways, with dementia: laboratory science, the medical clinic, literary scholarship and the human sciences more broadly, as well as cultural productions themselves. And every single domain is subject to economically driven deformations that skew the images and concepts of dementia that this domain projects. This chapter illustrates the increasing fluidity of present-day medico-scientific, healthcare and wider cultural concepts of dementia in this marketplace.

Immunization hope and hype: The patient as non-responder

Following the rise of immunology as a research discipline in the 1970s, a large number of long-identified conditions were redescribed in the 1980s as dysfunctions of the immunity.[9] The possibility that Alzheimer's disease may be caused by infectious agents like a virus or an abnormal regulation of the immune system was also explored. With the amyloid hypothesis steadily gaining ground, immunological reasoning led researchers to pursue the idea that amyloid could be tackled as if it were an infectious agent – a pharmacological approach that aimed to fulfil the vision of Selkoe and others, introduced in Chapter 5, namely that 'blocking of amyloid will be a therapy'.[10] Accordingly, immunological thinking explored the possibility of 'lowering the burden' of amyloid with a strategy against its production and accumulation.[11] This rationale introduced immunological notions of flexibility and spontaneity into the medico-scientific writing about Alzheimer's disease, which, I argue, caught up with budding cultural concepts of the patient as active. Disappointing drug trial outcomes, however, left patients all the more vulnerable, pushing them back into a position of passive objects of pity waiting for a cure. Pharmaceutical moves happening in the wake of these failed trials in turn reveal how scientific investigations themselves were increasingly driven by societal and cultural expectations.

First research results were based on the concept of what is termed 'active immunization', a strategy that challenges the patient's immune system to produce an active response against a pathological agent within the body (in this case amyloid). The results were collected in a murine model of neuropathological features of the condition and published in 1999.[12] The paper's language conveys the initial enthusiasm for the expected potential of this therapeutic approach. Moreover, its reception by the media bore witness to two developments: a transient conceptual shift in how the general public heard and read about the condition and an ever-stronger reach of scientific research into the mass media, met by the public's growing interest in the disease. The study's outcomes led the authors, based in the pharmaceutical industry, to present the new therapeutic agent as 'disrupting' plaque formation and blocking the pathology (175). The clinical recipient of this treatment was understood actively to develop a reaction against amyloid.

In other words, immunological strategies put amyloid on a par with infectious agents against which the body could mount an active response. They figured Alzheimer's – along with diseases such as cancer and infection – as a beatable condition. This transformed the image of the patient as victim of a killer disease to one enjoined to fight. Anthropologist Emily Martin observes how, during the 1990s, most scientists described the immune system in terms of 'the body at war' and a 'communication system' in which various elements 'orchestrate' cells and 'recognize' enemies; ideas of 'flexibility' and 'individuality' prevailed.[13] Such notions were contrary to earlier concepts of communication breakdown in Alzheimer's disease. They conveyed an understanding of the patient as active, just as Christine Boden fashioned herself. They credited the patient's body with characteristics that the wider cultural context linked to those of a postmodern self.

Ideas of a postmodern self include that of a self's fragmentation and dislocation as much as its spontaneity and flexibility. Like David B. Morris, I use the term 'postmodern' here to identify the comparatively coherent period in the development of rich societies in the West after the Second World War, 'a period of rapid transition when values and styles are losing their familiar shape', and a social force that evolved in negotiation with these changes.[14] As Arthur Frank argues, these changes also touched on illness. 'The postmodern experience of illness begins', Frank writes, 'when ill people recognize that more is involved in their experiences than the medical story can tell.' He then goes on to remark on the availability of 'rhetorical tools and cultural legitimacy' as necessary for the patient's successful expression.[15]

With Frank's argument in mind, we could take the increasing postmodern drift as providing both tool and legitimacy for patients credibly to tell their illness experience. The first patient narratives – such as Robert Davis's *My Journey into Alzheimer's Disease* (1989) and Diana Friel McGowin's *Living in the Labyrinth* (1993) – were coherently plotted. They were considered to be inauthentic because, as Anne Davis Basting argues, they described rather than performed the illness.[16] In support of her view, Jesse F. Ballenger expressly terms patient accounts postmodern when they tell their story not in a 'conventional' manner, that is, not in a traditional, well-structured chronological arrangement.[17] Terry Pratchett's television documentary *Living with Alzheimer's* (2009), directed by Charlie Russell, is a case in point. Shooting times could be adapted to the writer's moment-in-time acuity; and episodic reflections portray Pratchett as a spontaneous moderator and presenter.[18] Christine Bryden's *Dancing with Dementia: My Story of Living Positively with Dementia* (2005) articulates such a spontaneous postmodern identity more fully than her earlier account *Who Will I Be When I Die?* (1998). The combination of politically charged essay sections with accounts of her frequent travels acknowledges a short attention span and resulting fragmentation of thought. The same is true for Richard Taylor's collection of essays, *Alzheimer's from the Inside Out* (2007). Their length rarely exceeds two to three pages, but their poignancy regarding, for example, 'nude mice', 'a shell filled with Alzheimer's' or 'feelings of insecurity' made Taylor a prominent patient activist.[19] Finally, the general public's ideas of the immune system included images of 'dancing or playing', rendering Bryden's choice of title particularly meaningful.[20]

In summary, the pharmacological language related to medico-scientific ideas of Alzheimer's as a condition that could be tackled in immunological terms caught up with, perhaps even sanctioned, what the cultural realm had already worked out: that the patient had continued agency and was active and performing. I do not suggest that these research approaches redefined the condition's pathogenesis.[21] But immunological thinking in relation to therapeutic strategies introduced language which holds out the possibility of 'blocking' or 'counteracting' the threat from a terminal disease. This is shown by how the media reacted to the data presented by Schenk and colleagues. In the *New York Times*, for example, the results were 'hailed by scientists as a landmark achievement in the battle against Alzheimer's'.[22] In the same article, a professor of neurology considered immunization 'a radical innovation', and the chairman of the scientific and medical advisory board of the Alzheimer's Association judged the findings 'really exciting'.

So, immunological approaches to Alzheimer's disease shifted the condition away from its finality in pharmacological terms, and these concepts began to be picked up by the media. As already mentioned, with Bruno Latour, we should take the scientist's reach into the mass media as part of the strategies that champion the sale, belief in and spread of findings.[23] But such interaction becomes problematic when research results cannot deliver on the hopes held out – in this instance, the drug's safety, tolerability and efficacy in trials with human subjects. These trials, conducted between 2001 and 2003, revealed that cognitive testing scores remained unchanged, that a considerable number of subjects 'did not mount an immune response' to the drug.[24] These sobering outcomes left patients all the more vulnerable: they became increasingly exposed to the ups and downs of medico-scientific research and exaggerated reporting – a situation which was also influenced by distorted academic press releases as the pressure on science to find a cure continued to mount.[25] In addition, earlier notions of self-inflicted illness appeared to be enhanced by the patient's failure to respond.

Since the early to mid-2000s, a second line of investigation explored the benefits of passive immunization. This strategy does not rely on the body's active response, because an agent raised against the disease-related amyloid within the body is injected directly. In her novel *Still Alice* (2007), the trained neuroscientist Lisa Genova reflects on the patient's and caregiver's dilemma in relation to these drug developments.[26] Once an eminent professor in psycholinguistics, 50-year-old Alice Howland has to accept that the 'ability to use language, that thing that most separates humans from animals, was leaving her, and she was feeling less and less human as it departed' (270). As well as reducing mind to brain as other neuronovels, Genova's narrative additionally relies on socioculturally sanctioned stereotypes of science versus arts, male versus female, rational versus emotional, healthy versus sick. This polarization ultimately gears up towards a staging of the all-or-nothing promise of these new drugs. At the centre of her text, Genova introduces the fictional agent Amylix, which it is hoped will 'prevent the disease from progressing further' (139). Alice must decide whether to be included in a further trial of Amylix. Her husband John is not in favour of this option. A high-flying biomedical scientist himself, he has been living (just like Alice) with the conviction that 'anyone could be seduced by research when the results poured in. The trick was to love it when the results weren't forthcoming, and the reasons why were elusive' (28). That John now prefers an established combination of drugs, rather than helping the cause of science, tells of the helpless situation he is in as the caregiver who loses his partner in life. Alice, in contrast, tries to 'think analytically about her treatment options', arriving 'at a single, crisp image that made sense. A shotgun or a single bullet' (141). She opts for the trial. As she puts it to John, this 'isn't your one shot. You're brilliant, and you don't have Alzheimer's. You're going to have plenty of shots' (234). As a patient, Alice does not have anything to lose, because the diagnosis has already determined her end.

Genova acknowledges both 'Dr Rudy Tanzi and Dr Dennis Selkoe for an in-depth understanding of the molecular biology of this disease' (x). It is, therefore, unsurprising that her image should echo Tanzi's concept of an amyloid-directed inoculant as 'magic bullet' in his *Decoding Darkness* (246). This link, like Genova's novel itself, points to the increasing confluence of dementia representations in a growing marketplace – a marketplace of service users as well as for accounts about dementia. Medico-scientific, popular and cultural representations all closed in on the same concepts and disappointments. For John, the story ends in tears as he reads the lead article of the *New York Times* Health section, entitled 'Amylix fails trial' (287). Alice, by now looked after by a professional caregiver, continues to live at home together with one of her daughters. For the reader, the story ends on a redemptive note, as Alice cradles her newborn grandchild and we gather her continued ability to read other people's emotions. Sarah Falcus believes that the 'intimate experience of Alice's emotional, physical and intellectual states encourages the reader to construct and reconstruct Alice so that she remains not static, but still Alice'.[27] And Rebecca A. Bitenc finds that 'Genova highlights what frequently persists even in the last stages of dementia: the emotional and sensual aspects of relationships that do not depend on higher-order cognitive functioning', and identifies narrative devices in *Still Alice* which convey 'that no clear boundary exists between "self" and "loss of self"'.[28] But especially with the drug story in mind, I find it difficult not to agree with Stefan Merrill Block who argues that, 'considering Alice's formerly inquisitive, type-A, tough-as-nails personality, this does not feel like "still Alice". This feels like a transformed self'.[29]

Two years after Genova's fiction, the outcomes of first clinical trials on the passive immunization prototype drug bapineuzumab were published (its mechanism of action being comparable to the fictional Amylix). Again, research found no significant differences in cognitive testing outcomes.[30] Despite ever-intensifying research, Alzheimer's disease remained a condition with no cure in sight; in Genova's words, 'molecular murder or cellular suicide'

(1), with the patient its 'victim' (256). But, in comparison to the 1990s, the patient's perceived loss was amplified several fold: by the pharmaceutical industry's persuasive marketing of potential remedial interventions; by the mass media's exaggeration of press releases; and by the general public's so acquired common knowledge about the disease. I don't argue that science and industry were driving a condemnable profit-oriented campaign. Latour has made it abundantly clear that scientific research can only truly flourish in a marketplace in which different disease hypotheses compete for financial support.[31] The issue at stake was that Alzheimer's disease had reached an extremely value-laden prominence, and scientific research was expected to offer hope of deliverance from a condition which meanwhile had turned into a major public health issue and inspired fear in large parts of the population. This problematic situation is reflected in the claim made by William Thies, the chief medical and scientific officer of the Alzheimer's Association, at the association's International Conference in Vancouver in 2012, and the research carried out in its wake. Thies maintained that 'earlier detection of Alzheimer's, greater knowledge of dementia risk factors, and better treatments and prevention ... are critical in order to create a future where Alzheimer's disease is no longer a death sentence but a manageable, treatable, curable, or preventable disease'.[32] A key trial with bapineuzumab, and two trials with its sister compound, solanezumab, terminated within two months of the conference's end, with researchers having to acknowledge that the agents did not have an effect on cognitive performance.[33]

The film *Still Alice* (2014) does not broach the drug story at all, perhaps in view of the sobering appraisal of immunization-related strategies ten years after high hopes for an imminent cure had first been cherished.[34] Instead, the movie's energies – mise-en-scène, camera work and score – are all bent towards uncompromisingly depicting the disease progression of Julianne Moore's Alice: her loss of orientation, awareness and language, and her helpless subjection to the clinical diagnosis of memory loss (19:06). If journalist and writer Charles A. Riley is correct that the 'safest nomination bets for Oscar gold, year after year, are disability flicks', we cannot be surprised that Moore's performance earned her the eighty-seventh Academy Award for Best Actress. What I am more interested in, though, is Moore's acceptance speech on the occasion. Riley is convinced that such speeches zero in on the 'victim', who is 'momentarily blinded by flashbulbs', only to be forgotten weeks later.[35] Moore expressed the hope that the movie would 'shine a light on Alzheimer's disease', that it would make patients feel 'seen', because 'people with Alzheimer's deserve to be seen, so that we can find a cure'.[36]

Within two weeks of the film's appearance, news reported that fears 'drive the middle aged to seek checks', affirming how the big screen can help raise awareness (while creating an alarmist atmosphere).[37] Yet, the awareness raised was for what the film scholar Lucy Bolton describes as 'an incurable, humiliating nightmare through which we are transformed into a vision of the walking dead, locked into our own heads, unable to think or care for ourselves'.[38] The tie between the film's monolithic take on Alzheimer's disease and Moore's call for a cure adeptly condenses the patient's loss as it is societally perceived in the second decade of the twenty-first century. It is no longer only the medico-scientific or sociocultural understanding of the patient's loss of awareness, language and self. It is the absence of a cure that enhances this loss. This is especially the case in capitalist societies, where, according to Dana Lee Baker, we find an 'almost completely unquestioned assumption of superiority of cure over care', as both are treated as parts of one overarching effort: to increase funding for medico-scientific research.[39]

The political energy of the dementia patient's pharmacological loss becomes apparent from the fact that researchers on solanezumab revisited the clinical data collected up to 2012, and in 2015 published a paper on secondary outcomes. John Abraham's study on policy-related components to pharmaceutical research can help explain this move: specific outcomes may selectively be emphasized to keep drugs in trial for longer after large sums of money have already been invested in their development.[40] Reviewing clinical data for obtaining secondary outcomes also is (as seen earlier in the medico-scientific history of the condition) a further example of how specialist knowledge establishes the parameters according to which it wants to see its research subject defined. Stratification of subjects in this revision now revealed, in patients with mild Alzheimer's disease, less cognitive decline.[41] These new findings were effectively taken as justification for claims like Thies's that immunization might be beneficial especially in the preclinical stages of the condition.[42] By what Latour terms inventing new goals and interest groups, researchers began to turn to issues of prevention.[43] Overlapping with a slowly rising cultural visibility of the early-onset Alzheimer's patient, this move amplified concerns about health risks in the middle-aged.

Arguments are being made that an earlier diagnosis enables patients to take a more active role, for example, in making their wishes known.[44] But the film *Still Alice* depicts the deleterious consequences for the patient and her family of what it means to live with this diagnosis. The inheritable nature of the early-onset form of Alzheimer's disease challenges an entire family, and the prospect of inescapable decline leads the patient to consider a radical end-of-life decision.

Alice's conviction, in both book and film, is telling: 'She'd trade Alzheimer's for cancer in a heartbeat ... There was the chance that she could win' (117; 34:59). The continued absence of a cure only enhanced awareness of the patient's loss of memory, awareness and self, after immunological approaches and their representation in the media had begun picturing Alzheimer's as a beatable condition. Even during a period whose fascination with postmodern ways of life both granted and legitimized the narrative tools for fractured selves, a patient like Bryden eventually had to 'move away from the bright lights to a corner of the [dance] floor where the rhythm is slower and the music quieter, but still sweet. All I can do now is sit quietly and listen, and hope for a cure'.[45]

In the meantime, patients continue to rely on care. But, as Dana Lee Baker points out, 'public care is deliberately limited only to those in formally legitimated need ... Victim status is still often considered a necessary condition for provision of care'.[46] The 2014 UK initiative financially to reward general practitioners for every Alzheimer's diagnosis made has caused controversy across the media, political parties and healthcare associations.[47] Living with the disease and its label for an extended period can be acceptable for patients and their caregivers only when the label Alzheimer's no longer spoils the individual's social identity.[48] That is, once sociocultural attitudes to the condition change – under circumstances that Schweiger depicts (undoubtedly too simplistically) in a world ruled by love and empathy.[49] But reflected against a century-long narrative of dispossession, decline and loss, significant changes are needed politically and institutionally to deliver on the transformative potential of love. In a first instance, attitudes towards the patient must become as unprejudiced as Tilda's appreciation of her grandfather's continued identity outside memory loss and ageing as decline; and healthcare conditions must become more identity-affirming, if not celebratory. These are the issues I develop in the following two sections.

La guardiana di Ulisse: The patient beyond forgetting in children's literature and adult fiction of the new century

The most salient feature of *Honig im Kopf* is that the story is told from the perspective of a child. Moving the child's perspective from an underprivileged narrative space in a mother's account (as I discussed in Chapter 4) to the big screen during peak viewing season – both for that fact, and that it took thirty years to take this step – is noteworthy. First and foremost, this move draws attention to what partly facilitated it: a growing body of dementia narratives addressing children. In addition, that a dementia patient has become an increasingly prominent figure in books for children and young adults since the year 2000 is further evidence of the condition's popularization.[50]

To some extent, this rise indicates changing caregiver support strategies. Many such books, like those published by the American Psychological Association, were intended as pedagogical tools for caregivers, whose 'child may become stressed and anxious, especially as your family dynamics change'. But in providing support for the caregivers in the middle, they recounted the 1980s and 1990s story of the grandparent whose 'life and memories are leaving him'.[51] They also brought the grandmother as an object of scrutiny to the child, closing, for example, on the narrating child's observation that 'Nana and I can both wear striped shirts and flowered pants, and it will be okay'.[52] Such an account portrays the child being read about as an American 'improved self', as Lisa Diedrich would say.[53] But at the same time, this narrative strategy perpetuates the concept of the patient as child, underpinned by the striking visual resemblance between grandparent and grandchild, including facial expression, head shape and clothing. Such stereotypic representation, so Mareike Hachemer observes, reduces these narratives to didactic texts, without leaving children any room to negotiate with their own grandparent's otherness.[54] But narratives about dementia for children have also begun rewriting the cultural dementia narrative; they offer the chance to 'take a non-judgemental, unencumbered perspective onto a person, who is different'.[55] Tilda inverts the adult version of Alzheimer's disease, offering a vision that accepts the patient in all his peculiarity. For her, Amandus is on holiday.

It has variously been argued that applying literary criticism to children's literature demands an analytical consideration in its own right. Yet, I agree with the literary scholar Frank P. Riga that writing for children 'will appeal on a number of levels, eliciting different responses from different readers'.[56] This is the case, so Riga argues, because authors convey their message in both literal and figurative terms. Any supposedly simplified representation of patient and condition will tell the critical reader what is perceived to be the ultimate truth about illness and sufferer. I read Alessandro Borio's narrative with these genre presuppositions in mind. *La guardiana di Ulisse: Una malata di Alzheimer, un angelo e una domanda* (Ulysses's custodian: An Alzheimer's patient, an angel and a question; 2006) brings home its social criticism to all groups of readers in the form of an allegory.[57] It can, despite being addressed to children, readily be critiqued as a rejection of contemporary society's lack of reverence for patients – for their identity and personhood, awareness and needs.

Cynthia Marshall illustrates how a story's apparent resonance with a biblical passage will encourage readers to search for its deeper message.[58] Borio's allegory makes reference to a specific biblical source, alludes to Greek mythology and incorporates poetic symbolism. These heighten the reader's attention to imagery, especially because the narrative's meaning – the guardian angel Ulisse Mantovani's mission – is unclear from the outset. His name lets the reader foresee similarities to Homer's astute hero. But moved to a contemporary allegory the Ithacan king's plot-driving travels turn into Ulisse's (and the reader's) search for meaning. Ulisse is sent to Earth to watch over the Alzheimer's patient Maria Merlo who, he has been told, is a custodian. Ulisse has to figure out why Maria must be protected and what she is custodian of. Focalized through Ulisse, this search for meaning drives the plot. Maria's role on Earth is to preserve civilization and humanity that would be reflected in society's attitudes towards patients like herself.

Borio explicitly removes Maria's condition from its limited meaning as memory loss in the elderly, making it merely a vehicle for a larger truth. Maria stands in for 'individuals of all ages and origins' with a common characteristic: 'a deep suffering had signed them and now they lived in extreme, aggravated weakness. They were all ... "almost vegetables"' (78). Borio's invitation to revere those whose predicament obliterates any difference in age overlaps with Atul Gawande's observation that the current cultural climate has eclipsed the 'veneration of elders'. Gawande comments on the quality of the 'waning days of our lives ... spent in institutions – nursing homes and intensive care units – where regimented, anonymous routines cut us off from all the things that matter to us in life'.[59] Reflected against Gawande's deliberations, Borio's text revolves around the care for the patient. The power of Maria's status as patient – her name signalling an intercessionary role in the Roman Catholic tradition of Borio's Italy – is conveyed through the biblical source in Borio's text, Abraham's intercession for Sodom (Gen. 18.22-33). God will save Sodom from destruction, if only one righteous person can be found in the city (135–6). In Borio's tale, then, 'the only righteous people in the world ... are those whose condition outruns any possibility to sin, and on them rests the foundation of the world' (82). As Borio assigns this world-saving importance to patients like Maria, he invites us to consider how that world revolves around the patient.

Like Konrad Lang according to Miriam Seidler's critique, Maria herself is a flat character, and it is only through Ulisse's mind-reading (printed in italics) that we become aware of Maria's shrunken world. Still, the emotions and behaviour of all other protagonists develop in negotiation with her. This helps

Borio to develop his social critique, because less emotionally complex characters provoke complex feelings in other characters, and such mental triangulation can 'question existing social hierarchies'.[60] Maria's son Francesco feels that his mother has become 'her own tomb' (19). This view induces him to place her in a nursing home, and so remove her from any contact with her grandchildren. Maurizio, in particular, misses his grandmother, because for him Maria's memory loss has no negative meaning. It is around the reinstating of contact between grandchild and grandmother that the central message of Borio's text revolves. But how can the contact between two characters without apparent social agency come about? With the help of an angel and an act of faith: Ulisse's dream.

It is in Ulisse's dream that Maria and Maurizio are reunited; that Maurizio relishes his grandmother's hug; that Maria becomes active storyteller. And the story she writes down for Maurizio is the plot- and meaning-driving fable within Borio's allegory. *Il girasole immobile* (The static sunflower; 121–4) tells the reader how society should be thinking about the patient. Rebelling against the sun's lacking recognition of their undivided adoration of her, a field of sunflowers collectively decides that their gaze would no longer follow her rays. But in doing so, they become unpleasant and self-interested. In this climate of utter indifference, a new sunflower grows, who happily follows the sun's trajectory. Challenged by the older sunflowers, this new arrival tells them: 'For me, it is not important to know whether the sun is aware or not of my love, the fact alone of loving her makes me feel good, allows me to receive her vital energy that enlivens my colour and ripens my seeds' (124).

Literally becoming Borio's take-home message, Maria's letter is found, so arranged by Ulisse, in the armchair she occupied in Francesco's home. As a physically tangible object, Maria's story becomes real for those who doubt the truth of dreams, the views of children and the authority of patients.[61] On hearing this fable, Francesco adopts a perspective similar to Maurizio's. He sees his mother 'like a person and not like an Alzheimer's patient', and asserts that 'people are much more than what they manage or do not manage to do, understand, feel or remember' (129). In addition, Borio's choice to tell a tale featuring an angel encourages the reader's creativity.[62] Or, as sociologist Luigi Berzano suggests, with the return of angels in cinema and literature, the world dominated by technology and rationality becomes recoloured.[63] In line with such a reading, Borio lets Francesco learn. He allows Maurizio to see his grandmother again. And, having trusted the authority of an angel in an act of experiential reading, we have learnt our lesson: we are able to see 'something that resembled very much a smile' (141) on Maria's face when Maurizio finally hugs her.[64]

The angel's role in Borio's fiction obtains added meaning when read against Jens Brockmeier's reference to the Odyssey. For the scholar of psychology and philosophy, 'the fascination of Homer's epic poem stems ... from its conjuring up the idea of simultaneously moving forward and backward in time'.[65] Also Ulisse travels in time and space to collect memories of how Maria was in the past. These memories help Francesco to appreciate her in the present of her illness. But the reader does not access the past through Maria, and this helps Borio to remove dementia from notions of memory loss in the first place. Borio's representation of dementia is more in line with recent work that emphasizes the natural function of forgetting and demystifies the idea that the brain stores everything.[66] Such a change in perspective, what Brockmeier terms 'memory crisis', has developed since the 1990s, as archival notions of memory began to be questioned in social and cultural, literary and biological domains.[67] Here, I see parallels to the changes in identity-preserving concepts of forgetting which Nicholas Dames detects in the nineteenth century, exemplified in Timothy Forsyte. This dynamic suggests to me that a cultural shift in relation to memory is essential to, and anticipatory of, a cultural change in understanding dementia. This shift is not only seen in children's literature, but also in a growing body of adult fiction that portrays dementia as a mixed blessing: an opportunity for growth and fulfilment, or a condition that can help emotional healing because it entails forgetting.

One recent example that loosens decades-old depictions of dementia is Andrew O'Hagan's *The Illuminations* (2015).[68] O'Hagan juxtaposes the story of Anne Quirk, a 'wee lady ... [who] knows her own mind' (105), with a growing sense of spatial and temporal disorientation, to that of her grandson Luke, who is out on a mission in Afghanistan, which had turned 'their brains ... soft from months spent doing nothing' (38). Later in O'Hagan's narrative, it is the fighting and killing that 'wears away at you', so that 'there's less of you every day ... Less life. Less cause. Less morality. Less belief. Less judgement. Less energy. Less fucken hope. Just less.' (130). This language strongly echoes war-related imagery that became linked to the illness experience during the 1980s. However, O'Hagan strictly separates two narrative strands that are dominated by Anne and Luke, respectively, in their distant locations. In this way, he removes Alzheimer's disease in general and Anne's condition in particular from such imagery. Once Luke returns from a mission gone wrong the two narrative strands merge. In the conviction that she is 'losing parts of herself and gaining others' (166), Luke identifies 'a trace of something young' (252) in his grandmother. The flexibility with which Anne interprets her past illuminates his own mind, because she had 'preserved what she could of his young mind's entanglements' (255). Drawing on these insights, Luke eventually can 'imagine a future less taken up with loss' (265).

To appreciate that the essence of one's past is central to one's identity in the present has been fundamental to much adult-child caregiver life-writing since the 1990s. In O'Hagan's narrative the patient's regression into the past becomes the healing anchor for the younger generations who run the danger of losing themselves in the traumas inflicted by a fast-living society. Don DeLillo takes a comparable approach in *Falling Man* (2007). He interlaces an account of how individuals grapple with the aftermath of the 2001 terror attacks forever etched in their minds with Lianne's fears of having inherited her father's Alzheimer's disease. In this way, the reader is led to associate the condition with redemptive aspects of forgetting. In essence, narratives like Borio's and O'Hagan's suggest a view of Alzheimer's disease less intricately linked to negative meanings of forgetting, even partly removed from memory loss. They also remind the reader that ageing and old age can embody life experience, wisdom and love.

The concept of dementia as a mixed blessing evolved on the fringes of a cultural dementia narrative that continued to be dominated by notions of loss of self and life. It progressed alongside a rise in popular fiction whose success was at least partly carried by the unbroken fascination with all matters neurocognitive. Such fiction continues heavily to rely on archival concepts of memory. It develops stories of individuals who had been active in their past; and this past dominates both text time and narrative moves: the reader can only access it through and because of the patient's increasing memory loss. Debra Dean, for example, draws out Marina's 'slow erosion of self'; Samantha Harvey shows Jake as 'gone'; and Emma Healey characterizes Maud as 'cut loose'.[69] All three novels – Dean's *The Madonnas of Leningrad* (2006), Harvey's *The Wilderness* (2009) and Healey's *Elizabeth Is Missing* (2014) – play with the reader's expectation that the patient lives on a rich (but emptying) memory; all headed bestseller lists; all were the authors' debut novels. And this further illustrates how culturally current Alzheimer's disease has become during the past fifteen to twenty years – culminating in its exploitation in a big-screen comedy. But unlike *Honig im Kopf*, many of these narratives appear to propagate the story of the living dead on a mass audience level, writing, like Dean, that 'Marina herself has left ... It is all over but the waiting'.[70]

At the same time, however, this body of fiction has invited more ambivalent readings of dementia. There are gradually dualistic aspects to how the condition is viewed as a mixed blessing. Nicci Gerrard, for example, takes *The Wilderness* to illustrate that 'profound loss has a consolation', because Jake can forget his past.[71] In a similar fashion, Ian McEwan's move to let Briony succumb to dementia in *Atonement* (2001) can be read in binary terms. For Laura Salisbury, it serves to 'displace the question of the unreliability of subjective memory ... with a material form of loss'.[72] We could also read it as Briony's final atonement after lifelong efforts to come to terms with her deception. Or, she 'might not be unhappy – just a dim old biddy in a chair, knowing nothing, expecting nothing'.[73] Such an increasingly ambivalent perspective is also found in a number of recent adult-child caregiver accounts, and I will now turn to exploring how this growing ambivalence proves relevant for healthcare considerations.

Alzheimer mon amour: Healthcare changes and patient personality in contemporary caregiver memoirs

> On Monday I visited my godmother who has dementia. Her care home has flower-beds and seating at the front. But these are for staff and visitors. Residents have no way into the fresh air. They are imprisoned. It was a palaver to take my auntie through the security door to sit on the (too-low) bench. But outside the stifling day room circle of nodding souls, she brightened.
>
> Janice Turner, 'I Dreamt of Greek Olives as I Fumed in Munich'[74]

Janice Turner's frustration about the discrepancy between policy changes and their actual implementation strongly resonates with Scottish broadcaster and writer Sally Magnusson's account of her mother's dementia.[75] Magnusson particularly criticises 'an under-the-radar culture of poor practice in a system never designed for the longevity we have created for ourselves' (17) and brands 'the lack of empathy for confused patients and knowledge about their condition' (183). Magnusson, who together with her siblings cared for her mother at home, further explicates Alessandro Borio's idea of the patient as saviour of values, writing that 'the ultimate guarantee of civilisation for any society in any century ... is that every mind, every person, is regarded as equally precious' (375). In a similar fashion, the Italian adult-child caregiver, Elena De Dionigi, emphasizes her father's continued identity in the assisted living facility of Camelot. Her conjuring of associations with the legendary King Arthur locates Alzheimer's disease in the realm between the real and the imagined. It calls the reader to see in the patient all that remains rather than what is lost, an interpretation encouraged also by the narrative's title, *Prima di volare via* (Before flying away), which, like Maria Merlo's surname 'blackbird', alludes to the patient's flying away.[76]

But perceptions of loss continue to permeate caregiver life-writing. Their presence brings home how difficult it is to establish new standards in dementia discourse as much as healthcare practice. Giovanna Venturino, for example, comprehends her parent's condition as an expanding 'sea of nothingness'; Nucci A. Rota writes about her mother-turned-child; and Donatella Di Pietrantonio sees her mother's condition as one of 'lost control'.[77] Nicola Gardini, by comparison, describes the search for his father as a lifelong process that continues into his illness in *Lo sconosciuto* (The unknown; 2007). Such binary readings at least partly feed on, and also reinforce, celebratory aspects – and these have the potential to push healthcare approaches into new directions.[78] The relevance for healthcare considerations of this growing ambivalence is perhaps best put into words by Gawande himself: 'We've begun rejecting the institutionalized version of aging and death, but we've not established our new norm.'[79]

In this section, I explore what this new norm could be and what institutional and political changes need to happen to deliver on the transformative potential of love. I address how 'a policy bias in favor of a greater emphasis on caring medicine in old age', called for since the early 1990s, eventually began to manifest itself towards the end of the first decade of the new century.[80] As my example, I take the psychologist Cécile Huguenin's *Alzheimer mon amour* (Alzheimer my love; 2011) and read it in a cross-cultural way.[81] I do so because Huguenin herself thinks about differences in how various cultures understand and deal with Alzheimer's disease and ageing. In addition, the psychologist depicts the import of international research into the French context as influencing how she was advised on caregiving. Sociological research suggests that dementia-related sociopolitical movements in France echo the American model – a connection that points, among others, to the increasing globalization of dementia caregiving issues in the new century.[82]

Cécile Huguenin writes about her husband's illness in a short volume of not much more than one hundred pages. Divided into three parts, the account changes narrative form and perspective almost as often as the couple do their residence and living arrangements. These different forms mirror Huguenin's search for how to live with, and write about, the condition after having shared thirty years of life and love. In essence, it is a search for the discourse that would help her attitude towards patient and disease, at a time when care and cure policies continued to victimize the patient. As a product of this searching experience, Huguenin's narrative witnesses the life-changing effects of healthcare approaches that affirm the patient's continued identity and personhood, while leaving reflective space for the caregiver. This space allows for a view of the individual with dementia that is less prejudiced, as we have seen, for example, in the case of Borio's Maurizio.

The narrative's first part covers the years 2006 and 2007: 'Two long years to realize, powerless and lonely, his progressive and unavoidable drowning (*naufrage progressif et inéluctable*), as if every day a small piece of him was being swallowed by unknown depths' (17–18). It is told in the third person, and its simple language and uncomplicated syntax suggest the story belongs to a fictional world. These formal choices add to the reader's appreciation that Huguenin tells the tale of a couple wandering without hope or direction in the world of sickness. Huguenin describes this narrative choice as allowing her to distance herself from the illness experience.[83] More generally, this narrative strategy permits writers, like French TV director Françoise Laborde, in *Pourquoi ma mère me rend folle* (Why my mother drives me crazy; 2002), or the Swiss caregiver Ruth Schäubli-Meyer, in *Alzheimer: Wie will ich noch leben – wie sterben* (Alzheimer's: How will I continue to live – how will I die?; 2008), unflinchingly to reflect on problematic issues, such as challenging family constellations and difficult end-of-life decisions.

Initially escaping to the seaside, Huguenin ponders whether 'one can mourn somebody who is alive' (20). She particularly struggles with her reaction to ideas of self-infliction put forward by Daniel's neurologist: 'her anger, her desire to die, her obsessions to murder, to kill herself' (37). Eventually back in Paris, Huguenin seeks help from medical scientists, wanting to present 'this specimen of an addled (*décervelé*) man, this strong body deserted by the mind' (39). But Huguenin remains helpless in her 'absurd encounters' with the medical profession (43). The narrative's first part closes – in allusion to *Alice's Adventures in Wonderland* (1865) – with her impression that she only comes across 'crazy hatters, queens without a kingdom and rabbits in tails' (43). That Huguenin pictures herself and her husband as lost in an environment that resembles Carroll's nonsense world is perhaps best explained by the insights collected by Charlotte L. Clarke, professor of Nursing Practice Development Research, and her colleagues in a 2011 caregiver manual, namely that

> the reciprocity within a care relationship means that both the carer and the person with dementia are working towards keeping some stability, continuity and sense of the normal in their lives. Health and social care interventions, or others outside that caring relationship … throw a spotlight on the abnormal by focusing on the illness more than on getting on with life.[84]

With her husband's increasing dependence, it becomes Huguenin's responsibility to navigate the world of care; to find the couple's ideal way of getting on with life; to realize their dream 'of a future that would allow us still to live together' (60). Consequently, she tells the second part of her story in the first person.

Based on her experiences in Africa and India, Huguenin idealizes a life in a culture where Daniel's 'condition would be considered a blessing and he like a messenger from the hereafter' (61). Fieldwork in India has revealed that this culture interprets demented behaviour in relation to the patient's social status, and African traditions consider family support a moral norm and emphasize the role of elders. These value systems influence how family caregivers from different cultural backgrounds construct their idea of Alzheimer's disease.[85] Conversely, 'within an illness culture which emphasises the deficits of people', so observed by Clarke and colleagues, 'options for creative solutions to needs are reduced' (68).

In Madagascar, Daniel 'blossoms' (76). But the couple are forced to return to France, and Huguenin's renewed change of narrative form pinpoints her exhaustion. While continuing to tell her account in the first person, she now addresses her husband directly. Di Pietrantonio and Magnusson also use such a strategy in their stories. But in recounting the mother's past to the mother in the second person, their approach is an attempted 'transfusion of memory', a strategy aimed at writing the parent back to life and identity.[86] This strategy gains all the more thrust when considering that even patients like Wendy Mitchell use it for holding on to acquired capabilities for as long as possible in the present of their illness. In *Alzheimer mon amour*, by comparison, this narrative choice emphasizes the demands Daniel places on Huguenin, his needs that make her 'prisoner' (82) – that lead to 'the crash that forces [her] to land' (83). But Huguenin lands on her feet. The clinician, who takes care of her husband after a severe fall, points Huguenin to an assisted living facility that enables 'a life put in place (*aménagée*) to take care of those exiled by a society that can no more or does not know yet how to accept them' (93). The narrative's final part is set in that place. A third of the entire text tells about how Huguenin experiences the way of life in this nursing home. Its calm tone and settled narrative perspective bring out the peace that comes with having found this place. The couple have reached their destination – in discourse practised, care received and narrative form enabled.

The place's caring philosophy centres around the understanding that the professional caregivers 'have to adapt to [the patients]' (94). This approach resembles the values pushed by a working group around the geriatrician and president of the francophone association for the rights of the elderly, François Blanchard. Similar to those set down in the 2007 UK personalization agenda and the National Dementia Strategy in England in 2009, these values centre on the 'imperative of the respect for the individual'.[87] In the French context, for example, Nicolas Philibert's *La moindre des choses* (Every little thing; 1996) demonstrates that such values – to credit the knowledge of service users – were practised in institutions prior to policy changes. In Germany, policy changes are equally recent, even though initiatives like Demenz Support Stuttgart have promoted the ideal of caregivers as 'companions' of people with dementia since 2002.[88] But the hands-on implementation of such concepts continues to be slow. Contemporary healthcare keeps privileging, as Clarke and colleagues noted in 2011, 'the knowledge held by health and social care practitioners and underplays the importance of knowledge held by service users' (16).

However, the power of appropriate care is far reaching. Daniel's carers take the view that 'conversation between persons with dementia and their interlocutors is a privileged site for ongoing cognitive engagement'.[89] Huguenin notices that 'for the first time, every one of his interlocutors has addressed him. Directly looking him in the eyes, without passing via myself as an intermediate and without ignoring him' (99). Her move to mention the caregiver is noteworthy for the fact that scholars of ageing like Kathleen Woodward continue to lament the narrative absence of the frail elderly's professional caregiver.[90] That said, De Dionigi praises Marco in particular, and professional caregivers in general; and Diane Keaton ponders hospice care and explicitly acknowledges her mother's caregivers.[91] I take their choice as a significant step towards change in prosperous societies, especially because Julia Neuberger, in 2008, called for more respect as well as improved training for healthcare staff, 'who are paid abominably, and are themselves bullied by penny-pinching, technocratic management regimes'.[92]

The attitude of professional caregivers patterns Huguenin's own narrative approach and her perception of her husband. After she had deprived Daniel of a narrative voice throughout parts one and two of the text, she now includes several of his short poems. These assert his continued identity and personality, as Huguenin sees it (91). Huguenin's choice is encouraged by how professional caregivers themselves 'amass treasures', recording for themselves 'a word, a gesture, a mimic' of individual patients (111). During the first decade of the new century, such person-centred healthcare practices suggest a shift from person with dementia to personality within dementia. Resulting also from such celebratory sociolinguistic approaches, the views and opinions of people with dementia began to be heard. An international journal of social research and practice, *Dementia*, was launched in 2002, inviting people with dementia to contribute or join the editorial board.[93] Plays enacting dementia were developed with educational purpose, with some of these plays originating from interviews with patients and their families; and the patient's perspective was becoming constitutively integrated in the education and training of healthcare staff and students of health-related disciplines.[94] In the same vein, recent dementia guides address separate chapters to caregivers and the individual with dementia, or include the patient perspective among articles on medical, legal and ethical aspects of dementia.[95]

Also Huguenin herself considers Daniel her teacher. In a short epilogue she thanks her husband for 'having made me go through the hell of your illness so that I would become a different person' (131). Having crossed this hell, she can now – perhaps reminding us of how Borio envisaged his creation of Maria Merlo as custodian of love and empathy – appreciate in other human beings 'the endless power of being' (131). Or, as Magnusson puts it with respect to her mother's continued identity:

> You are still you ... You will be you when the last of the fight has left you. You will be you when you no longer know who I am. You will be you even if I can recognise nothing of what you are and connect it with nothing of what you used to be. You will be you, I really do believe, until the last breath leaves your body. (329)

This perspective matches the view patients like Christine Bryden and Claude Couturier increasingly offered about themselves in new forms of patient life-writing.

To assign wisdom and life experience to an individual with dementia reminds us of the positive images that collaborative health humanities projects began to uncover for ageing and old age during the 1980s.[96] Since the early 2000s, a comparable gradual shift seems to be underway for dementia.[97] Meanwhile, the popular view that pharmacological research could be closing in on what Huguenin describes as 'this molecule of forgetting that corrodes him inexorably' (39) partly enabled such a change. More obviously and effectively, such a change was supported by narrative medicine-based healthcare approaches to, and a health humanities interest in, ageing, which furthered research projects, for example, on creativity and dementia, and explicitly included elderly people.[98] Nevertheless, a truly lasting cultural transformation is likely to develop only by incremental steps, because the concepts at stake – loss, decline, passivity, dependence – are particularly resilient as they are embedded within a variety of mutually enhancing discourses.

Take the health humanities as an example. Over the past thirty years, they have been of significant service to people with dementia and their caregivers. But scholarly research has, to a large extent, bought into and further amplified the cultural imperatives about dementia, in this way exacerbating the notion of

Alzheimer's disease as a condition of old age. In the field of literary scholarship, this phenomenon is at least partly explained by a focus on caregiver narratives which, by nature, predominantly feature elderly patients. As I already explained, I believe this focus to be conditioned by an (until recently) almost monolithic attention to John Bayley's account and – maybe unavoidable – its prominent filmic adaptation. But it is also true that, even though films like *Still Alice* (2014) feature the early-onset patient, entertainment culture more generally continues to instil the idea of dementia as a condition of old age. Prominent examples include Sarah Polley's frequently critiqued *Away from Her* (2006) and Phyllida Lloyd's *The Iron Lady* (2012). The latter is a special case in point, because it portrays Margaret Thatcher as young and powerful as long as she is in office; the only 'Thatcher' outside office that the film shows is the former politician in a state of hallucinating dementia in old age.

Revealingly, identity-affirming fiction like Borio's or personality-centred life-writing like Huguenin's leave very little impact on the language used in relation to Alzheimer's disease, regardless of the patient's age. Notions of loss reflect society's deepest fear about the condition: the death of the self and the fear of how a person who is losing their self will be treated by that society. The fear of dying, which Mary Mothersill describes as central to the experience of old age, is aggravated in the experience of Alzheimer's disease – for both patient and caregiver.[99] Celebratory approaches can do very little to change this, especially in rich societies in which the perceived value of individuals is tied to their productivity. In these societies, so Lucy Burke recently observed, the 'rhetoric of crisis' regarding dementia is closely knitted to financial gain and capitalist marketization.[100] I do not argue that the number of dementia diagnoses has increased with the rise of new capitalism, but I believe that principles of individualism and an atmosphere of entrepreneurialism have contributed to marginalizing people living with dementia even more.[101] In the final section of this chapter, I explore more fully the reasons for the persistence of notions of loss and decline in a nearly twenty-year period that has seen tentative shifts in how the person with dementia is viewed in a variety of disciplines and realms.

We Are Not Ourselves: The cultural image of Alzheimer's disease in the twenty-first-century bildungsroman

Edmund Leary's ailment in Matthew Thomas's novel *We Are Not Ourselves* (2014) is Alzheimer's disease as it is perceived and dealt with in current

American culture: a condition of dependence, decline and loss, but most of all of dispossession in both intellectual and economic terms.[102] A century ago, John Galsworthy deployed dementia as a micro-cultural image of macro-cultural concepts of degeneration in order to portray a particular culture. Thomas's early-twenty-first-century narrative, by comparison, tells the story of a character's disease trajectory as degeneration literally rather than symbolically.

It is by way of Thomas's genre-typical scrutiny of family conflict in the bildungsroman that we learn about the culture at large, in which this illness is experienced. I argue that, first, the qualities assigned to Edmund and his illness, and, second, the ways in which this character fits into the narrative form and frame suggest that the condition stands conceptually opposed to all the values held high in twenty-first-century enterprise societies. The account's overt emphasis on loss highlights how an economic dementia discourse may well obliterate the tentative conceptual changes in healthcare and cultural realms.

Thomas neither offers any new images nor describes signature events that would characterize the condition in radically different ways. Instead, it is the sheer number of well-established images and stock scenes stretching out over 620 pages that make it a meticulous study of the condition's insidious nature. For the first time in the literary history of dementia, they make the reader experience the patient's decline in real time. It is ultimately the epic length and consequent detail of Thomas's account that lead reviewers like Nicci Gerrard to assign it a central place in recent dementia fiction.[103] Stefan Merrill Block, in particular, considers it 'the truest and most harrowing account of a descent into dementia', and ascribes it 'a "Buddenbrooks"-like quality, offering a vision of a society in flux that is at once expansive and tangibly detailed'.[104]

I take Merrill Block's association as an invitation to read Thomas's account against sinologist Yi-Ling Ru's work on the family novel, whose first prototype she considers to be Thomas Mann's *Buddenbrooks* (1901). Thomas uses the genre of the family novel to characterize Alzheimer's as a disease. The generic similarities as well as obvious differences bring out the meaning he ascribes to the condition: a disease of decline and degeneration. This is especially true when we look at how Ru, who bases her study, among others, on Galsworthy's saga, sees the family novel defined by four salient features, namely,

> first, it deals realistically with a family's evolution through several generations; second, family rites play an important role and are faithfully recreated in both their familial and communal contexts; third, the primary theme of the novel always focuses on the decline of a family; and fourth, such a novel has a peculiar

narrative form which is woven vertically along the chronological order through time and horizontally among the family relationships.[105]

For Thomas, Alzheimer's disease impinges on exactly these four elements. I will initially attend to features two and four to show the ways in which *We Are Not Ourselves* methodically aligns with concepts of the family novel. Exploring Thomas's breaking with characteristics one and three, thereafter, will bring out how his novel questions particular conceptualizations of Alzheimer's disease.

Alzheimer's disease disrupts family rituals and the continuity and stability they symbolize. Most obviously, Christmas celebrations change over the years. In 1991, Edmund's wife Eileen has to remind her husband to buy a present for his son; in 1994, she exhorts everyone to 'please have a nice Christmas' (403). One year later, Christmas 'saw no enthusiastic ripping open of presents', and Eileen resolves 'to wring enough perfection out of next Christmas to last her the rest of her years on earth' (430). But throughout 1996, Edmund is 'declining so quickly' (441) that he has to enter a nursing home. The momentousness of this change is signalled by the length of part five that covers the events of a single year, but is nearly as long as part four about the preceding five years. Connell's plan to bring his father home for Christmas fails dismally, because his father 'was gone, gone' (542). Eileen's efforts are reminiscent of Enid Lambert's in *The Corrections* (2001). In light of her husband Alfred's progressing Parkinson's dementia, Enid wishes to bring the entire family together for one last Christmas, but it 'wasn't the Christmas [she'd] hoped for', because Alfred is rapidly 'failing' in body and mind.[106]

Change becomes most overtly clear from Thomas's focus on family relationships, as he reverses the roles of husband and wife, and father and son. At first sight, this role reversal merely reflects the actual experience of caregiving, with the account's length allowing for all the excruciating details of Eileen's burden. For Merrill Block, these are 'mostly intimate, domestic scenes, but Thomas understands how the numbing, repetitive minutiae of caretaking are often freighted with unspeakable pain'.[107] That Eileen 'would have to play nursemaid to him' (391) is both subtle and critical given that she is a nurse by profession. And her realization that Edmund 'would have to become something like a child to her' (337) obtains additional meaning, because it is Eileen through whom the narrative is mainly focalized.

The reader, in fact, does not experience Edmund Leary's mind. This move ensures that Edmund does not turn to living in the past, and it matches the genre's concept of driving the plot towards the younger generation's life and choices. It also means that Thomas does not perpetuate the image of Alzheimer's disease as a condition of ageing and old age. Edmund Leary is only 50 years old, in keeping with the condition (one hundred years after Alois Alzheimer's case of 51-year-old Auguste D.) as it was originally described, as an illness of middle age. Except that anxieties about ageing particularly pervade middle age.[108] In this respect, Thomas's narrative is truly about ageing.

Edmund's age documents the increasing societal awareness of early-onset Alzheimer's disease, as seen with *Still Alice*. But most of all it is an important narrative move. *We Are Not Ourselves* tells of Alzheimer's disease as attacking the middle of life, the time of an individual's prime productivity. Or, as Eileen puts it: '"early-onset is the most virulent kind … It's the true Alzheimer's", she said, with something like pride at the thought that if her husband were to be destroyed by a degenerative neurological disorder, it would be the undiluted article, the aristocrat of brain diseases' (341). Thomas does exactly this: depicting Alzheimer's as the aristocrat of brain diseases – a powerful player outside the common rules, a condition that is beyond the reach of science and medicine, that destroys everything the middle classes have built for themselves. He shows Alzheimer's disease as attacking all that 'capital' has come to embody in rich societies according to Foucault's definition: 'everything that in one way or another can be a source of future income'.[109]

In a first step, of course, it is Edmund's capital that is destroyed: his mind. As an 'expert on the brain' (73), an enthusiastic lecturer and dedicated scientist, it is Edmund's profession that makes the horror of his disintegration exceptionally vivid. But Thomas's vision of Alzheimer's disease as the killer of Foucauldian capital goes deeper. In a second, even more powerful way, Thomas uses the theme of role reversal in relation to the narrative scheme of the family novel. To pursue this argument further, consider Ru's definition of family constellations within such a novel:

> The husband, just like the father, cares for the family honor and power … He tries to possess and dominate his wife … while the wife … strives for an individual form of freedom just as the rebellious son does. Such a husband-wife conflict is, then, characterized by the opposition between power and will … there is a clash between tradition and individuality.[110]

The individualist is Leary. Edmund's name already signals his doom. Notions of *King Lear* as *the* classical cultural imprint of Alzheimer's disease continue to persist in scholarship, narrative and on stage.[111] As expounded in the introduction, I disagree with interpretations of *King Lear* as demented, rather seeing him as an ageing parent gone mad over the consequences of his acting. But here also lies the point of comparison. Edmund's values in life make him unfit for the narrative purpose of a strong father figure, because 'his ambition had never been for fancier titles and fatter pay-checks; he was after something unquantifiable, philosophical, the kind of aim never properly rewarded in earthly terms' (96). These values symbolize 'the grand objectives of the Enlightenment (including Truth, Justice, Reason and Equality)' that, among others, Anthony Elliott describes as dissolving or becoming irrelevant in the dominant presence of corporate capitalism and mass popular culture.[112] Thomas brings out precisely this conflict of values through the focus on Eileen as the prototype American middle-class climber. He traces the rise and Alzheimer's-related fall of the Tumulty-Leary clan, and thereby illustrates all that the condition embodies in the rich West.

As a bildungsroman, *We Are Not Ourselves* follows Eileen Tumulty's development and traces her change, as she adjusts her system of values in the confrontation with Edmund's dementia.[113] The novel's first part prepares for an understanding of how Eileen's life – her relationships to her private property, her family, household and retirement – makes her, as Foucault would express it, 'a sort of permanent and multiple enterprise'. It narrates Eileen's harsh childhood in an Irish American household, and portrays her as someone who learns that life was 'what you made of it' (45). It is towards the end of part one that we reach the peak of the Tumulty family cycle, stabilized – again in Foucault's words – by the 'economic rationalization constituted by marriage'.[114] Eileen's choice had fallen on Edmund, because the 'men that stirred her [were] reliable ones, predictable ones' (51). But the scene is set for the family's decline, once Eileen realizes Edmund's 'excessive vigilance about the effects of capitalism' (78) that opposes her own strong sense of property.[115]

The conflict between Eileen and Edmund arises over Eileen's desire to comply with her father's wish to own a house (66). Edmund's opposition is initially related to his disinterest in property and later to his disease-imposed need for environmental stability. This makes it increasingly difficult for Eileen to see Edmund 'as a fully vested partner in her future' (96–7). As Eileen recognizes that Edmund cannot be her capital in marriage, her capital becomes investing in a house. In letting Eileen realize in retrospect that this house contained her 'former *future* life' (575; emphasis original), Thomas very obviously characterizes Alzheimer's disease as severely restricting her entrepreneurial aspirations. What was a minor event in Timothy's final days – the Forsytes' by then eldest member losing the ability to manage his accounts – moves centre stage in this family's life, creating (as Eileen's surname suggests) disorder and agitation. The novel's two central and longest parts cover the purchase of this house, and illustrate Edmund's rapid decline in it. In keeping with earlier considerations regarding its symbolic nature, Eileen's house embodies her sense of achievement as well as her world of values. Yet, the illness prevents the family from truly owning this house: the rooms remain 'empty spaces' (279). And living in the house – the state of the house itself – becomes one with living with the disease, as 'a lot of her house looked like a waiting room' (314) and Eileen is unable to remedy its 'state of disrepair' (339).

Like the husband–wife relationship, so the father–son relationship takes a special place in the family novel. To quote Ru again, the antagonisms explored are 'the father's thirst for power versus his love for his children, the sons' need for authority versus their rebellion against their fathers, the fathers' dominance versus the sons' disobedience, and the fathers' defense of the old ways versus the sons' quest for the new'.[116] Revisiting themes covered in the husband–wife relationship, Alzheimer's disease plays its part in the workings of all these conflicts. Thomas particularly explores the consequences of 'losing a father early, being given all that responsibility' (491). What I want to look at here in more detail is the role of the youngest generation as such in Thomas's family novel, because, as Ru puts it, by 'defying and denying the father, [the sons] are reborn in the end'.[117] Connell defies the disease. And in this defiance, he covers the roles of both rebellious outcast and loving son.

Edmund Leary dies in March 1999, the last year of the old century; and the narrative's final part ends with the turn of the millennium. But Edmund's death is not a landmark event. While *The Forsyte Saga* features several significant deaths over time, Thomas's narrative only truly includes Leary's. Its understated singularity within the family novel – Connell misses his father's last few living hours – underpins notions of Alzheimer's disease as a condition of slow dying alive. Shortly after his diagnosis Leary himself claims, 'I *am* dying' (325; emphasis original).

The depiction suggests that the death of the father does not really mark a liberation from the old: Alzheimer's disease lives on in the epilogue set in the year 2011 and remains the challenge of the new century. *We Are Not Ourselves* uses the apparent dissolution of the Tumulty-Leary family to show how family life in the West has been subjected to Alzheimer's disease as one of 'the conditions of postmodern capitalist life'.[118] Thomas explicitly lifts Eileen's individual story onto the level of collective, national relevance, when he lets her ponder that we 'move around too much in this country' (575), a thought which evokes a feature Foucault assigns to capitalist thinking, namely an increasing mobility to improve income.[119] Characteristic of capitalist societies, such increased mobility for financial reasons destabilizes the individual as well as the family as a whole. Underlining this argument, Thomas develops the fate of the Tumulty-Leary clan against the backdrop of the intact and extended family of Italian immigrants who make their fortune within the underprivileged environment from which Eileen wants to escape.

It is, therefore, significant that Thomas offers a possible answer in the final ten pages of his account. Again outside the narrative time and frame, the epilogue eventually brings into play the theme of change embodied in the youngest generation: Connell's courage and choice to continue the family line despite a risky gene. Or, as he puts it:

> He could honor his father by loving the kid the way his father had loved him. And if he had to be vulnerable in front of the kid, if he had to be defenseless and useless and pathetic, if he had to forget things and piss himself and get lost on the way home, then so be it. If the kid didn't handle it well – well, that was what kids did. (619)

With Connell realizing that '*he* was his father's real estate' (588; emphasis original), this epilogue can be read as an ode to the continuation of the family. As the nineteenth century turned into the twentieth, dementia was an image of degeneration. Two decades into the twenty-first century, Alzheimer's disease is the predominant trope for the decline and loss of the thriving individual as the smallest unit of capitalist societies.[120] It pictures the fractured lives led in such societies, but concurrently represents the kind of disease for which exactly these societies have neither space nor place.[121]

Where *The Forsyte Saga* revolves around the dying out of the clan because of the young generation's individuality, Thomas's account emphasizes the continuity of life granted by that very individuality. This individuality entails an attitude to dementia like that of the child Maurizio who has grown up without having lost his capacity for empathy in favour of the will to 'profit, selfishness, and the drive for ever more bigness'.[122] Such loss of empathy (already in childhood), according to scholar of psychology Arno Gruen, is particularly encouraged in cultures 'that make it difficult for people to develop the self-esteem that comes from a sense of one's inner worth, which can evolve only if people learn to accept and share their suffering, pain, and adversity'.[123] In Borio's allegory, the values of this materialistic culture are embodied by the demon Giancarlo, who seeks to undermine the power of custodians like Maria. Giancarlo focuses his attention on Maurizio, taking inspiration from a book that instructs how to rob children of their carefree years by letting them watch television programmes where 'children portray the behaviour of adults' (90).

More generally, some recent healthcare initiatives attempt to foster continued empathy from an early age. Sally Chivers, for example, expands on the effects of South Korea's 'interactive and intergenerational social program' directed at dementia care. This social agenda explicitly includes contact between children and people with dementia. As a consequence, Chivers continues, 'alarmist news stories will not be the only source of knowledge these children receive about dementia in late life'.[124] In reverse, fiction for young adults may contribute, as Clarke and colleagues observe, to 'frequent (mis)representations of dementia that escalate people's concerns'. The authors particularly mention the 'dementors' in the *Harry Potter* series of books, 'which suck the soul out of people and cause a sense of terror in anyone they are near' (87). Borio's book for children genuinely counters this. The careful fabrication of the book – including the use of deckle-edged paper and sewn binding – emblematizes a niche publisher's traditional craftsmanship.[125] It opposes the fast-lived consumerism encouraged by the story's anti-hero, Giancarlo.

Still, especially in the absence of a cure, the calm acceptance of suffering, pain and adversity appears to be a sheer impossibility; and not enough has changed politically, institutionally and societally that would support such acceptance. The condition will not lose its traumatic connotations – not least if we can read the title of Huguenin's narrative as alluding to *Hiroshima mon amour* (Hiroshima my love; 1959). This French-Japanese co-production tells a story about memory, forgetting and lost relationships during and after the Second World War. Revolving around the harrowing consequences of the first atomic bombing, it explores, as Cathy Caruth contends, the productivity of communicating across cultural boundaries and demands 'a different kind of listening and ... a speaking literally of the body'.[126] Caruth's perceptions of *Hiroshima mon amour* open an additional perspective on Huguenin's caregiver experience and, I find, they also suggest how Arno Gruen's insights can be implemented in changing healthcare approaches: such approaches acknowledge the patient's inner worth beyond his ability to remember, and the caregiver shares her story in the acceptance of her own risk of dementia. Connell Leary eventually adopts this perspective. But, I reiterate, it is a perspective difficult to imagine in a cultural climate that equates dementia with the death of the self.[127]

How can attitudes evolve in societies that face sharp socio-economic and demographic challenges? This is a question worth asking, especially when media coverage on 'dementia levels' as 'stabilising' may easily endanger slowly budding changes, even if this coverage were merely aimed at toning down the populist alarmism we have been witnessing in recent years.[128] Where families are far flung due to economically imposed mobility, quality healthcare is no longer a matter of choice. It becomes a service many cannot afford. Choice is often about inexpensive deals. Low-cost healthcare is provided by global care chains; yet, as Woodward has remarked, these labour markets 'move love as surplus value, as a commodity, and as an export from one country to another'.[129] In addition, quality care is undermined by the profit-seeking exploitation of staff: poor communication skills, high workloads and insufficient training leave little time and space for compassionate attention and dedication.[130]

Compassionate attention and dedication are essential to the caregiving attitude which, during what Gawande describes as a time of transition, Schweiger's film portrays. The very product of consumerist culture – I especially think about its product placement, its comedy form and its entrepreneurial setting (as well as the fact that Schweiger has filmed a US remake, *Head Full of Honey*, with Nick Nolte as lead character) – it still gives the answer to how society must deal with people with dementia in these circumstances: 'Amandus' means 'to be loved'.[131] But love and empathy can only become transformative on a larger societal scale once political, institutional and cultural values and beliefs about ageing take a momentous turn. For the perceivedly productive part of the population, this means accepting ageing as a natural part of life, and appreciating growth, development and success in ageing and old age. For the elderly and perceivedly less productive, this means bringing their wisdom to bear on value systems that are overly determined by individualism and financial gain.

8

Conclusion

This book has identified the path taken by the presentation of dementia in scientific, medical and literary texts across the twentieth century. It has illustrated that, since the late nineteenth century, when the condition was first described as an organic brain disease, to the present day, when it is viewed as a disorder of cognition, the literary representation of the condition has progressed alongside a continuously evolving medico-scientific discourse about dementia.

The biological psychologies of the late eighteenth century, the somaticization of diseases a century later and the biomedicalization of illness with its pharmacological redefinition of health towards the end of the twentieth century have all led to the objectification of patients, a process not confined to dementia. The memory sciences instigated the systematic neurological study of the dementias at the fin de siècle, thereby creating the patient with a condition of memory loss; the attention of the biomedical sciences, in turn, amplified by sociocultural, political and economic concerns, focused on Alzheimer's disease towards the end of the twentieth century. Yet I argue that the medical sciences' main influence in dementia discourse developments has been of a reinforcing nature. Rather than introducing new metaphorical concepts, on several occasions, medico-scientific language paralleled or followed the wider sociocultural discourse, and was taken to strengthen already existing images. By the beginning of the twenty-first century, this enhancing dynamics had placed the patient in a position from which it was difficult to reclaim a social identity and achieve recognition of their continued self and authority in contributing to healthcare planning.

Scientific, medical and literary iterations have all contributed to the current cultural dementia narrative; and, indeed, it would be wrong to search out an antithetic dualism between the psycho-sciences and the neurosciences in its development. Rather, I see these different disciplines connected by way of a dialogical space occupied by discourses of ageing and memory, and it is in this space that the narrative has developed. As the layout of this book indicates, at different times the discourses of ageing and memory each held sway over twentieth-century developments. The scientization of memory and its loss and the biomedicalization of Alzheimer's disease have created an object of knowledge which – prior to its naming – had already existed but was dealt with in non-scientific terms in the sociocultural realm. Given that the role of science in discourse is strongly linked to wider cultural changes, the discourse that developed in the medico-scientific field around this new object of knowledge, dementia, fed on the already existing sociocultural images. As the medical sciences developed these images further, their discourse – in the continued exchange with the sociocultural realm – strengthened and reinforced the original images. The images and concepts so developed impinged on the newly created object of knowledge: the dementia patient.

One of my central concerns has been to explore how these dialogical processes affected individuals with neurodegeneration. I wanted to understand why specific notions of Alzheimer's disease are so persistent; why particular themed metaphors continue to reign supreme; and why some research concepts continue to dominate the medical sciences and also the health humanities at the expense of recognizing the patient's individuality and personhood. The foregoing chapters have already answered these questions in much more detail than I can do here.

Concepts of old age and cognitive decline are deeply engrained in the cultural dementia narrative. They are so durable because they formed in a process of reinforcement. The medical sciences adopted images from the sociocultural realm and developed them further; and these images were then taken to corroborate the wider sociocultural ideas about the patient. In the process individuals with memory loss became embedded in a new system of norms and standards in which they could be objectively assessed – a system revolving around antonyms of growth and development versus loss and degeneration, productive versus dependent, young versus old, performing versus passive. The individual turned into an object of knowledge. I argue that Kraepelin's move to focus on neuropathological features and the condition's cognitive component can be taken as activating this process (while at the same time being a representative example of the period's fascination with materialist approaches). The medico-scientific definition of Alzheimer's disease centred on objectively assessable parameters. The patient's thinking mind and feeling psyche were excluded from this focus and as a consequence concepts of passivity and lost agency became manifest in wider sociocultural thinking.

The patient's mind entered the literary arena much earlier and more explicitly as compared to when (only in the 1950s) and how old-age psychiatrists and social medics began to take an interest in this area. The true achievement of modernist narratives, like Hoult's *There Were No Windows*, in relation to dementia discourse developments was their calling into question a 'Cartesian geography of mind'.[1] They challenged old-age psychiatry to aim for an understanding of the patient's mental thought processes and a characterization of the illness from within. My point is not that the psycho-disciplines had a more positive or constructive influence on metaphorical language than the neurosciences; they each emphasized different aspects of the same problem. Rather, meaningful shifts in discourse developments occurred where the human sciences managed to spotlight the patient's subjectivity, where interest was roused in the patient's illness experience and appreciation was able to foster better communication with the patient.

With a steadily growing attention to the patient's narrative – first on the part of old-age psychiatrists, then social scientists, psycho- and sociolinguists, and eventually health humanists – we have witnessed a slow change in discourse. Patients have begun to be seen as individuals with agency, who can contribute to the knowledge to be had about their condition. With a move, in dementia discourse analysis, from a focus on language deterioration to an emphasis on remaining linguistic capabilities, ideas of continued identity and personhood replaced those of communication breakdown. The passive patient turned interlocutor and entered the stage – their own voice being heard in the caregiver's narrative, at the conference podium, and in bookshops and policy discussions. This is a shift with parallels in fiction. Think of Claire Temple, who is being perceived as tedious by her visitors; Faith Gadny not being listened to; later on, Rosemary Travis leads the plot and Maria Merlo challenges how the patient is viewed. Concurrently, we observe that increased value is placed upon the child's perspective on the patient: from Timothy's childlessness and the narrative absence but inferred hostility of Faith Gadny's grandchildren, via Maurizio's valuing the patient outside of memory loss, to Connell Leary's defiance.

Tracing a continuous dynamic between the serious and the comic, the idealist and the derogatory in the history of the novel, Thomas G. Pavel perceives the tension between depiction of 'human isolation' and a 'possible reconciliation between the individual and the world'.[2] The literary history of dementia appears to be no exception: Faulkner, Bailey, Beckett and Bernlef home in on aspects of isolation, while Hoult, Dibdin, Genova and Thomas see scope for reconciliation with the caregiver and a potential for discourse developments in favour of the patient's humanity. In this regard, the literary discourse is characterized by poles similar to those observed in the medico-scientific realm. Cure, at the beginning of the twenty-first century, is not only a pharmacological need; it is also a necessity in terms of a 'cure narrative' of and for the self – the ageing self as well as the remembering self.[3]

I opened this book with a quotation by Claude Couturier from her 2004 diary, in which she advocates the organic paradigm as saving the patient's personhood and agency. A return to the organic paradigm may well restore patients their individuality as elderly citizens with a diseased brain, as long as this does not imply the objectification of the patient. But similarly, to consider dementia as a condition of the failing mind supports the patient, as long as notions regarding, for example, disorientation in space and time solicit supportive healthcare thinking rather than dissatisfaction with the patient's performance and efficiency. Both approaches can turn out to be effective in changing century-old ideas of the patient as cognitively under-performing and the disease as one of old age – especially as long as ageing itself continues to be considered in predominantly negative terms.

Throughout this book, we have encountered sociocultural perceptions of the healthy middle-aged as the only truly productive stratum of the population. Yet, during the past decade or so popular awareness of Alzheimer's disease as a condition that also afflicts the middle-aged has arisen. This is explained by two factors. First, by the decision of scientists to shift both the goal and the interest group of pharmacological investigations towards disease prevention in the middle-aged, and how this move has been reported on in the media. And second, by Richard Glatzer's prominent motion picture *Still Alice* (2014), whose popularizing effect cannot be overestimated, particularly when thinking back to the effect of Richard Eyre's *Iris* thirteen years earlier. This ever-increasing visibility of Alzheimer's disease as a possible condition of middle age has just begun to affect how this younger age group anticipates it will be perceived in the future.

This could prove a turning point in discourse developments, because – as we have seen all through the century-long period under consideration – new ways of classifying may, as Ian Hacking convincingly articulates, 'systematically affect the people who are so classified, or the people themselves may rebel against the knowers, the classifiers, the science that classifies them'.[4] In exchange with technological and biological domains, the literary and wider cultural realms have questioned archival notions of memory.[5] This has led to Alzheimer's disease at times being perceived of as a mixed blessing, as possibilities for human fulfilment begin to be considered possible within or despite memory loss. The response of the human sciences to a growing medico-scientific emphasis on prediction and prevention in the middle-aged will contribute to the conceptual reconfiguration of the future dementia patient.

Caregivers, especially the women (and men) in the middle, have become militant in their writing, with respect to the financial and healthcare support they need, but also regarding their intention to change the dementia narrative itself. With the very recent trends in cultural representations as well as medico-scientific research endeavours that include the middle-aged subject more generally, advocating change is no longer the business of the overburdened healthcare provider and the still underprivileged patient alone. It has become of concern to those middle-aged individuals whose choices define and eventually reinforce the values and beliefs of enterprise societies which continue to have neither space nor place for the condition. But as still productive members are increasingly at risk of the condition, the outlook of these societies will have to change – although this is hard to imagine in practice at the time of writing, when renewed nationalist thinking driven by economic interests seriously damages the slowly rising culture of care and empathy. As those increasingly perceived to be at risk begin to articulate their anxieties about marginalization and illness-related preconceptions, the human sciences will begin to challenge notions of ageing and cognition as the only determinants in the cultural dementia narrative.

Glossary

This glossary offers lay-term explanations of specialist terminology. Honouring a readership from within and beyond the literary, scientific and medical disciplines, it includes medical and scientific as well as literary terminology. Explanations referring to current medical and scientific understandings cannot be free of the metaphorical language and conceptualizations it is the focus of this study to investigate. The literary terms included are those I experienced as challenging in discipline-crossing conversations and teaching. Terms printed in bold have their own entry within this glossary.

Acetylcholine	The **Neurotransmitter** produced and used by a specific set of brain cells termed 'cholinergic'. Cholinergic brain cells, among others, are located in the **Brain area** called basal forebrain. See also **Cholinergic hypothesis**.
Acetylcholinesterase	Terminates the lifetime of action of an individual **Acetylcholine** molecule. In **Alzheimer's disease**, acetylcholine levels are lower than physiologically normal. Some drugs can increase the lifetime of still available acetylcholine molecules by inhibiting acetylcholinesterase. These drugs are called pro-cholinergic; they are acetylcholinesterase inhibitors. See also **Drug treatment**.
Alzheimer's disease	The most frequently diagnosed form of **Dementia** in older people; neuropsychological symptoms include memory loss (see also **Amnesia**), the loss of speech (see also **Aphasia**) and the inability to recognize things or people. Further symptoms include behavioural changes and delusional symptoms such as hallucinations. The condition is named after Alois Alzheimer, a German psychiatrist (1864–1915).
Amnesia	The partial or total loss of memory; one of the neuropsychological symptoms of **Alzheimer's disease** and other (but not all) forms of **Dementia**.
Amyloid (hypothesis, peptide and plaques)	Amyloid plaques are one of the morphological features detected (after death) in the brains of people diagnosed with **Alzheimer's disease**. These plaques consist of agglomerated amyloid. The amyloid hypothesis considers these plaques to be the cause of Alzheimer's disease and the associated brain cell death. The amyloid hypothesis drives pharmacological

	research that aims at removing amyloid plaques from the brain or preventing these plaques from forming in the first place. To date, such research has not led to the formulation of clinically effective drugs.
Amyotrophic lateral sclerosis	A form of motor neuron disease; a usually fast progressing terminal neurological disorder that exclusively affects voluntary muscle control, including the ability to speak, swallow and breathe. It leaves mental acuity intact until the very final stage of the condition.
Aphasia	The inability to understand or produce speech; one of the neuropsychological symptoms of **Alzheimer's disease**.
Bildungsroman	A novel of development, usually featuring a younger character and tracing their development into mature adulthood. A form further developed into the **Reifungsroman**, in which the main character is of middle or older age.
Biomarker	Molecules identified in peripheral (not brain) tissue, including skin or blood cells and cerebrospinal fluid. Metabolic events involving the molecules detected in these compartments are understood to mirror pathological alterations in the living patient's brain. Biomarkers are used in diagnostic tests, and alterations in their levels are also taken as indicators of treatment effects. To date, there is no single biomarker available that would confirm a diagnosis of **Alzheimer's disease**. See also **Pathology**.
Brain areas and Alzheimer's disease	Brain cell death in **Alzheimer's disease** predominantly affects an area at the bottom front of the brain, the basal forebrain (specifically, the Nucleus basalis of Meynert). The basal forebrain is connected to higher brain areas via elongated nerve cells; these higher brain areas are the cerebral **Cortex** (which harbours our ability to think and reason) and the **Hippocampus** (which is fundamentally involved in memory and learning). When input from the basal forebrain is lost, the function of the cortex and hippocampus gradually declines due to decreased connectivity. Over time, the function of the cortex and hippocampus decreases further also because of cell death in these brain areas. See also **Cholinergic hypothesis**.
Cholinergic hypothesis	The first brain cells to die in **Alzheimer's disease** are usually those of the basal forebrain (see **Brain areas**); they are cholinergic nerve cells, because they use **Acetylcholine** as their **Neurotransmitter**. The cholinergic hypothesis argues that the death of cholinergic nerve cells most accurately

	correlates with the cognitive deficits noted in those diagnosed with **Alzheimer's disease**. The cholinergic hypothesis forms the rationale for currently available **Drug treatment**.
Chromosomes	The molecular structure within the **Nucleus** of a cell holding, in a tightly coiled up and packed form, the genetic instructions of an organism, the **DNA**. The information contained in the DNA is encoded in the unique sequence of a small number of specific molecules. Their 'transcription' and 'translation' yield **Proteins** which are the basis of cellular structures and components as well as functional units of the cell.
Computer tomography (CT)	An imaging method building on X-ray technology, offering anatomical, but not functional, information. In comparison to single-plane X-ray recordings, CT collects a series of planes (comparable to virtual brain slices) which then can be taken together to yield three-dimensional information.
Cortex	The area of the brain which harbours our ability to think and reason, receives sensory information and initiates movement. One of the **Brain areas** impaired in **Alzheimer's disease** and other forms of **Dementia**.
Dementia	From Latin, meaning 'out of mind', the term refers to a broad range of mental conditions, including **Alzheimer's disease**. Not every form of dementia presents with severe memory loss: fronto-temporal dementia, for example, is predominantly characterized by personality changes and hallucinations.
DNA	The macromolecule in the **Nucleus** of the cell that contains the genetic information of an organism. The large DNA molecules are tightly coiled and packed into **Chromosomes**. One such chromosome contains many different **Genes**, which, in turn, consist of varying lengths of DNA.
Drug treatment in Alzheimer's disease	The current principal treatment strategy aims at correcting the cognitive deficit that originates from the gradual loss of **Acetylcholine** (see also **Cholinergic hypothesis**). The treatment is termed pro-cholinergic, because it increases the lifetime of acetylcholine molecules in their action. This is achieved by inhibiting **Acetylcholinesterase**, which is responsible for the physiological breakdown of acetylcholine. The drugs are called acetylcholinesterase inhibitors and include donepezil and tacrine.
Epidemiology	The branch of medical research trying to answer the questions why, and how frequently, a particular disease

occurs in a specific population. Research projects (like the Nun Study) usually focus on a well-defined study sample which shares many characteristics with the target population.

Family novel A particular form of the **Bildungsroman**, tracing the decline of a family clan, habitually over three generations. Decline is often embodied in individual family members' health, well-being and financial stability. The family novel had its heyday during the final third of the nineteenth century, Thomas Mann's *Buddenbrooks* being considered its epitome.

Focalization Referring to the limiting of information or knowledge given in a narrative where the events are told through the perspective, or from within the world of experience, of a character. Where the narrative is told in **Free indirect style** the reader feels particularly drawn into the perspective of the character through whom the narrative is focalized. This device is somewhat characteristic of literary **Modernism**.

Free indirect style A narrative strategy forcing the reader into the character's world of experience; a character's thinking and feeling control the third-person narration. This makes it difficult for the reader to separate out what is part of the perceivably objective account and what is subjective experience. See also **Focalization** and **Modernism**.

Genes Representing units of inheritable information, and located in a defined region on a **Chromosome**, composed of varying lengths of **DNA**.

Hippocampus The area of the brain which is fundamentally involved in memory and learning. One of the **Brain areas** impaired in **Alzheimer's disease** and other forms of **Dementia** that are characterized by **Amnesia**.

Histology The study of tissue at the microscopic level, for the purpose of physiological or pathological characterization, in the latter case termed 'histopathology'. This can involve the staining of tissue (using, e.g. Golgi silver stain) to make particular areas visible. See also **Pathology**.

Huntington's disease A familial, incurable and degenerative neurological disorder that affects both motor function (especially movement control) and cognition; the only one of the most common neurodegenerative disorders exclusively linked to a specific **Chromosome**.

Inward turn Referring to the modernist novel's narrative strategy, through which the reader becomes privy to the lived experience of fictional characters. See also **Free indirect style** and **Modernism**.

Lewy body dementia   A type of **Dementia** that often co-presents with **Parkinson's disease**. Cognitive decline is usually fast progressing and more severe than in other types of dementia. The condition is named after Fritz H. Lewy (1885–1950), who described small inclusion bodies in the brain, comparable to **Amyloid** plaques in **Alzheimer's disease**.

Life-writing   Referring to writing concerned with the autobiographical or biographical, it can, for example, present in the form of a diary, comic, picture or photo book, but also as a blog.

Magnetic resonance imaging (MRI)   An imaging methodology that assesses the environment of hydrogen atoms in the body, specifically depending on the inherent magnetism of the nuclei at the centre of atoms. The methodology yields information on the anatomy of an organ and is often complementary to information obtained using **Computer tomography**. MRI can also yield functional information by detecting alterations in the environment of hydrogen atoms following *external* stimuli (compared to **Positron emission tomography**, which relies on administration of specifically labelled drugs into the blood stream).

Mild cognitive impairment   Defining the transitional stages of gradually decreasing cognitive performance that lead up to a diagnosis with **Alzheimer's disease** or another form of **Dementia**. The term is also used to describe a less-than-normal cognitive state or performance that will not necessarily lead to a diagnosis with dementia.

Mise-en-scène   In the context of film, a term referring to the arrangement of objects within a single frame during filming, including lighting and background.

Modernism   Referring to the movement arising from the profound industrial, societal and cultural shifts during the final decades of the nineteenth century – a period additionally marked by the feeling of loss of certainty in the wake of the Great War. The modernist novel leaves behind the tradition of the all-knowing narrator typical, for example, of the classical **Family novel**. Instead, it works with mimetic approaches, that is, text mimicking or imitating confusion or fragmentation. It also uses narrative strategies such as **Free indirect style** to admit the reader into the lived experience of its characters. See also **Inward turn** and, by comparison, **Postmodernism**.

Multiple sclerosis	A progressive, degenerative neurological condition in which the immune system attacks the casing that is wrapped around nerve fibres. This myelin sheathing supports signal transmission, and its loss can impair motor control, vision or speech; neuropsychological symptoms such as memory loss are less frequent.
Narrator	The character telling the story in a piece of writing or a film. Taking two films considered in this book as examples: an extradiegetic narrator does not feature within the story or action of a film, like the voiceover in *Black Daisies for the Bride*; an intradiegetic narrator has her own role in the story told, like Tilda in *Honig im Kopf*.
Neurofibrillary tangles	The appearance of the aggregated **Tau protein** in **Alzheimer's disease**. Particular elements of normal tau are chemically altered (specifically, they are hyperphosphorylated) such that the resulting protein molecules link together, which impairs their function.
Neuronovel	Novels in which mind is reduced to brain. In the case of narratives featuring **Dementia** or **Alzheimer's disease**, the narrative strategy is often defined by disease characteristics.
Neurotransmitter	Molecules (like **Acetylcholine**) typically produced in specific sets of brain cells characteristic of distinct **Brain areas** (like cholinergic nerve cells in the basal forebrain). They are released in response to electrical impulses and propagate information to surrounding cells. See also **Cholinergic hypothesis**.
Nosology	The classification of diseases; also referring to the understanding of a specific disease and its characteristics.
Nucleus	The structure within the cell that contains the **DNA**. The term is also used in chemistry and physics to describe the centre of an atom. See **Magnetic resonance imaging**.
Ontogeny	Referring to the development of an individual organism, from fertilization to its mature form. Ernst Haeckel (1834–1919) hypothesized the recapitulation of **Phylogeny** in the foetus's ontogeny, with the foetus, over the nine-month gestation period, passing through the different evolutionary developmental stages of the human species.
Parkinson's disease	One of the most frequently diagnosed neurodegenerative diseases in older age. The condition predominantly presents with impaired motor function. Like **Alzheimer's disease**, Parkinson's disease continues to be incurable, but there is better **Drug treatment** available that can alleviate symptoms

over a longer time period. Parkinson's disease can co-present with a particular form of **Dementia** termed **Lewy body dementia**.

Pathogenesis Concerning the origin and development of a disease.

Pathology The medical discipline concerned with the description of disease processes. A disease is usually diagnosed based on the analysis of tissue samples (e.g. skin or brain tissue) and the detection of specific **Biomarkers**. The term also refers to the progression of a disease itself. See also **Histology**.

Peptide A short **Protein**; for example, **Amyloid** peptide.

Phylogeny Describing the evolutionary path of an entire species. Ernst Haeckel (1834–1919) hypothesized the recapitulation of phylogeny in the foetus's **Ontogeny**, with the foetus, over the nine-month gestation period, passing through the different evolutionary developmental stages of the human species.

Plot Referring to how the events in a story are arranged and told within the account as presented to the reader; not necessarily organized in logical or chronological order.

Positron emission tomography (PET) An imaging methodology that traces the distribution of specifically labelled compounds, commonly a radiopharmaceutical form of glucose, fludeoxyglucose (abbreviated FDG). Glucose is essential for cellular activity, which means that data collected by means of PET yields information about cellular activity. Using **Computer tomography**, PET images of several individual planes are combined to yield three-dimensional information. See, by comparison, **Single photon emission computed tomography**, which uses a tracer that is not taken up by the cell.

Postmodernism Identifies the period of rapid changes after the Second World War; also used to describe a social force evolving in negotiation with these transitions; a period demanding flexibility and spontaneity in lifestyle choices, but also leading to the experience of fragmentation and dislocation – notions reflected in postmodern literary experimentation. See, by comparison, **Modernism**.

Protein The basis of cellular structures and components, such as **Tau protein**. Proteins are also essential for the physiological functioning of the body, like, for example, **Acetylcholinesterase** terminating the action of **Acetylcholine**.

Psycholinguistics	In the context of research related to **Alzheimer's disease** and other forms of **Dementia**, referring to the study of the patient's gradual language breakdown. The discipline emphasizes parallels between deteriorating linguistic skills and cognitive decline. See, by comparison, **Sociolinguistics**, which focuses on remaining linguistic abilities.
Reifungsroman	Also termed 'midlife novel'; the novel's main character is often of middle or older age, and experiences the awareness of their own ageing as a process of becoming wiser. Considered a particular type of the **Bildungsroman**.
Single photon emission computed tomography (SPECT)	An imaging methodology that measures blood flow, and blood supply to an organ. A radioactive tracer is injected into the blood stream, and its journey through the blood is detected by recording the emitted gamma-ray radiation. The methodology is combined with **Computer tomography** to allow for recording of a series of images which are then combined to yield three-dimensional information. Data collected is understood as informing about organ function rather than organ anatomy. Unlike glucose in **Positron emission tomography**, the tracer is not taken up into cells. This means that SPECT does not inform about cellular activity itself.
Sociolinguistics	In the context of research related to **Alzheimer's disease** and other forms of **Dementia**, referring to the study of the patient's remaining linguistic abilities with an emphasis on context and engagement. See, by comparison, **Psycholinguistics**, which focuses on language breakdown.
Tau protein and tau hypothesis	Tau usually features as an essential cytoskeletal **Protein** in the brain, giving structural stability to the architecture of individual cells. In **Alzheimer's disease**, this tau protein is aggregated. It is hyperphosphorylated, that is, chemically altered, such that the protein is structurally and functionally changed, appearing in a fibrillary and tangled form. The tau hypothesis claims that characteristic distribution patterns of **Neurofibrillary tangles** in the brain allow for the differentiation of various stages in the **Pathology** of **Alzheimer's disease**.
Theory of mind	An individual's awareness of the existence of another person's mind; the knowledge of how to interpret one's own as well as the other person's thought processes.

Notes

1 Introduction

1 Andrés Barba, *Ahora tocad música de baile* [Now strike a dance tune] (Barcelona: Editorial Anagrama, 2004), 85; Martina Zimmermann, 'Deliver Us from Evil: Carer Burden in Alzheimer's Disease', *Medical Humanities* 36 (2010): 101–7.
2 Claude Couturier, *Puzzle, Journal d'une Alzheimer* [Jigsaw, diary of an Alzheimer's patient] (Paris: Éditions Josette Lyon, 2004), 143.
3 Lars-Christer Hydén, Hilde Lindemann and Jens Brockmeier, introduction to *Beyond Loss: Dementia, Identity, Personhood*, ed. Lars-Christer Hydén, Hilde Lindemann and Jens Brockmeier (New York: Oxford University Press, 2014), 1–7.
4 'Alzheimer's Disease Information Page', Society for Neuroscience, last modified 27 March 2019, https://www.ninds.nih.gov/Disorders/All-Disorders/Alzheimers-Disease-Information-Page.
5 Nikolas Rose and Joelle M. Abi-Rached, *Neuro: The New Brain Sciences and the Management of the Mind* (Princeton, NJ: Princeton University Press, 2013), 43.
6 Ludmilla J. Jordanova, introduction to *Languages of Nature: Critical Essays on Science and Literature*, ed. Ludmilla J. Jordanova (London: Free Association Books, 1986), 20; consider also Paola Spinozzi, 'Representing and Narrativizing Science', in *Discourses and Narrations in the Biosciences*, ed. Paola Spinozzi and Brian Hurwitz (Göttingen: V&R Unipress, 2011), 31–60.
7 Lilian R. Furst, *Between Doctors and Patients: The Changing Balance of Power* (Charlottesville: University Press of Virginia, 1998).
8 Marlene Goldman, for example, asserts that '"Gothic" biomedical models rely on a metonymic process of substitution of the person for increasingly smaller cellular and ultra-cellular units'. Concurrently, she describes her work as in 'contrast to the social scientific and biomedical approaches', intended to bring 'an alternative perspective to bear on the experiences of aging and age-related dementia'; Marlene Goldman, *Forgotten: Narratives of Age-Related Dementia and Alzheimer's Disease in Canada* (Montreal: McGill-Queen's University Press, 2017), 178, 7.
9 Anne Whitehead and Angela Woods, introduction to *The Edinburgh Companion to the Critical Medical Humanities*, ed. Anne Whitehead and Angela Woods (Edinburgh: Edinburgh University Press, 2016), 15.
10 Jesse F. Ballenger, *Self, Senility, and Alzheimer's Disease in Modern America: A History* (Baltimore: Johns Hopkins University Press, 2006), 3.

11 Goldman, *Forgotten*, 47, 317.
12 Martina Zimmermann, 'Dementia in Life Writing: Our Health Care System in the Words of the Sufferer', *Neurological Sciences* 32 (2011): 1233–8; and *The Poetics and Politics of Alzheimer's Disease Life-Writing* (Basingstoke: Palgrave Macmillan, 2017).
13 Consider, for example, Christopher A. Vassilas, 'Dementia and Literature', *Advances in Psychiatric Treatment* 9 (2003): 439–45; Robert E. Yahnke, 'Old Age and Loss in Feature-Length Films', *The Gerontologist* 43 (2003): 426–8; Kurt Segers, 'Degenerative Dementias and Their Medical Care in the Movies', *Alzheimer Disease and Associated Disorders* 21 (2007): 55–9; Desmond O'Neill, 'Ageing with Style', *The Lancet* 387 (2016): 639. Also Alzheimer's Research UK, formerly Alzheimer's Research Trust and exclusively focused on biomedical research, increasingly invests in caregiver support; 'Support for People Affected by Dementia', Alzheimer's Research UK, accessed 25 October 2019, https://www.alzheimersresearchuk.org/about-dementia/helpful-information/support-for-carers/.
14 Oliver Sacks, *The Man Who Mistook His Wife for a Hat* (1985; London: Picador, 2007), 3.
15 Laura Otis, *Membranes: Metaphors of Invasion in Nineteenth-Century Literature, Science, and Politics* (Baltimore: Johns Hopkins University Press, 1999), 8, 36.
16 William R. Clark, *Sex and the Origins of Death* (Oxford: Oxford University Press, 1996), 175.
17 See, inter alia, Paul John Eakin, *How Our Lives Become Stories: Making Selves* (Ithaca, NY: Cornell University Press, 1999), 46; Jerome Bruner, *Making Stories: Law, Literature, Life* (Cambridge, MA: Harvard University Press, 2002), 86; Howard Brody, *Stories of Sickness*, 2nd edn (New York: Oxford University Press, 2003), 67–9.
18 A. Ralph Barlow, 'Senile Dementia: Metaphor for Our Time. The Effect of Senile Dementia on Relationships with Others Echoes the Sense of Discontinuity in American Society', *Rhode Island Medical Journal* 66 (1983): 359–60; Norm O'Rourke, 'Alzheimer's Disease as a Metaphor for Contemporary Fears of Aging', *Journal of the American Geriatrics Society* 44 (1996): 220–1.
19 Wayne Booth, *The Art of Growing Older: Writers on Living and Aging* (1992; Chicago: University of Chicago Press, 1996), 41–97.
20 Ronald A. Carson, foreword to *What Does It Mean to Grow Old? Reflections from the Humanities*, ed. Thomas R. Cole and Sally A. Gadow (Durham, NC: Duke University Press, 1986), xiv.
21 S. Kay Toombs, 'The Meaning of Illness: A Phenomenological Approach to the Patient–Physician Relationship', *Journal of Medicine and Philosophy* 12 (1987): 229.
22 Hannah Zeilig, 'Dementia as a Cultural Metaphor', *The Gerontologist* 54 (2014): 260; Martina Zimmermann, 'Alzheimer's Disease Metaphors as Mirror and Lens to the Stigma of Dementia', *Literature and Medicine* 35 (2017): 90–1; Cristina Douglas,

'When Memory Gets Political: Dementia and the Project of Democracy in Post-Communist Romania' (paper, Dementia and Cultural Narrative Symposium, Aston University, 9 December 2017).

23 Susan Sontag, *Illness as Metaphor* and *AIDS and Its Metaphors* (1977 and 1988; London: Penguin, 2002), 130–1.

24 Margaret Lock, *The Alzheimer Conundrum: Entanglements of Dementia and Aging* (Princeton, NJ: Princeton University Press, 2013), 5.

25 Stephen Bertman, 'The Ashes and the Flame: Passion and Aging in Classical Poetry', 157–71; Carol Clemeau Esler, 'Horace's Old Girls: Evolution of a Topos', 172–82; both in *Old Age in Greek and Latin Literature*, ed. Thomas M. Falkner and Judith de Luce (Albany: State University of New York Press, 1989).

26 Louis Roberts, 'Portrayal of the Elderly in Classical Greek and Roman Literature', in *Perceptions of Aging in Literature: A Cross-Cultural Study*, ed. Prisca von Dorotka Bagnell and Patricia Spencer Soper (New York: Greenwood, 1989), 22, 21.

27 Pat Thane, *Old Age in English History: Past Experiences, Present Issues* (Oxford: Oxford University Press, 2000), 34.

28 M. I. Finley, introduction to Falkner and de Luce, *Old Age in Greek and Latin Literature*, 17.

29 Ferdinand P. Moog and Daniel Schäfer, 'Aspekte der Altersdemenz im antiken Rom. Literarische Fiktion und faktische Lebenswirklichkeit' [Aspects of senile dementia in ancient Rome: literary fiction and factual reality], *Sudhoffs Archiv* 91 (2007): 73–81.

30 Helen Small, *The Long Life* (Oxford: Oxford University Press, 2007), 2.

31 Consider, for example, Sara Munson Deats, 'The Dialectic of Aging in Shakespeare's *King Lear* and *The Tempest*', in *Aging and Identity: A Humanities Perspective*, ed. Sara Munson Deats and Lagretta Tallent Lenker (Westport, CT: Praeger, 1999), 25–7.

32 Nina Taunton, *Fictions of Old Age in Early Modern Literature and Culture* (New York: Routledge, 2007), 71.

33 Kirk Combe and Kenneth Schmader, 'Shakespeare Teaching Geriatrics: Lear and Prospero as Case Studies in Aged Heterogeneity', in Deats and Lenker, *Aging and Identity*, 39, 40.

34 Richard Horton, 'Editorial: Shakespeare: the Bard at the Bedside', *The Lancet* 387 (2016): 1693; Jonathan Bate, '"The Infirmity of His Age": Shakespeare's 400th Anniversary', *The Lancet* 387 (2016): 1715–16.

35 *Oxford English Dictionary Online* (June 2014); G. E. Berrios and H. L. Freeman, 'Dementia before the Twentieth Century', in *Alzheimer and the Dementias*, ed. G. E. Berrios and H. L. Freeman (London: Royal Society of Medicine, 1991), 9–27.

36 Alan Richardson, *British Romanticism and the Science of the Mind* (Cambridge: Cambridge University Press, 2001), xiv, 1. For comprehensive accounts of

how nineteenth-century biology used cerebral localization to relate mind to brain, see Robert M. Young, *Mind, Brain and Adaptation in the Nineteenth Century: Cerebral Localization and Its Biological Context from Gall to Ferrier* (Oxford: Clarendon, 1970); and Edwin Clarke and L. S. Jacyna, *Nineteenth-Century Origins of Neuroscientific Concepts* (Berkeley: University of California Press, 1987).

37 Jenny Bourne Taylor and Sally Shuttleworth, introduction to *Embodied Selves: An Anthology of Psychological Texts, 1830–1890*, ed. Jenny Bourne Taylor and Sally Shuttleworth (Oxford: Clarendon, 1998), xiii.

38 Ian Hacking, *Rewriting the Soul: Multiple Personality and the Sciences of Memory* (Princeton, NJ: Princeton University Press, 1995), 4.

39 Michel Foucault, *The Birth of the Clinic: An Archaeology of Medical Perception*, trans. A. M. Sheridan (1973; London: Routledge, 2005), 205, 206.

40 Richard M. Torack, 'The Early History of Senile Dementia', in *Alzheimer's Disease: The Standard Reference*, ed. Barry Reisberg (New York: Free Press, 1983), 23–8; Stanley Finger, 'The Neuropathology of Memory', in *Origins of Neuroscience: A History of Explorations into Brain Function* (New York: Oxford University Press, 1994), 349–68; François Boller and Margaret M. Forbes, 'History of Dementia and Dementia in History: An Overview', *Journal of the Neurological Sciences* 158 (1998): 125–33; François Boller, 'History of Dementia', in *Handbook of Clinical Neurology: Dementias*, ed. Charles Duyckaerts and Irene Litvan (New York: Elsevier, 2008), 3–13; Douwe Draaisma, 'A Labyrinth of Tangles: Alzheimer's Disease', in *Disturbances of the Mind*, trans. Barbara Fasting (New York: Cambridge University Press, 2009), 199–228.

41 Thomas S. Kuhn, *The Structure of Scientific Revolutions*, 4th edn (Chicago: University of Chicago Press, 2012), 102, 180.

42 Paolo Mazzarello, *Golgi: A Biography of the Founder of Modern Neuroscience*, trans. Aldo Badiani and Henry A. Buchtel (New York: Oxford University Press, 2010), 3–6, ch. 19.

43 Roger Luckhurst and Josephine McDonagh, introduction to *Transactions and Encounters: Science and Culture in the Nineteenth Century*, ed. Roger Luckhurst and Josephine McDonagh (Manchester: Manchester University Press, 2002), 11.

44 Anne Stiles, introduction to *Neurology and Literature, 1860–1920*, ed. Anne Stiles (Basingstoke: Palgrave Macmillan, 2007), 1; Laura Salisbury and Andrew Shail, introduction to *Neurology and Modernity: A Cultural History of Nervous Systems, 1800–1950*, ed. Laura Salisbury and Andrew Shail (Basingstoke: Palgrave Macmillan, 2010), 28.

45 Nicholas Dames, *Amnesiac Selves: Nostalgia, Forgetting, and British Fiction, 1810–1870* (New York: Oxford University Press, 2001); Sally Shuttleworth, '"The Malady of Thought": Embodied Memory in Victorian Psychology and the Novel', in *Memory and Memorials: From the French Revolution to World War One*, ed.

Matthew Campbell, Jacqueline M. Labbe and Sally Shuttleworth (New Brunswick, NJ: Transaction, 2004), 46–59.

46 Lucy Burke, 'Thinking about Cognitive Impairment', *Journal of Literary and Cultural Disability Studies* 2 (2008): iii; Goldman, *Forgotten*, 245–51.

47 Patrick McDonagh, *Idiocy: A Cultural History* (Liverpool: Liverpool University Press, 2008).

48 Daniel Pick, *Faces of Degeneration: A European Disorder, c. 1848–c. 1918* (Cambridge: Cambridge University Press, 1989).

49 Thomas R. Cole, 'Part One: The Tattered Web of Cultural Meanings', in Cole and Gadow, *What Does It Mean to Grow Old?*, 4.

50 Small, *The Long Life*, 2, 215; emphasis original.

51 Miriam Solomon, 'Epistemological Reflections on the Art of Medicine and Narrative Medicine', *Perspectives in Biology and Medicine* 51 (2008): 409.

52 Bruno Latour, *Science in Action: How to Follow Scientists and Engineers through Society* (Cambridge, MA: Harvard University Press, 1987), 174–5.

53 P. W. Anderson, 'More Is Different: Broken Symmetry and the Nature of the Hierarchical Structure of Science', *Science* 177 (1972): 393; consider also Michael J. Joyner, Laszlo G. Boros and Gregory Fink, 'Biological Reductionism versus Redundancy in a Degenerate World', *Perspectives in Biology and Medicine* 61 (2018): 519–20.

54 Carroll L. Estes and Elizabeth A. Binney, 'The Biomedicalization of Aging: Dangers and Dilemmas', *The Gerontologist* 29 (1989): 590.

55 Anderson, 'More Is Different', 393.

56 Latour, *Science in Action*, 15.

57 Andrew S. Reynolds, *The Third Lens: Metaphor and the Creation of Modern Cell Biology* (Chicago: University of Chicago Press, 2018), 11.

58 Martyn Evans and Jane Macnaughton, 'Should Medical Humanities Be a Multidisciplinary or Interdisciplinary Study?' *Medical Humanities* 30 (2004): 1–4.

59 José Van Dijck, *Imagenation: Popular Images of Genetics* (Basingstoke: Palgrave Macmillan, 1998), 197.

60 Anne Davis Basting, 'Looking Back from Loss: Views of the Self in Alzheimer's Disease', *Journal of Aging Studies* 17 (2003): 87–99; Lucy Burke, 'Alzheimer's Disease: Personhood and First Person Testimony' (paper, Inaugural Conference of the Cultural Disability Studies Research Network, Liverpool John Moores University, May 2007), accessed 22 August 2011, http://www.cdsrn.org.uk/Burke_CDSRN_2007.pdf; Ellen Bouchard Ryan, Karen A. Bannister and Ann P. Anas, 'The Dementia Narrative: Writing to Reclaim Social Identity', *Journal of Aging Studies* 23 (2009): 145–57.

61 G. S. Rousseau, 'The Discourses of Literature and Medicine: Theory and Practice (1)', in *Enlightenment Borders: Pre- and Post-Modern Discourses, Medical, Scientific* (Manchester: Manchester University Press, 1991), 5.

62 Furst, *Between Doctors and Patients*, 14.
63 Stefan Merrill Block, 'A Place beyond Words: The Literature of Alzheimer's', *New Yorker*, 20 August 2014; emphasis original; https://www.newyorker.com/books/page-turner/place-beyond-words-literature-alzheimers.
64 David Lodge, *Consciousness and the Novel: Connected Essays* (Cambridge, MA: Harvard University Press, 2002), 31; emphasis original.
65 David Herman, introduction to *The Emergence of Mind: Representations of Consciousness in Narrative Discourse in English*, ed. David Herman (Lincoln: University of Nebraska Press, 2011), 3.
66 John Frow, *Genre* (Abingdon: Routledge, 2015), 81–3.
67 Angela Woods, 'The Limits of Narrative: Provocations for the Medical Humanities', *Medical Humanities* 37 (2011): 74; Sarah Falcus and Katsura Sako, *Contemporary Narratives of Dementia: Ethics, Ageing, Politics* (Abingdon: Routledge, 2019), 205.
68 W. Andrew Achenbaum, *Crossing Frontiers: Gerontology Emerges as a Science* (New York: Cambridge University Press, 1995); Ballenger, *Self, Senility, and Alzheimer's Disease*.
69 Raquel Medina, *Cinematic Representations of Alzheimer's Disease* (London: Palgrave Macmillan, 2018); 'Alzheimer's Disease, a Shifting Paradigm in Spanish Film: ¿Y tú quién eres? and Amanecer de un sueño', *Hispanic Research Journal* 14 (2013): 356–72; and 'From the Medicalisation of Dementia to the Politics of Memory and Identity in Three Spanish Documentary Films: *Bicicleta, cullera, poma, Las voces de la memoria* and *Bucarest: la memòria perduda*', *Ageing and Society* 34 (2014): 1688–710.
70 Laëtitia Ngatcha-Ribert, *Alzheimer: la construction sociale d'une maladie* [Alzheimer's: the social construction of a disease] (Paris: Dunod, 2012), 31–7.
71 Kavita Sivaramakrishnan, *As the World Ages: Rethinking a Demographic Crisis* (Cambridge, MA: Harvard University Press, 2018), 198.
72 Charles A. Riley II, *Disability and the Media: Prescriptions for Change* (Lebanon: University Press of New England, 2005), 219–23.
73 Lennard J. Davis, *Enforcing Normalcy: Disability, Deafness, and the Body* (London: Verso, 1995), xiii.
74 Zimmermann, *The Poetics and Politics*, 18–19.

2 From brain inspection to cell death

1 Geoffrey Harvey, introduction to *The Forsyte Saga*, by John Galsworthy (Oxford: Oxford University Press, 2008), xxi; all references to Galsworthy's saga are from this edition and incorporated in the text.

2 Earl Eugene Stevens, 'John Galsworthy: An Annotated Bibliography of Writings about Him. Supplement I', *English Literature in Transition, 1880–1920*, 7 (1964): 93–110; see therein, for example, Stevens's summary of Ione Dodson Young, 'The Social Conscience of John Galsworthy' (PhD thesis, University of Texas, 1955); Tracy Hargreaves, '"We Other Victorians": Literary Victorian Afterlives', *Journal of Victorian Culture* 13 (2008): 278–86.
3 *The Complete Forsyte Saga*, directed by Andy Wilson (2002; ITV Studios Home Entertainment, 2003), DVD.
4 On notions of decay and degeneration in England, see Pick, *Faces of Degeneration*, 153–221.
5 A specific 'battle' centres on the consequences of Timothy's nephew Soames's failed love life. As a 'Man of Property', he is unable to consider his wife anything other than a further possession. Irene's consequent adultery, later divorce from Soames and marriage to his cousin Young Jolyon are central to the Forsytes' fragmentation.
6 William Greenslade, *Degeneration, Culture and the Novel, 1880–1940* (New York: Cambridge University Press, 1994), 2.
7 Harvey, introduction, xvii.
8 Galsworthy specifically alludes to Darwin's theories on *The Descent of Man, and Selection in Relation to Sex* (1871), which had, among others, emphasized the female's dominant role in selecting mating partners. This further accentuates Soames's perceived failure to maintain his family line.
9 Karen Chase, *The Victorians and Old Age* (Oxford: Oxford University Press, 2009); on the 'consciousness of senescence' at the time of Queen Victoria's Diamond Jubilee, see pp. 197–8.
10 Charles Booth, *Pauperism: A Picture, and the Endowment of Old Age: An Argument* (London: Macmillan, 1892), 167–8. Booth's reference to 'a more perfect nerve' suggests the growing popular awareness of research that related neurological health to bodily performance.
11 Thomas R. Cole, 'The "Enlightened" View of Aging: Victorian Morality in a New Key', in Cole and Gadow, *What Does It Mean to Grow Old?*, 125.
12 Philippa Moylan, 'The Nervous Economies of John Galsworthy's Forsyte Chronicles', *English Literature in Transition, 1880–1920* 54 (2011): 56–78; Yi-Ling Ru, *The Family Novel: Toward a Generic Definition* (New York: Peter Lang, 1992).
13 Harvey, introduction, xviii.
14 Harold V. Marrot, *The Life and Letters of John Galsworthy* (London: William Heinemann, 1935), 169; emphasis original.
15 Helen Small, 'The Unquiet Limit: Old Age and Memory in Victorian Narrative', in Campbell, Labbe and Shuttleworth, *Memory and Memorials*, 62, 63–4.
16 Laura Otis, *Organic Memory: History and the Body in the Late Nineteenth and Early Twentieth Centuries* (Lincoln: University of Nebraska Press, 1994), 34.

17 Elaine Scarry, *The Body in Pain: The Making and Unmaking of the World* (New York: Oxford University Press, 1985), 32–3; William F. May, 'The Virtues and Vices of the Elderly', in Cole and Gadow, *What Does It Mean to Grow Old?*, 46. The 1960s screen adaptation brings Timothy to life faithfully in Galsworthy's narrative up to this last appearance. However, John Baskcomb's Timothy dies lying in bed in his last appearance, rather than performing his cognitive state; *The Forsyte Saga*, directed by David Giles and James C. Jones (1967; BBC Worldwide, 2004), DVD.

18 Timothy has remained excluded from the Forsytes' gossiping throughout. Gossip has been described – in relation to George Eliot's earlier social realist novel *Middlemarch* (1871–2) – as 'the fundamental linking force', the tool with which 'characters articulate both their individual and communal identity'. Timothy's silence induces the reader to question his identity: we perceive him as an outsider, if not a stranger throughout. As such, Timothy has always been a symbol for the Forsyte clan's undermined situation, a living dead. Soames seems to imply this. When he asks his sister whether she did 'ever know a publisher', Winifred suggests 'Uncle Timothy' and Soames retorts with 'Alive, I mean' (803); Sally Shuttleworth, *George Eliot and Nineteenth-Century Science: The Make-Believe of a Beginning* (Cambridge: Cambridge University Press, 1984), 147–8; see also Laura Otis, *Networking: Communicating with Bodies and Machines in the Nineteenth Century* (Ann Arbor: University of Michigan Press, 2011), 81.

19 Heike Hartung, *Ageing, Gender and Illness in Anglophone Literature: Narrating Age in the Bildungsroman* (Abingdon: Routledge, 2016), 23; consider also Thomas R. Cole, *The Journey of Life: A Cultural History of Aging in America* (Cambridge: Cambridge University Press, 1992), ch. 4.

20 Havelock Ellis, *The Criminal* (New York: Scribner and Welford, 1890), 212, https://archive.org/details/criminal00elli/page/212. Sally Shuttleworth illustrates how nineteenth-century research projects 'locate the human infant in the pre-human sphere'; Sally Shuttleworth, *The Mind of the Child: Child Development in Literature, Science, and Medicine, 1840–1900* (Oxford: Oxford University Press, 2010), 255, 181–206.

21 F. B. Smith, 'Health', in *The Working Class in England, 1875–1914*, ed. John Benson (Beckenham: Croom Helm, 1985), 38.

22 Dames, *Amnesiac Selves*, 163.

23 Greenslade, *Degeneration, Culture and the Novel*, 238.

24 Jill Matus, 'Emergent Theories of Victorian Mind Shock: From War and Railway Accident to Nerves, Electricity and Emotion', in Stiles, *Neurology and Literature*, 163–83.

25 Marrot, *The Life and Letters of John Galsworthy*, 426.

26 McDonagh, *Idiocy*, 79–101.

27 Philippe Pinel, *Nosographie philosophique ou méthode de l'analyse appliquée à la médecine*, 6th edn (Paris: Brosson, 1818), 3:127, https://archive.org/details/nosographiephilo03pine/page/126; Jean Étienne Esquirol, *Mental Maladies: A Treatise on Insanity*, trans. E. K. Hunt (Philadelphia: Lea and Blanchard, 1845), 422; originally published in French in 1838. The Scottish physician William Cullen (1710–1790) first defined 'amentia' as a loss of memory and intellectual faculties; Berrios and Freeman, 'Dementia before the Twentieth Century', 12.

28 James C. Prichard, *A Treatise on Insanity and Other Disorders Affecting the Mind* (Philadelphia: Haswell, Barrington and Haswell, 1837), 17, https://wellcomelibrary.org/item/b21007597#?c=0&m=0&s=0&cv=0&z=- 1.0448%2C-0.0907%2C3.0895%2C1.8137.

29 Ibid., 76.

30 Bénédict-Auguste Morel, *Traité des Maladies Mentales* [Treatise on mental illnesses] (Paris: Librairie Victor Masson, 1860); Samuel G. Howe had expounded theories of moral degeneration in the United States before Morel advanced his treatise, but his work appeared only the year after Morel's publication in the United Kingdom; McDonagh, *Idiocy*, 264–5.

31 Gillian Beer, *Darwin's Plots: Evolutionary Narrative in Darwin, George Eliot and Nineteenth-Century Fiction*, 3rd edn (Cambridge: Cambridge University Press, 2009), 119–20.

32 *Entartung* appeared in Italy as *Degenerazione* in 1893, reached France as *Dégénérescence* in the same year and was published in the United Kingdom and United States as *Degeneration* in 1895. For a detailed analysis of Morel's as compared to Darwin's theories and their societal assimilations, see Ian Dowbiggin, *The Quest for Mental Health: A Tale of Science, Medicine, Scandal, Sorrow, and Mass Society* (New York: Cambridge University Press, 2011), 56–60.

33 Berrios and Freeman, 'Dementia before the Twentieth Century', 22. The case of Phineas Gage brings notions of damage to the frontal lobe together with a consideration of the patient's marked change in personality and behaviour, which was also described as 'demented'; John M. Harlow, *Recovery from the Passage of an Iron Bar through the Head* (Boston: David Clapp, 1869), 10, https://archive.org/details/66210360R.nlm.nih.gov/page/n13.

34 John Hughlings Jackson, 'On Affections of Speech from Disease of the Brain', *Brain* 1 (1879): 309; Salisbury and Shail, introduction, 32.

35 Insanity represented, for much of the nineteenth century, an umbrella term also for mania and melancholia; consider, for example, James Kennaway, 'Singing the Body Electric: Nervous Music and Sexuality in *Fin-de-Siècle* Literature', 141–60; and Mark S. Micale, 'Medical and Literary Discourses of Trauma in the Age of the American Civil War', 184–206; both in Stiles, *Neurology and Literature*.

36 On clinical observations and experimental work supporting localization theories carried out, inter alia, by Franz Josef Gall (1758–1828), Paul Broca (1824–1880),

Gustav Fritsch (1838–1927) and Eduard Hitzig (1838–1907), David Ferrier (1843–1928) and Carl Wernicke (1848–1905), see, as previously referenced Young, *Mind, Brain and Adaptation*, and Clarke and Jacyna, *Nineteenth-Century Origins of Neuroscientific Concepts*.

37 Stiles, introduction, 6.
38 John Hughlings Jackson, *On Convulsive Seizures* (London: British Medical Association, 1890), 7; Michael J. Clark, '"A Plastic Power Ministering to Organisation": Interpretations of the Mind–Body Relation in Late Nineteenth-Century British Psychiatry', *Psychological Medicine* 13 (1983): 489.
39 Michael J. Clark, 'The Rejection of Psychological Approaches to Mental Disorder in Late Nineteenth-Century British Psychiatry', in *Madhouses, Mad-Doctors, and Madmen: The Social History of Psychiatry in the Victorian Era*, ed. Andrew Scull (Philadelphia: University of Pennsylvania Press, 1981), 287, 298.
40 Otis, *Organic Memory*, 192. The French psychologist Théodule A. Ribot (1839–1916) already in 1881 claimed that 'memory is a general function of the nervous system'; Théodule A. Ribot, *The Diseases of Memory*, trans. J. Fitzgerald (New York: Humboldt, 1883), 47.
41 For a full account of Freud's move from materialism to psychology and a detailed reading of 'Project for a Scientific Psychology', see Katja Guenther, *Localization and Its Discontents: A Genealogy of Psychoanalysis and the Neuro Disciplines* (Chicago: University of Chicago Press, 2015), ch. 3, 72, 73, 75, 81, 84.
42 Clark, 'The Rejection', 271.
43 David Shenk, *The Forgetting. Alzheimer's: Portrait of an Epidemic* (2001; New York: Anchor Books, 2003), 77.
44 The lectures on brain pathology, which Alzheimer attended at the Charité Hospital in Berlin, had been devised by Griesinger and further developed by Westphal. Griesinger's textbook *Die Pathologie und Therapie der psychischen Krankheiten für Aerzte und Studierende* (Pathology and therapy of psychiatric diseases for practitioners and students; 1845) was a cornerstone in the emerging scientific epoch of psychiatry which sought to characterize brain damage in psychotic conditions; Clarke and Jacyna, *Nineteenth-Century Origins of Neuroscientific Concepts*, 133–8.
45 For a detailed account of the circumstances outlined in this paragraph, see W. F. Bynum, *Science and the Practice of Medicine in the Nineteenth Century* (New York: Cambridge University Press, 1994), 95–103.
46 Thomas Schlich, for example, shows how international German surgery was in its approach to innovation and how British and German research communicated; Thomas Schlich, 'Farmer to Industrialist: Lister's Antisepsis and the Making of Modern Surgery in Germany', *Notes and Records of the Royal Society* 67 (2013): 245–60.

47 P. Hoff, 'Alzheimer and His Time', in Berrios and Freeman, *Alzheimer and the Dementias*, 49.
48 Otis, *Organic Memory*, 113; Silvia Acocella, *Effetto Nordau: Figure della degenerazione nella letteratura Italiana tra Ottocento e Novecento* [Nordau effect: Figures of degeneration in Italian literature between the nineteenth and twentieth centuries] (Naples: Liguori Editore, 2012), 105–24.
49 Samuel Wilks, 'Clinical Notes on Atrophy of the Brain', *Journal of Mental Science* 10 (1864): 382–3; Torack, 'The Early History of Senile Dementia', 26.
50 James Crichton-Browne, 'Clinical Lectures on Mental and Cerebral Diseases', *British Medical Journal* 1 (1874): 601, 603.
51 Alois Alzheimer, 'Neuere Arbeiten über die Dementia senilis und die auf atheromatöser Gefässerkrankung basierenden Gehirnkrankheiten' [Recent work on senile dementia and brain diseases based on atheroma], *Monatsschrift für Psychiatrie und Neurologie* 3 (1898): 101–15.
52 Lauren Berlant, 'On the Case', *Critical Inquiry* 33 (2007): 666; Alois Alzheimer, 'Über eine eigenartige Erkrankung der Hirnrinde' [On a peculiar illness of the cortex], *Allgemeine Zeitschrift für Psychiatrie und Psychisch-gerichtliche Medizin* 64 (1907): 146–8; all references from this edition are incorporated in the text. Consider also the paper's two present-day translations: 'A Characteristic Disease of the Cerebral Cortex', in *The Early Story of Alzheimer's Disease: Translation of the Historical Papers by Alois Alzheimer, Oskar Fischer, Francesco Bonfiglio, Emil Kraepelin, Gaetano Perusini*, ed. Katherine Bick, Luigi Amaducci and Giancarlo Pepeu (Padua: Liviana, 1987), 1–3; and Alois Alzheimer et al., 'An English Translation of Alzheimer's 1907 Paper, "Über eine eigenartige Erkrankung der Hirnrinde"', *Clinical Anatomy* 8 (1995): 429–31.
53 The Swiss psychiatrist and neurologist Otto Ludwig Binswanger (1852–1929) used the term 'presenile dementia' already in 1894; Otto Ludwig Binswanger, 'Die Abgrenzung der allgemeinen progressiven Paralyse' [Differentiation of general progressive paralysis], *Berliner klinische Wochenschrift* 31 (1894): 1181.
54 Michael Lynch, 'Discipline and the Material Form of Images: An Analysis of Scientific Visibility', *Social Studies of Science* 15 (1985): 43; emphasis original.
55 Melissa M. Littlefield, 'Matter for Thought: The Psychon in Neurology, Psychology and American Culture, 1927–1943', in Salisbury and Shail, *Neurology and Modernity*, 269.
56 Together with the neurologist Max Lewandowsky (1876–1918), Alzheimer later co-edited the newly established specialist journal *Zeitschrift für die gesamte Neurologie und Psychiatrie* with the aim of bringing the fields of psychiatry and neurology closer together.
57 Konrad Maurer and Ulrike Maurer, *Alzheimer: Das Leben eines Arztes und die Karriere einer Krankheit* [Alzheimer: The life of a physician and the career of a disease] (Munich: Piper, 1998), 73, 128.

58 Already the nineteenth-century clinical case study had seen remarkable textual and formal changes: the autopsy report refocused the case history on a 'narrative of causality instead of analogy', and statistics gained in value; Meegan Kennedy, *Revising the Clinic: Vision and Representation in Victorian Medical Narrative and the Novel* (Columbus: Ohio State University Press, 2010), 57, 76.

59 Francesco Bonfiglio, 'Di speciali reperti in un caso di probabile sifilide cerebrale' [Concerning special findings in a case of probable cerebral syphilis], *Rivista Sperimentale di Freniatria e Medicina Legale delle Alienazioni Mentali* 34 (1908): 196–206; all references incorporated in the text. Translations of Bonfiglio's as well as Perusini's cases are included in Bick, Amaducci and Pepeu, *The Early Story of Alzheimer's Disease*, 19–31, 82–128.

60 The pathologist Rudolf Virchow (1821–1902) used the term on several occasions in his textbook on *Die Cellularpathologie in ihrer Begründung auf physiologische und pathologische Gewebelehre* (1858); translated into English in 1860 with the title *Cellular Pathology as Based upon Physiological and Pathological Histology*. Also Alzheimer had observed 'degenerative alterations' in relation to nerve cell morphology in a patient with atrophy of a muscle at the base of the thumb in 1892; Alois Alzheimer, 'Über einen Fall von spinaler progressiver Muskelatrophie und hinzutretender Erkrankung bulbärer Kerne und der Rinde' [On a case of spinal progressive muscular atrophy with additional affliction of bulbar marrow and cortex], *Archiv für Pychiatrie und Nervenkrankheiten* 23 (1892): 465 et passim.

61 Foucault, *The Birth of the Clinic*, 146.

62 Gaetano Perusini, 'Über klinisch und histologisch eigenartige psychische Erkrankungen des späteren Lebensalters' [On clinically and histologically peculiar psychiatric illnesses of later life], in *Histopathologische Arbeiten über die Großhirnrinde unter besonderer Berücksichtigung der pathologischen Anatomie der Geisteskrankheiten* [Histopathological studies on the cerebral cortex with particular reference to the pathological anatomy of mental illnesses], ed. Franz Nissl and Alois Alzheimer (Jena: Gustav Fischer, 1910), 3: 297–352.

63 G. Macchi, C. Brahe and M. Pomponi, 'Alois Alzheimer and Gaetano Perusini: Should Man Divide What Fate United?' *European Journal of Neurology* 4 (1997): 210–13.

64 Emil Kraepelin, 'VII. Das senile and präsenile Irresein' [Senile and presenile dementia], in *Psychiatrie: Ein Lehrbuch für Studierende und Ärzte* [Psychiatry: A textbook for students and practitioners] (Leipzig: Johann Ambrosius Barth, 1909–10), 627; all further references from this edition are incorporated in the text. Chapter VII (533–632) contains three sections: 'A. Das präsenile Irresein' (presenile dementia; 534–54), 'B. Das arteriosklerotische Irresein' (arteriosclerotic dementia; 554–93) and 'C. Der Altersblödsinn' (senile dementia; 593–632).

65 In this section, Kraepelin elaborated on 'depressive madness' particularly in women between 45 and 50 years of age. He described these presenile cases as 'very

considerably mentally deficient, poor in thought, without judgement, unreasonable, dull and weak in will' (548).
66 Note that *Rückbildung* can be translated as 'regression' or 'involution'.
67 Thomas S. Kuhn has illustrated the role of textbooks in establishing a 'paradigm for granted'; Kuhn, *The Structure of Scientific Revolutions*, 20.
68 Mary Boyle, 'The Non-discovery of Schizophrenia? Kraepelin and Bleuler Reconsidered', in *Reconstructing Schizophrenia*, ed. Richard P. Bentall (London: Routledge, 1990), 10, 5, 13.
69 G. E. Berrios, 'Alzheimer's Disease: A Conceptual History', *International Journal of Geriatric Psychiatry* 5 (1990): 363; Lawrence Cohen, *No Aging in India: Alzheimer's, the Bad Family, and Other Modern Things* (1998; Berkeley: University of California Press, 1999), 79–80. Kraepelin treated 'Das arteriosklerotische Irresein' (arteriosclerotic dementia; 554–93) in a separate section, even though vascular phenomena had in fact been described, but with less detail, by Alzheimer, Bonfiglio and Perusini.
70 Hacking, *Rewriting the Soul*, ch. 5.
71 Kraepelin's haste has given rise to a series of historical hypotheses that could account for the selective emphasis on characteristics when establishing Alzheimer's disease in these terms. The medical historian G. E. Berrios convincingly discards all of them, including those concerning rivalry with Arnold Pick (1851–1924), who researched on dementia in Prague, and Sigmund Freud, who was known for his non-materialist views; Berrios, 'Alzheimer's Disease', 362–3.
72 Guenther, *Localization and Its Discontents*, 65–7.
73 Ibid., 81; Berrios, 'Alzheimer's Disease', 362. Also Sander L. Gilman notes, inter alia, Freud's work on neurological diseases like multiple sclerosis as related to 'the endogenous decay of the cell'; Sander L. Gilman, *The Case of Sigmund Freud: Medicine and Identity at the Fin de Siècle* (Baltimore: Johns Hopkins University Press, 1993), 157–8.
74 Sacks, *The Man Who Mistook*, 91. Neurobiological concepts of 'excess' – as with increased levels of the neurotransmitter glutamate or genetic explanations of loss of function through gain of function – have, as far as I am aware, not reached the popular dementia narrative. I note that the justification for the pharmacological inhibition of acetylcholinesterase (which breaks down the neurotransmitter acetylcholine) I discuss in Chapter 6 is not its excess, but rather insufficient substrate (i.e. acetylcholine).
75 Alois Alzheimer, 'Die diagnostischen Schwierigkeiten in der Psychiatrie' [Diagnostic difficulties in psychiatry], *Zeitschrift für die gesamte Neurologie und Psychiatrie* 1 (1910): 1–19; references incorporated in the text.
76 Nordau took inspiration from England's Henry Maudsley (1835–1918), Italy's Cesare Lombroso (1835–1909), France's Auguste Morel and Germany's Richard von Krafft-Ebing (1840–1902); Dowbiggin, *The Quest for Mental Health*, 59. At least

on three occasions, Alzheimer set himself apart from the kind of degenerationism that fed into criminal anthropology and, later, eugenicist healthcare politics. In the publication 'Ein "geborener Verbrecher"' (A 'born criminal'), Alzheimer asserted himself against the contemporary conceptualization of the genetically predisposed as criminal, as propagated in Lombroso's 1876 *L'Uomo Delinquente* (The criminal). In 1907, he argued against increasing eugenicist tendencies in Germany, and, supporting notions of *regeneration*, Alzheimer's stance echoed diatribes like Egmont Hake's, published in response to Nordau's *Entartung*. Finally, in his speech 'Der Krieg und die Nerven' (War and nerves), Alzheimer vehemently opposed himself to views that modernity in all its facets would lead to degeneration; Alois Alzheimer, 'Ein "geborener Verbrecher"', *Archiv für Psychiatrie und Nervenkrankheiten* 28 (1896): 327–53; 'Über die Indikationen für eine künstliche Schwangerschaftsunterbrechung bei Geisteskranken' [On the indication for induced abortion in the mentally ill], *Münchener medizinische Wochenschrift* 54 (1907): 1617–22, cited in Maurer and Maurer, *Alzheimer*, 185; Egmont Hake, *Regeneration: A Reply to Max Nordau* (Westminster: Archibald Constable, 1895), cited in *The Fin de Siècle: A Reader in Cultural History, c. 1880–1900*, ed. Sally Ledger and Roger Luckhurst (Oxford: Oxford University Press, 2000), 17–19; Alois Alzheimer, *Der Krieg und die Nerven* (Breslau: Verlag von Preuß und Jünger, 1915), https://digital.staatsbibliothek-berlin.de/werkansicht?PPN=PPN716104202&PHYSID=PHYS_0001&DMDID=DMDLOG_0001.

77 Alois Alzheimer, 'Über eigenartige Krankheitsfälle des späteren Alters' [On peculiar illness cases in later life], *Zeitschrift für die gesamte Neurologie und Psychiatrie* 4 (1911): 356–85; references incorporated in the text.

78 Gonzalo R. Lafora, 'Beitrag zur Kenntnis der Alzheimerschen Krankheit oder präsenilen Demenz mit Herdsymptomen' [Contribution to the understanding of Alzheimer's disease or presenile dementia with focal symptoms], *Zeitschrift für die gesamte Neurologie und Psychiatrie* 6 (1911): 16.

79 Ernst Grünthal, 'Klinisch-anatomisch vergleichende Untersuchungen über den Greisenblödsinn' [Comparative clinico-anatomical examinations on senile dementia], *Zeitschrift für die gesamte Neurologie und Psychiatrie* 111 (1927): 774.

80 Johannes Schottky, 'Über präsenile Verblödungen' [On cases of presenile stultification], *Zeitschrift für die gesamte Neurologie und Psychiatrie* 140 (1932): 342, 361.

81 The photographs were shown at the conference of the Society of German Neurologists and Psychiatrists in 1936; Maurer and Maurer, *Alzheimer*, 276–9; see p. 277 for representative images (of course, their selection in 1998 from the archived slides could also be discourse influenced).

82 Ernst Grünthal, 'Über die Alzheimersche Krankheit: Eine histopathologisch-klinische Studie' [On Alzheimer's disease: A histopathological-clinical study], *Zeitschrift für die gesamte Neurologie und Psychiatrie* 101 (1926): 128–57; references incorporated in the text.

83 Schottky, 'Über präsenile Verblödungen', 396.
84 A full-text search of the *Asylum Journal* (later the *Journal of Mental Science* and today the *British Journal of Psychiatry*) and *Brain: A Journal of Neurology* (today *Brain*) suggests that British and French interest in Alzheimer's disease surged only after the publication of Alzheimer's second case in 1911 and Kraepelin's publicity, with discussion focused on histopathological phenomena and their use for differential diagnosis; consider, for example, Samuel T. Orton, 'A Study of the Satellite Cells in Fifty Selected Cases of Mental Disease', *Brain* 36 (1914): 525–42; G. Mingazzini, 'On Aphasia due to Atrophy of the Cerebral Convolutions', *Brain* 36 (1914): 493–524; L. Bouman, 'Senile Plaques', *Brain* 57 (1934): 128–42; A. A. W. Petrie, 'Differential Diagnosis of Organic and Functional Nervous Disorders', *British Medical Journal* 2 (1934): 503–6.
85 W. H. McMenemey (also McMenemy), 'Alzheimer's Disease: A Report of Six Cases', *Journal of Neurology and Psychiatry* 3 (1940): 211–40.
86 M. A. Green et al., 'Cerebral Biopsy in Patients with Presenile Dementia', *Diseases of the Nervous System* 13 (1952): 303–7; Myre Sim and W. Thomas Smith, 'Alzheimer's Disease Confirmed by Cerebral Biopsy: A Therapeutic Trial with Cortisone and ACTH', *Journal of Mental Science* 101 (1955): 604–9.
87 Consider, for example, the following two connected papers: Myre Sim, Eric Turner and W. Thomas Smith, 'Cerebral Biopsy in the Investigation of Presenile Dementia: I. Clinical Aspects', *British Journal of Psychiatry* 112 (1966): 119–25; and W. Thomas Smith, Eric Turner and Myre Sim, 'Cerebral Biopsy in the Investigation of Presenile Dementia: II. Pathological Aspects', *British Journal of Psychiatry* 112 (1966): 127–33.
88 Smith, Turner and Sim, 'Cerebral Biopsy', 129.
89 A further conference on 'Alzheimer's Disease and Related Conditions' was held in London in 1969, inter alia, drawing attention to familial cases of Alzheimer's disease.
90 Ch. Müller and L. Ciompi, eds, *Senile Dementia: Clinical and Therapeutic Aspects* (Bern: Hans Huber, 1968), quoted from the sleeve text of the conference book.
91 J. A. N. Corsellis and P. H. Evans, 'The Relation of Stenosis of the Extracranial Cerebral Arteries to Mental Disorder and Cerebral Degeneration in Old Age', *Proceedings of the 5th International Congress of Neuropathology* (1965): 546; B. E. Tomlinson, G. Blessed and M. Roth, 'Observations on the Brains of Demented Old People', *Journal of the Neurological Sciences* 11 (1970): 205–42; V. C. Hachinski, N. A. Lassen and J. Marshall, 'Multi-infarct Dementia: A Cause of Mental Deterioration in the Elderly', *The Lancet* 2 (1974): 207–10.
92 Ballenger, *Self, Senility, and Alzheimer's Disease*, 109; Furst, *Between Doctors and Patients*, 6.
93 G. Stanley Hall, *Senescence: The Last Half of Life* (New York: D. Appleton, 1922), 199, 201, 366–438.

94 Consider also Schottky, 'Über präsenile Verblödungen', and note that McMenemey's publication discusses female cases only.
95 Jeanette King, *Discourses of Ageing in Fiction and Feminism: The Invisible Woman* (Basingstoke: Palgrave Macmillan, 2013), 39–42.
96 Norah Hoult, *There Were No Windows* (London: Persephone Books, 2005); references incorporated in the text.
97 Lisa Appignanesi, *Mad, Bad and Sad: A History of Women and the Mind Doctors from 1800 to the Present* (2008; London: Virago, 2009).
98 Julia Briggs, afterword to Hoult, *There Were No Windows*, 329. According to Briggs, the writer Violet Hunt (1862–1942) was Claire Temple's real-life model.
99 Nick Hubble and Philip Tew, *Ageing, Narrative and Identity: New Qualitative Social Research* (Basingstoke: Palgrave Macmillan, 2013), 185.
100 McDonagh, *Idiocy*, 102–28.
101 Suzanne Nalbantian, *Memory in Literature: From Rousseau to Neuroscience* (Basingstoke: Palgrave Macmillan, 2003), 93.
102 Briggs, afterword, 333.
103 Greenslade, *Degeneration, Culture and the Novel*, 38–40; Hoult makes Claire Temple particularly aware of the slum courts in London's East End.
104 Jesse F. Ballenger, 'Disappearing in Plain Sight: Public Roles of People with Dementia in the Meaning and Politics of Alzheimer's Disease', in *The Neurological Patient in History*, ed. L. Stephen Jacyna and Stephen T. Casper (Rochester: University of Rochester Press, 2012), 109–28.
105 Shuttleworth, *The Mind of the Child*, 33.
106 Joanna Moncrieff, 'The Politics of a New Mental Health Act', *British Journal of Psychiatry* 183 (2003): 8.
107 Joan Lane, *A Social History of Medicine: Health, Healing and Disease in England, 1750–1950* (London: Routledge, 2001), 96–119.
108 This was the case also outside Britain. For the asylum situation in the United States, consult Ballenger, *Self, Senility, and Alzheimer's Disease*; on the relationship between criminalization and madness in France, see Robert A. Nye, *Crime, Madness, and Politics in Modern France: The Medical Concept of National Decline* (Princeton, NJ: Princeton University Press, 1984); and for the German context, consider Paul Weindling, *Health, Race and German Politics between National Unification and Nazism, 1870–1945* (1989; Cambridge: Cambridge University Press, 1993).
109 Michel Foucault, *Psychiatric Power: Lectures at the Collège de France, 1973–1974*, trans. Graham Burchell (2006; Basingstoke: Palgrave Macmillan, 2008), 108; Shuttleworth, *The Mind of the Child*, 183.
110 Berrios and Freeman, 'Dementia before the Twentieth Century', 26.
111 Greenslade agrees on this with Nye, who scrutinizes the period of the French Third Republic (1870–1940); however, Pick brings the period's end forward to

1918; Weindling considers 1945 to be its end point in the German context of his cultural enquiry.
112 Bynum, *Science and the Practice of Medicine*, 194.
113 Emil Kraepelin, 'Altersblödsinn' [senile dementia], in *Einführung in die psychiatrische Klinik: Dreissig Vorlesungen* [Introduction into clinical psychiatry: Thirty lectures] (Leipzig: Johann Ambrosius Barth, 1901), 237, https://archive.org/details/einfhrungindiep01kraegoog/page/n249.
114 Thane, *Old Age in English History*, 436–8.
115 Chase, *The Victorians and Old Age*, 61.
116 Also the passing appearance of an elderly woman, who suffers from memory loss, in *Lady Audley's Secret* seems to me to be a clear manifestation of this phenomenon; Mary E. Braddon, *Lady Audley's Secret* (London: Penguin, 2012), 456.
117 Lodge, *Consciousness and the Novel*, 58.
118 Stiles, introduction, 1.
119 Thomas G. Pavel, *The Lives of the Novel: A History* (Princeton, NJ: Princeton University Press, 2013), 267.
120 I borrow this term from Cary Smith Henderson, *Partial View: An Alzheimer's Journal* (Dallas, TX: Southern Methodist University Press, 1998).
121 Kennedy, *Revising the Clinic*, 200.
122 Lodge, *Consciousness and the Novel*, 17.

3 Culture shapes politics shapes science

1 Robert N. Butler, *Why Survive? Being Old in America* (Baltimore: Johns Hopkins University Press, 2002), xi.
2 Emma Brown, 'Robert N. Butler Dies, "Father of Modern Gerontology" Was 83', *Washington Post*, 7 July 2010.
3 Stephen Katz, *Disciplining Old Age: The Formation of Gerontological Knowledge* (Charlottesville: University Press of Virginia, 1996), 41.
4 Elie Metchnikoff, 'Old Age', in *Annual Report of the Smithsonian Institution for the Year Ending June 30, 1904* (Washington, DC: Smithsonian Institution, 1903–4), 537, 541, 548. Metchnikoff's definition signalled the epidemiologic transition from infectious to degenerative diseases around the turn of the century, as he appreciated the role of infectious diseases on health in ageing, but did not yet clearly separate infectious from chronic diseases; Abdel R. Omran, 'The Epidemiologic Transition: A Theory of the Epidemiology of Population Change', *Milbank Quarterly* 83 (2005): 731–57; reissue of 49 (1971): 509–38; cf. ch. 2, 29; Achenbaum, *Crossing Frontiers*, 30 (for a fuller account of Metchnikoff's scientific research on ageing, see pp. 23–33).

5 Achenbaum, *Crossing Frontiers*, 37.
6 Pat Thane, 'Geriatrics', in *Companion Encyclopedia of the History of Medicine*, ed. W. F. Bynum and Roy Porter (Abingdon: Routledge, 1993), 1103–4.
7 Georges Canguilhem, *The Normal and the Pathological*, trans. Carolyn R. Fawcett (1978; New York: Zone Books, 1991), 160. For a concise account of the medicalization of the aged body, see Katz, *Disciplining Old Age*, 40–8.
8 Achenbaum, *Crossing Frontiers*, 86. Conferences on geriatrics and gerontology became more frequent, and pioneering textbooks emerging from these conferences marked a growing collaboration among researchers from different disciplines – a collaboration that promoted new professional organizations; Ballenger, *Self, Senility, and Alzheimer's Disease*, 57; see also Achenbaum, *Crossing Frontiers*, chs 2 and 3, on the role of Edmund Cowdry, Edward Sieglitz and Oscar Kaplan for discipline building in relation to gerontology and geriatrics.
9 The International Association of Gerontology formed in Liège in 1950, and medical journals devoted to geriatrics were being established across Europe, including *Giornale di Gerontologia* (Italy, 1958), *The Gerontologist* (United Kingdom, 1961) and, slightly later, *Age and Aging* (United Kingdom, 1972), *Gerontology* (Switzerland, 1976) and *Age* (Netherlands, 1978). The first journal to carry 'Alzheimer's' in its title was the US journal *Alzheimer Disease and Associated Disorders*, launched in 1987.
10 Thane, *Old Age in English History*, 437; see ch. 22 on the rise of geriatric medicine in Britain.
11 Marjorie Warren, 'Care of the Chronic Aged Sick', *The Lancet* 1 (1946): 841.
12 Thane, *Old Age in English History*, 385.
13 Joseph H. Sheldon, *The Social Medicine of Old Age: Report of an Inquiry in Wolverhampton* (London: Nuffield Foundation, 1948), references incorporated in the text. On Peter Townsend's (1928–2009) socio-medical work in London around mid-century, see Charlotte Greenhalgh, *Aging in Twentieth-Century Britain* (Oakland: University of California Press, 2018), ch. 2.
14 Kenneth Hazell, *Social and Medical Problems of the Elderly* (London: Hutchinson, 1973), 19 (preface to 1960 edition); all further references incorporated in the text with each edition's year of publication.
15 Thane, *Old Age in English History*, 385.
16 Trevor H. Howell, *Our Advancing Years: An Essay on Modern Problems of Old Age* (London: Phoenix House, 1976), 13.
17 Butler, *Why Survive?*, 12.
18 Barbara Pym, *Quartet in Autumn* (London: Bello, 2013), 92, 103; all further references incorporated in the text.
19 H. L. Freeman, 'Social Aspects of Alzheimer's Disease', in Berrios and Freeman, *Alzheimer and the Dementias*, 60; Ngatcha-Ribert, *Alzheimer*, 128–49.

20 J. De Ajuriaguerra and R. Tissot, 'Some Aspects of Psycho-neurologic Disintegration in Senile Dementia'; and R. Tissot, 'Discussion Remarks on Senile Disintegration'; both in Müller and Ciompi, *Senile Dementia*, 69 and 83, respectively.
21 De Ajuriaguerra and Tissot, 'Some Aspects', 74.
22 Ibid., 76, 77.
23 Ch. Müller, foreword to Müller and Ciompi, *Senile Dementia*, 12.
24 Ballenger, *Self, Senility, and Alzheimer's Disease*, 56.
25 Robert Katzman, 'Editorial. The Prevalence and Malignancy of Alzheimer Disease: A Major Killer', *Archives of Neurology* 33 (1976): 217.
26 Martin S. Brander, 'The Scientist and the News Media', *New England Journal of Medicine* 308 (1983): 1170–3; Van Dijck, *Imagenation*, 72.
27 Ballenger, *Self, Senility, and Alzheimer's Disease*, 115.
28 Constance Rooke lists thirty-nine works, fifteen and sixteen of which were published in the 1970s and 1980s, respectively; Constance Rooke, 'Old Age in Contemporary Fiction: A New Paradigm of Hope', in *Handbook of the Humanities and Aging*, ed. Thomas R. Cole, David D. Van Tassel and Robert Kastenbaum (New York: Springer, 1992), 256–7.
29 Sally A. Gadow, 'Part Two. Subjectivity: Literature, Imagination, and Frailty. Introduction', in Cole and Gadow, *What Does It Mean to Grow Old?*, 132.
30 The epigraph is from the 1973 edition, pp. 23–4.
31 Paul Bailey, *At The Jerusalem* (London: Penguin, 1982); all references incorporated in the text.
32 Thane, *Old Age in English History*, 385.
33 Personal conversation with the author subsequent to: Paul Bailey, 'Contemporary Narrative and the Coming of Ageing' (Literary dialogue with Courttia Newland, Senate House, University of London, 30 October 2013).
34 Mike Hepworth, *Stories of Ageing* (Buckingham: Open University Press, 2000), 77.
35 Robert N. Butler, 'The Life Review: An Interpretation of Reminiscence in the Aged', in *New Thoughts on Old Age*, ed. Robert Kastenbaum (New York: Springer, 1964), 266.
36 Thane, *Old Age in English History*, 452.
37 Charles L. Rose, 'Social Correlates of Longevity', in Kastenbaum, *New Thoughts on Old Age*, 87; see also in the same volume: M. Elaine Cumming, 'New Thoughts on the Theory of Disengagement', 3–18. Compare this with Barbara Pym's poignant vignette of the theme of social disengagement in *Quartet in Autumn*. Both Letty and Marcia retire at the same time, but only Letty keeps 'making an effort with her clothes' (7) and going to 'the library for her own pleasure and possible edification' (3). Marcia, in contrast, does 'not bother[ ] to touch up the roots of her hair' (102)

and spends most of her time doing 'a good deal of classifying and sorting' of tinned foods in her cupboard (54).
38 Small, *The Long Life*, 21.
39 Sidney Levin, 'Depression in the Aged: The Importance of External Factors', in Kastenbaum, *New Thoughts on Old Age*, 179–85.
40 It is perhaps only rhetorically correct to suggest that workhouses were 'not intended for the very aged and frail' because almost all the aged poor were destitute since unable to work; F. B. Smith, 'Old Age', in *The People's Health, 1830–1910* (1979; London: Weidenfeld and Nicolson, 1990).
41 Erving Goffman, *Asylums: Essays on the Social Situation of Mental Patients and Other Inmates* (1961; London: Penguin, 1991), 43, 25; Michel Foucault, *Madness and Civilization: A History of Insanity in the Age of Reason*, trans. Richard Howard (1965; London: Vintage, 1988).
42 On grooming in old age in Britain between the 1940s and 1970s, see Greenhalgh, *Aging in Twentieth-Century Britain*, ch. 4.
43 Hepworth, *Stories of Ageing*, 77.
44 Kathleen Woodward, *Aging and Its Discontents: Freud and Other Fictions* (Bloomington: Indiana University Press, 1991), 70.
45 Butler, *Why Survive?*, 188, 189.
46 Bailey, 'Contemporary Narrative'; R. D. Laing, *The Politics of Experience* and *The Bird of Paradise* (1967; Harmondsworth: Penguin, 1987), 110.
47 R. D. Laing, *The Divided Self: An Existential Study in Sanity and Madness* (1960; London: Penguin, 2010), 25–6 *et passim*; see also Sander L. Gilman, *Disease and Representation: Images of Illness from Madness to AIDS* (1988; Ithaca, NY: Cornell University Press, 1991), 98–111.
48 Irving Wardle, 'A Worthy Guest', *Times*, 12 June 1974.
49 Herrman L. Blumgart, 'Caring for the Patient', *New England Journal of Medicine* 270 (1964): 452.
50 Small, *The Long Life*, 28.
51 J. L. Villa and L. Ciompi, 'Therapeutic Problems of Senile Dementia', in Müller and Ciompi, *Senile Dementia*, 135.
52 Helen Dunmore, introduction to *Ending Up*, by Kingsley Amis (London: Penguin, 2011), ix, xi; all references from this edition are incorporated in the text.
53 Paul Bailey, 'Author, Author: Paul Bailey', *Guardian*, 15 January 2011, https://www.theguardian.com/books/2011/jan/15/paul-bailey-author-author.
54 John Updike, *The Poorhouse Fair* (London: Penguin, 2006), 5.
55 Robert H. Binstock, Stephen G. Post and Peter J. Whitehouse, 'The Challenges of Dementia', in *Dementia and Aging: Ethics, Values, and Policy Choices*, ed. Robert H. Binstock, Stephen G. Post and Peter J. Whitehouse (Baltimore: Johns Hopkins University Press, 1992), 1.

4 The loss of self in healthcare and cultural discourse

1. Nancy L. Mace and Peter V. Rabins, *The 36-Hour Day: A Family Guide to Caring for Persons with Alzheimer Disease, Related Dementing Illnesses, and Memory Loss in Later Life* (Baltimore: Johns Hopkins University Press, 1981), 139; all further references incorporated in the text are from this first of (to date) six editions, unless noted otherwise.
2. Rosalie Walsh Honel, *Journey with Grandpa: Our Family's Struggle with Alzheimer's Disease* (Baltimore: Johns Hopkins University Press, 1988); all references incorporated in the text. Yasushi Inoue, *Chronicle of My Mother* (New York: Kodansha International, 1985), originally published in 1975, was translated from Japanese in 1982, but had not received the same endorsement.
3. Jane Lewis and Barbara Meredith, *Daughters Who Care: Daughters Caring for Mothers at Home* (London: Routledge, 1988), 10.
4. May, 'The Virtues and Vices', 46.
5. Margaret Forster, *Have the Men Had Enough?* (London: Vintage, 2004), 94–114, 177.
6. Elaine M. Brody, '"Women in the Middle" and Family Help to Older People', *The Gerontologist* 21 (1981): 471–80.
7. Peter V. Rabins, foreword to Honel, *Journey with Grandpa*, ix.
8. Thane, *Old Age in English History*, 430.
9. This motif reflects the impact of population changes in Italy, where since the 1980s, demographic figures have caused serious concern; Rita C. Cavigioli, *Women of a Certain Age: Contemporary Italian Fictions of Female Aging* (Madison, NJ: Fairleigh Dickinson University Press, 2005), 50.
10. Sawako Ariyoshi, *The Twilight Years*, trans. Mildred Tahara (London: Peter Owen, 1984).
11. The first time I encountered this term was in Richard C. Adelman's questioning of the NIA's political strategies in 'The Alzheimerization of Aging', *The Gerontologist* 35 (1995): 526–32.
12. Binstock, Post and Whitehouse, 'The Challenges of Dementia', 5. Honel became active in response to Ronald Reagan's 1982 call for November to be the National Alzheimer's Disease Awareness Month. She provided the local newspaper with factual information (166), participated in a Chicago television news show (173) and made 'a documentary film on the problems of aging' (193). She also published a poem about her experience in a prominent medical journal. This, by itself, I take as a first and significant step that forged a way for caregivers as members of the general public to feed back into, and thus contribute to shaping, how the medical profession thought about the condition; *My Mother, My Father*, directed by James Vanden Bosch (1984; Concord Media, n.d.), DVD; Rosalie Walsh Honel, 'Alzheimer's Disease', *New England Journal of Medicine* 309 (1983): 1524. On

the popularizing effect of Alzheimer's Awareness Week, consider, for example, Nadine Brozan, 'Coping with Travail of Alzheimer's Disease', *New York Times*, 29 November 1982. On the central role of the Alzheimer's Disease and Related Disorders Association (ADRDA) in the Alzheimer's movement, see Patrick Fox, 'From Senility to Alzheimer's Disease: The Rise of the Alzheimer's Disease Movement', *Milbank Quarterly* 67 (1989): 58–102. In a chapter developed around oral narratives, Marlene Goldman beautifully illustrates the personal involvement of caregivers which led to the creation of the Alzheimer's Society of Canada; Goldman, *Forgotten*, 193–215.

13 Honel's use of the term 'Grandpa' could be seen as perpetuating the image of the individual with memory loss as elderly, but her choice might be inspired by Ariyoshi's Akiko referring to *Ojii-chan* (Grandpa), a term the Japanese use endearingly for a male elderly; also Inoue writes about his mother as *Obā-chan* (Granny).

14 Fox, 'From Senility to Alzheimer's Disease', 73–4.

15 Cohen, *No Aging in India*, 58, 54.

16 Richard J. Martin and Stephen G. Post, 'Human Dignity, Dementia, and the Moral Basis of Caregiving', in Binstock, Post and Whitehouse, *Dementia and Aging*, 66.

17 Andrea Fontana and Ronald W. Smith, 'Alzheimer's Disease Victims: The "Unbecoming" of Self and the Normalization of Competence', *Sociological Perspectives* 32 (1989): 35, 36.

18 Susan M. Behuniak, 'The Living Dead? The Construction of People with Alzheimer's Disease as Zombies', *Ageing and Society* 31 (2011): 70–92.

19 Ballenger, *Self, Senility, and Alzheimer's Disease*, 113.

20 Rabins, foreword, ix; Shenk, *The Forgetting*, 137.

21 Anne Davis Basting, *Forget Memory: Creating Better Lives for People with Dementia* (Baltimore: Johns Hopkins University Press, 2009), 35–45; Peter J. Whitehouse with Daniel George, *The Myth of Alzheimer's: What You Aren't Being Told about Today's Most Dreaded Diagnosis* (New York: St. Martin's Griffin, 2008), 97–102.

22 A feature article of the period, for example, opened with accounts of famous stars such as Rita Hayworth turned *Schreckbild* (image of horror); 'Das Hirn wird brüchig wie ein alter Stiefel' [The brain turns crumbly like an old boot], *Der Spiegel*, 19 June 1989.

23 Lawrence Meyer, 'A Cure Is Sought for Disease of Aged', *Washington Post*, 8 January 1982; Katzman, 'Editorial'.

24 M. Clark, M. Gosnell and D. Witherspoon, 'A Slow Death of the Mind', *Newsweek*, 3 December 1984.

25 Donna Cohen and Carl Eisdorfer, *The Loss of Self: A Family Resource for the Care of Alzheimer's Disease and Related Disorders* (New York: W. W. Norton, 1986), 64; all further references incorporated in the text.

26 King's Fund, *Living Well into Old Age: Applying Principles of Good Practice to Services for People with Dementia* (London: King's Fund, 1986); references incorporated in the text.
27 Philip M. Boffey, 'Alzheimer's Disease: Families Are Bitter', *New York Times*, 7 May 1985.
28 Note that objectifying terms such as the 'impaired person' have been replaced in later editions of *The 36-Hour Day*. The 1999 edition changed it to 'person with dementia', and the 2011 edition wrote about 'the person who has dementia'. Consider also the focus of Jaber F. Gubrium, 'Structuring and Destructuring the Course of Illness: The Alzheimer's Disease Experience', *Sociology of Health and Illness* 9 (1987): 1–24.
29 Larry Rose, *Show Me the Way to Go Home* (Forest Knolls, CA: Elder Books, 1996), 26 (manual title *sic*).
30 Jill Manthorpe, 'A Child's Eye View: Dementia in Children's Literature', *British Journal of Social Work* 35 (2005): 316; Stuart Murray, *Representing Autism: Culture, Narrative, Fascination* (Liverpool: Liverpool University Press, 2008), 68–70.
31 Martina Zimmermann, '"Journeys" in the Life-Writing of Adult-Child Dementia Caregivers', *Journal of Medical Humanities* 34 (2013): 385–97.
32 Arthur Kleinman, *The Illness Narratives: Suffering, Healing, and the Human Condition* (New York: Basic Books, 1988), 183.
33 Compare to this Kleinman's later call for 'practical and emotional support' for caregivers; Arthur Kleinman, 'Caregiving: The Odyssey of Becoming More Human', *The Lancet* 373 (2009): 293.
34 Samuel Beckett, 'Company', in *Company, Ill Seen Ill Said, Worstward Ho, Stirrings Still* (London: Faber and Faber, 2009), 4; all further references from this edition are incorporated in the text.
35 Hartung, *Ageing, Gender and Illness*, 180.
36 Ballenger, *Self, Senility, and Alzheimer's Disease*, 153, 173, 179.
37 Anthony Elliott highlights 'fragmentation, dislocation and decomposition of identity and everyday working life' as central to postmodern theory; Anthony Elliott, *Concepts of the Self*, 2nd edn (Cambridge: Polity, 2008), 141.
38 Peter Tinniswood, '"Company" by Samuel Beckett', *Times*, 26 June 1980. I could not find any reference to Alzheimer's disease in connection with Beckett's *Company* in the *Times*.
39 Joseph M. Foley, 'The Experience of Being Demented', in Binstock, Post and Whitehouse, *Dementia and Aging*, 30.
40 J. Bernlef [Hendrik Jan Marsman], *Out of Mind*, trans. Adrienne Dixon (London: Faber and Faber, 1988); references incorporated in the text.
41 Alexander Zweers, 'The Narrator's Position in Selected Novels by J. Bernlef', *Canadian Journal of Netherlandic Studies* 19 (1998): 35; Hartung, *Ageing, Gender and Illness*, 187.

42 Steven Connor, *Postmodernist Culture: An Introduction to Theories of the Contemporary*, 2nd edn (Oxford: Blackwell, 1997), 127.
43 Ellipses are very important in this novel; to distinguish them from editorial omissions, original suspension points are reported without space on either side of the ellipsis.
44 Hannah Zeilig, 'Gaps and Spaces: Representations of Dementia in Contemporary British Poetry', *Dementia* 13 (2014): 167.
45 Luce Irigaray, *Le Langage des Déments* [The manner of speaking of the demented] (The Hague: Mouton, 1973).
46 Barry Reisberg, 'Clinical Presentation, Diagnosis, and Symptomatology of Age-Associated Cognitive Decline and Alzheimer's Disease', in Reisberg, *Alzheimer's Disease*, 174–5.
47 W. G. Sebald, *Austerlitz* (Frankfurt am Main: Fischer, 2013), 173, 269.
48 Andreas Huyssen, *Present Pasts: Urban Palimpsests and the Politics of Memory* (Stanford, CA: Stanford University Press, 2003), 4.
49 Rick Crownshaw, 'Theoretical Anticipations: Memory and Photography in W. G. Sebald's *Austerlitz*', in *The Politics of Cultural Memory*, ed. Lucy Burke, Simon Faulkner and Jim Aulich (Newcastle upon Tyne: Cambridge Scholars, 2010), 60.
50 Saul Bellow, *Mr. Sammler's Planet* (1969; London: Penguin, 1995), 73.
51 Harry Mulisch, *The Assault*, trans. Claire N. White (New York: Pantheon, 1985), originally published in Dutch in 1982; references incorporated in the text.
52 Thomas DeBaggio, *Losing My Mind: An Intimate Look at Life with Alzheimer's* (2002; New York: Free Press, 2003), 1, 184; Zimmermann, 'Alzheimer's Disease Metaphors', 82–3.
53 Barbara Ehrenreich, *Smile or Die: How Positive Thinking Fooled America and the World* (2009; London: Granta Books, 2010); Zimmermann, *The Poetics and Politics*, ch. 4.
54 Stuart Taberner, *Aging and Old-Age Style in Günter Grass, Ruth Klüger, Christa Wolf, and Martin Walser: The Mannerism of a Late Period* (Rochester: Camden House, 2013), 40. Helen Small makes a similar observation on Bellow's writing, considering it 'a hyper-self-reflexive prose, forever revising its own claims, refining, sharpening, justifying, retracting, softening, debunking'; Small, *The Long Life*, 108.
55 Barlow, 'Senile Dementia', 359.
56 Woodward, *Aging and Its Discontents*, 89; emphasis original.
57 Hartung, *Ageing, Gender and Illness*, 86.
58 Zimmermann, *The Poetics and Politics*, ch. 5.
59 Consider, for example, Lawrence Meyer, 'A Family Stranger', *Washington Post*, 9 January 1982.
60 J. Wuest, P. K. Ericson and P. N. Stern, 'Becoming Strangers: The Changing Family Caregiving Relationship in Alzheimer's Disease', *Journal of Advanced Nursing* 20 (1994): 437.

61 Anita Desai, 'The Narrator Has Alzheimer's', *New York Times*, 17 September 1989.
62 Eakin, *How Our Lives Become Stories*, 46.
63 Ruth M. Tappen, 'Awareness of Alzheimer Patients', *American Journal of Public Health* 78 (1988): 987–8.
64 Zweers, 'The Narrator's Position', 40.
65 Desai, 'The Narrator Has Alzheimer's'.
66 Foley, 'The Experience of Being Demented', 40, 42.
67 Basting, *Forget Memory*, 39.
68 *Black Daisies for the Bride*, directed by Peter Symes, aired 30 June 1993, on BBC Two, https://www.youtube.com/watch?v=c8YxHk7yMo8 and https://www.youtube.com/watch?v=6Z2r1brcBrY; Tony Harrison, *Black Daisies for the Bride* (London: Faber and Faber, 1993), 1, 10, 14, 9.
69 Lucy Burke, 'The Poetry of Dementia: Art, Ethics and Alzheimer's Disease in Tony Harrison's *Black Daisies for the Bride*', *Journal of Literary and Cultural Disability Studies* 1 (2007): 70.
70 Basting, 'Looking Back from Loss'; Ballenger, *Self, Senility, and Alzheimer's Disease*, ch. 6; Burke, 'Alzheimer's Disease'; and, slightly later, Ryan, Bannister and Anas, 'The Dementia Narrative'.
71 Ann Robertson, 'The Politics of Alzheimer's Disease: A Case Study in Apocalyptic Demography', *International Journal of Health Services* 20 (1990): 439; Jeroen Spijker and John MacInnes, 'Population Ageing: The Timebomb That Isn't?', *Biomedical Journal* 347 (2013): f6598.
72 David Callahan, 'Dementia and Appropriate Care: Allocating Scarce Resources', in Binstock, Post and Whitehouse, *Dementia and Aging*, 141–52.
73 On efforts of the 'gray lobby' in the United States, see Achenbaum, *Crossing Frontiers*, chs 4 and 6.
74 Estes and Binney, 'The Biomedicalization of Aging', 593.
75 Ronald Sahyouni, Aradhana Verma and Jefferson Chen, *Alzheimer's Disease Decoded: The History, Present, and Future of Alzheimer's Disease and Dementia* (New Jersey: World Scientific, 2017), 236.

5 The narrative of loss in a growing biomedical and literary marketplace of Alzheimer's disease

1 Bick, Amaducci and Pepeu, *The Early Story of Alzheimer's Disease*; consider also Alzheimer et al., 'An English Translation'.
2 A 'topic' Web of Science™ (http://wok.mimas.ac.uk/) search indicates that the number of studies naming 'Alzheimer' rose from twenty-eight in the 1940s to forty-six and ninety per decade in the 1950s and 1960s, respectively. From the 1980s to

the 1990s, publication numbers on 'Alzheimer' increased 5-fold, while those on 'dementia' (though much larger in absolute terms) increased by only 2.7 times.
3   Ballenger, *Self, Senility, and Alzheimer's Disease*, 87.
4   Arne Brun, 'An Overview of Light and Electron Microscopic Changes', in Reisberg, *Alzheimer's Disease*, 37, 38, 40, 43.
5   Rose and Abi-Rached, *Neuro*, 43.
6   Robert N. Butler and Marian Emr, 'An American Perspective', in Reisberg, *Alzheimer's Disease*, 464.
7   David B. Morris, *Illness and Culture in the Postmodern Age* (Berkeley: University of California Press, 1998), 2.
8   Otis, *Networking*, 3.
9   Reynolds, *The Third Lens*, 115–17.
10 P. Davies and A. J. F. Maloney, 'Selective Loss of Central Cholinergic Neurons in Alzheimer's Disease', *The Lancet* 2 (1976): 1403. Similar discoveries were made contemporaneously in two other British laboratories: David M. Bowen et al., 'Neurotransmitter-Related Enzymes and Indices of Hypoxia in Senile Dementia and Other Abiotrophies', *Brain* 99 (1976): 459–96; and E. K. Perry et al., 'Correlation of Cholinergic Abnormalities with Senile Plaques and Mental Test Scores in Senile Dementia', *British Medical Journal* 2 (1978): 1457–9.
11 Peter J. Whitehouse et al., 'Alzheimer's Disease and Senile Dementia: Loss of Neurons in the Basal Forebrain', *Science* 215 (1982): 1237–9; references incorporated in the text.
12 Otis, *Networking*, 80.
13 Ibid., 220.
14 David A. Drachman, 'The Pharmacological Basis for Cholinergic Investigations of Alzheimer's Disease: Evidence and Implications', in Reisberg, *Alzheimer's Disease*, 342; emphasis original.
15 Michael Ignatieff, *Scar Tissue* (London: Vintage, 1994); references incorporated in the text.
16 Dementia occurs in the final stages of ALS, but goes largely unnoticed due to the rapid progression of the condition; Majid Fotuhi, Vladimir Hachinski and Peter J. Whitehouse, 'Changing Perspectives Regarding Late-Life Dementia', *Nature Reviews Neurology* 5 (2009): 649–58.
17 Also Otis underlines the conceptual link between connectedness and the ability to convey intelligence and identity when referring to the late astrophysicist and ALS patient Stephen Hawking as 'an extreme case of merged organo-technical communications systems'; Otis, *Networking*, 221.
18 Michelle Wildgen, *You're Not You* (New York: Picador, 2007), 117, 228.
19 Joel Havemann, *A Life Shaken: My Encounter with Parkinson's Disease* (Baltimore: Johns Hopkins University Press, 2002); references incorporated in the text.

20 The neuromolecular model of Parkinson's disease ascribes loss of movement control to a loss of the neurotransmitter dopamine in the brain region named substantia nigra.
21 Sandi Gordon, *Parkinson's: A Personal Story of Acceptance* (Boston: Branden, 1992), 21.
22 Jonathan Franzen, *The Corrections* (London: Fourth Estate, 2010), 642.
23 C. G. Rasool, C. N. Svendsen and Dennis J. Selkoe, 'Neurofibrillary Degeneration of Cholinergic and Noncholinergic Neurons of the Basal Forebrain in Alzheimer's Disease', *Annals of Neurology* 20 (1986): 482–8.
24 Michael Kidd, 'Paired Helical Filaments in Electron Microscopy of Alzheimer's Disease', *Nature* 197 (1963): 192–3.
25 I. Grundke-Iqbal et al., 'Microtubule-Associated Protein Tau: A Component of Alzheimer Paired Helical Filaments', *Journal of Biological Chemistry* 261 (1986): 6084.
26 Kenneth S. Kosik, Catherine L. Joachim and Dennis J. Selkoe, 'Microtubule Associated Protein Tau (τ) Is a Major Antigenic Component of Paired Helical Filaments in Alzheimer Disease', *Proceedings of the National Academy of Sciences* 83 (1986): 4044; Heiko Braak and Eva Braak, 'Neuropathological Stageing of Alzheimer-Related Changes', *Acta Neuropathologica* 82 (1991): 239–59.
27 Rudolph E. Tanzi and Ann B. Parson, *Decoding Darkness: The Search for the Genetic Causes of Alzheimer's Disease* (Cambridge, MA: Perseus, 2000), 103; all further references incorporated in the text.
28 David A. Snowdon, *Aging with Grace: The Nun Study and the Science of Old Age: How We Can All Live Longer, Healthier and More Vital Lives* (2001; London: Fourth Estate, 2002), 92; all further references incorporated in the text.
29 Henryk M. Wisniewski, 'Neuritic (Senile) and Amyloid Plaques', in Reisberg, *Alzheimer's Disease*, 57.
30 Dennis J. Selkoe, 'The Deposition of Amyloid Proteins in the Aging Mammalian Brain: Implications for Alzheimer's Disease', *Annals of Medicine* 21 (1989): 73–6.
31 George G. Glenner and Caine W. Wong, 'Alzheimer's Disease: Initial Report of the Purification and Characterization of a Novel Cerebrovascular Amyloid Protein', *Biochemical and Biophysical Research Communications* 120 (1984): 885–90; Jie Kang et al., 'The Precursor of Alzheimer's Disease Amyloid A4 Protein Resembles a Cell-Surface Receptor', *Nature* 325 (1987): 733–6.
32 Lock, *The Alzheimer Conundrum*, 155.
33 Peter H. St. George-Hyslop et al., 'The Genetic Defect Causing Familial Alzheimer's Disease Maps on Chromosome 21', *Science* 235 (1987): 885–90.
34 Alison Goate et al., 'Segregation of a Missense Mutation in the Amyloid Precursor Protein Gene with Familial Alzheimer's Disease', *Nature* 349 (1991): 704–6.
35 John A. Hardy and Gerald A. Higgins, 'Alzheimer's Disease: The Amyloid Cascade Hypothesis', *Science* 256 (1992): 184.

36 Gina Kolata, 'Alzheimer's Researchers Close In on Causes', *New York Times*, 26 February 1991.
37 Ronald Kotulak, 'Stalking "Demons of the Mind"', *Chicago Tribune*, 4 June 1980.
38 Van Dijck, *Imagenation*, 37. The exaggerated importance ascribed to genetic research became ironically manifest in the genotyping of Auguste D.'s condition in 1998. The general public's belief in the importance of genetics, in turn, was conveyed in the related newspaper report which conceded that 'Alois Alzheimer had not erred'; M. B. Graeber et al., 'Histopathology and APOE Genotype of the First Alzheimer Disease Patient, Auguste D.', *Neurogenetics* 1 (1998): 223–8; 'Alois Alzheimer hat sich nicht geirrt', *Frankfurter Allgemeine Zeitung*, 15 April 1998.
39 Dennis J. Selkoe, 'Missense on the Membrane', *Nature* 375 (1995): 734. The presenilin 1 and 2 genes were identified on chromosomes 14 and 1, respectively.
40 Van Dijck, *Imagenation*, 39.
41 Daniel A. Pollen, *Hannah's Heirs: The Quest for the Genetic Origins of Alzheimer's Disease*, exp. edn (New York: Oxford University Press, 1996); all references incorporated in the text.
42 Van Dijck, *Imagenation*, 122.
43 Latour, *Science in Action*, 164.
44 Foucault, *The Birth of the Clinic*, 166–7; as also cited in Nikolas Rose, *The Politics of Life Itself: Biomedicine, Power, and Subjectivity in the Twenty-First Century* (Princeton, NJ: Princeton University Press, 2007), 193.
45 Rose, *The Politics of Life Itself*, 93.
46 Charles P. Pierce, *Hard to Forget: An Alzheimer's Story* (New York: Random House, 2000); references incorporated in the text.
47 Lucy Burke, '"The Country of My Disease": Genes and Genealogy in Alzheimer's Life-Writing', *Journal of Literary and Cultural Disability Studies* 2 (2008): 63, 66. Pierce's book remains, to my knowledge, the only Alzheimer's narrative that explicitly centres around genetic testing. But consider Lisa Snyder's account of 33-year-old Consuelo, who underwent (positive) genetic testing; Lisa Snyder, *Speaking Our Minds: What It's Like to Have Alzheimer's*, rev. edn (Baltimore: Health Professions, 2009), 133–50. Tracy Mobley tells of disease onset at age 38 in *Young Hope: The Broken Road* (2007); however, genetic testing does not play a major role in her narration.
48 Burke, '"The Country of My Disease"', 71.
49 Hacking, *Rewriting the Soul*, 111.
50 On a powerful exploration of how genes function 'within culture and within the body in culture as condensed arguments and contests', see Valerie Hartouni, *Cultural Conceptions: On Reproductive Technologies and the Remaking of Life* (Minneapolis: University of Minnesota Press, 1997), 128–9. Consider also Ana M. Soto and Carlos Sonnenschein, 'Reductionism, Organicism, and Causality

in the Biomedical Sciences. A Critique', *Perspectives in Biology and Medicine* 61 (2018): 493–4.
51 Jean Baréma, *The Test: Living in the Shadow of Huntington's Disease* (New York: Franklin Square, 2005); references incorporated in the text.
52 Anne Hunsaker Hawkins, *Reconstructing Illness: Studies in Pathography*, 2nd edn (West Lafayette: Purdue University Press, 1999), 5.
53 Franziska Draeger, 'Da war ich tot' [Then I was dead], *Die Zeit*, 27 March 2013. Also Sandy Sulaiman's *Learning to Live with Huntington's Disease: One Family's Story* (2007) evolves around each family member's risk and testing dilemma. It remains to be demonstrated whether the recent gene silencing success in Huntington's disease research can slow or prevent the disease before symptoms develop; James Gallagher, 'Huntington's Breakthrough May Stop Disease', *BBC News*, 11 December 2017, http://www.bbc.co.uk/news/health-42308341. The pharmaceutical company Roche launched a Phase 3 clinical trial in December 2018 and started enrolling patients in early 2019.
54 E. H. Corder et al., 'Gene Dose of Apolipoprotein E Type 4 Allele and the Risk of Alzheimer's Disease in Late Onset Families', *Science* 261 (1993): 921–3.
55 David A. Snowdon et al., 'Linguistic Ability in Early Life and Cognitive Function and Alzheimer's Disease in Late Life: Findings from the Nun Study', *Journal of the American Medical Association* 275 (1996): 528–32; references incorporated in the text.
56 Ann E. Vandenberg, 'What We Can Say about Cognition in Aging: Arguments for and against Cognitive Health Promotion' (PhD thesis, Emory University, 2012), 66, 174, 210.
57 Whitehouse, *The Myth of Alzheimer's*, xix, 244–9; Vandenberg, 'What We Can Say', 136. For a more recent contribution that closes on a chapter on 'where you live, what you eat', consider Jay Ingram, *The End of Memory: A Natural History of Aging and Alzheimer's* (2014; London: Rider Books, 2016), 241–52.
58 John A. Hardy, 'ApoE, Amyloid, and Alzheimer's Disease', *Science* 263 (1994): 454.
59 Whitehouse, *The Myth of Alzheimer's*, xvii.
60 Gayatri Devi, *The Spectrum of Hope: An Optimistic and New Approach to Alzheimer's Disease and Other Dementias* (New York: Workman, 2017), viii, ix.
61 Daniel R. George and Peter J. Whitehouse, 'Marketplace of Memory: What the Brain Fitness Technology Industry Says about Us and How We Can Do Better', *The Gerontologist* 51 (2011): 595.
62 Stephen Katz and Barbara L. Marshall, 'Tracked and Fit: FitBits, Brain Games, and the Quantified Aging Body', *Journal of Aging Studies* 45 (2018): 63, 66.
63 Huyssen, *Present Pasts*, 16.
64 Annie Ernaux, *Une femme* [A woman] (1987; Paris: Éditions Gallimard, 2002), 42; Mary Gordon, *The Shadow Man: A Daughter's Search for Her Father* (1996;

New York: Vintage, 1997), 56; Lillian B. Rubin, *Tangled Lives: Daughters, Mothers, and the Crucible of Aging* (Boston: Beacon, 2000), viii. I am not aware of a similar body of work in the Spanish or Italian culture.

65 Lisa Appignanesi, *Losing the Dead: A Family Memoir* (London: Chatto and Windus, 1999), 167.

66 Linda Grant, *Remind Me Who I Am, Again* (London: Granta Books, 1999), 31; all further references incorporated in the text.

67 Nicola King, *Memory, Narrative, Identity: Remembering the Self* (Edinburgh: Edinburgh University Press, 2000), 2–3, 40; emphasis original.

68 Huyssen, *Present Pasts*, 8.

69 Eleanor Cooney, *Death in Slow Motion: A Memoir of a Daughter, Her Mother, and the Beast Called Alzheimer's* (New York: Harper Perennial, 2004), 210; all further references incorporated in the text.

70 Michael Bury, 'Chronic Illness as Biographical Disruption', *Sociology of Health and Illness* 4 (1982): 167–82; Gay Becker, *Disrupted Lives: How People Create Meaning in a Chaotic World* (Berkeley: University of California Press, 1997).

71 Rooke, 'Old Age in Contemporary Fiction', 255.

72 Ehrenreich, *Smile or Die*, 59.

73 Kathlyn Conway, *Illness and the Limits of Expression* (Ann Arbor: University of Michigan Press, 2007), 1.

74 G. Thomas Couser, *Recovering Bodies: Illness, Disability, and Life Writing* (Madison: University of Wisconsin Press, 1997), 5.

75 Appignanesi, *Losing the Dead*, 81.

76 Margarita Retuerto Buades, *Mi vida junto a un enfermo de Alzheimer* (Madrid: La Esfera de los Libros, 2003); references incorporated in the text.

77 Booth, *The Art of Growing Older*, 40.

78 Flora Sáez, 'Mi marido tiene Alzheimer' [My husband has Alzheimer's], *El Mundo*, 22 September 2002, https://www.elmundo.es/cronica/2002/362/1032772217.html.

79 Ann Burack-Weiss, *The Caregiver's Tale: Loss and Renewal in Memoirs of Family Life* (New York: Columbia University Press, 2006), 69.

80 Bradley T. Hyman, 'Alzheimer's Disease or Alzheimer's Diseases? Clues from Molecular Epidemiology', *Annals of Neurology* 40 (1996): 136.

81 John Bayley, *The Iris Trilogy: A Memoir of Iris Murdoch* (London: Abacus, 2003); all references from this edition are incorporated in the text.

82 Among the narratives referring to Bayley's account are Gillies's *Keeper*, Tilman Jens's *Demenz: Abschied von meinem Vater* (Dementia: Farewell to my father; 2009), Nucci A. Rota's *La bimbamamma: Cosa vuol dire convivere con l'Alzheimer. Il diario di una figlia* (The mommy girl: What it means to share your life with Alzheimer's. A daughter's diary; 2009), and David Sieveking's *Vergiss mein nicht: Wie meine Mutter ihr Gedächtnis verlor und ich meine Eltern neu entdeckte* (Forget me

not: How my mother lost her memory and I discovered my parents anew; 2012); Zimmermann, *The Poetics and Politics*, 51.
83 *Iris*, directed by Richard Eyre (2001; Miramax Home Entertainment, 2002), DVD.
84 Sally Chivers, *The Silvering Screen: Old Age and Disability in Cinema* (Toronto: University of Toronto Press, 2011), 82.
85 For a broader discussion, see David M. R. Orr and Yugin Teo, 'Carers' Responses to Shifting Identity in Dementia in *Iris* and *Away from Her*: Cultivating Stability or Embracing Change?', *Medical Humanities* 41 (2015): 81–5.
86 Consider, for example, Vassilas, 'Dementia and Literature'; Yahnke, 'Old Age and Loss'.

6 Neurotechnologies and narrative examine the failing mind

1 G. E. Berrios and H. L. Freeman, introduction to Berrios and Freeman, *Alzheimer and the Dementias*, 5–6.
2 'Project on the Decade of the Brain', Library of Congress, last modified 1 March 2000, http://www.loc.gov/loc/brain/.
3 R. P. Friedland et al., 'The Diagnosis of Alzheimer-Type Dementia: A Preliminary Comparison of Positron Emission Tomography and Proton Magnetic Resonance', *Journal of the American Medical Association* 252 (1984): 2752.
4 Kelly A. Joyce, *Magnetic Appeal: MRI and the Myth of Transparency* (Ithaca, NY: Cornell University Press, 2008), 2; emphasis original.
5 J. P. Seab et al., 'Quantitative NMR Measurements of Hippocampal Atrophy in Alzheimer's Disease', *Magnetic Resonance in Medicine* 8 (1988): 200–8.
6 William Bondareff et al., 'Magnetic Resonance Imaging and the Severity of Dementia in Older Adults', *Archives of General Psychiatry* 47 (1990): 47; all further references incorporated in the text.
7 Barry F. Saunders, *CT Suite: The Work of Diagnosis in the Age of Noninvasive Cutting* (Durham, NC: Duke University Press, 2008), 5.
8 Bruno Latour and Steve Woolgar, *Laboratory Life: The Construction of Scientific Facts* (1979; Princeton, NJ: Princeton University Press, 1986), 51.
9 Hacking, *Rewriting the Soul*, ch. 7.
10 G. Blessed, B. E. Tomlinson and M. Roth, 'The Association between Quantitative Measures of Dementia and of Senile Changes in the Cerebral Grey Matter of Elderly Subjects', *British Journal of Psychiatry* 114 (1968): 799; Marshal F. Folstein, Susan E. Folstein and Paul R. McHugh, 'Mini-Mental State: A Practical Method for Grading the Cognitive State of Patients for the Clinician', *Journal of Psychiatric Research* 12 (1975): 189–98.

11 S. L. Rogers et al., 'A 24-Week, Double-Blind, Placebo-Controlled Trial of Donepezil in Patients with Alzheimer's Disease', *Neurology* 50 (1998): 136–45; references incorporated in the text.
12 The Diagnostic and Statistical Manual DSM-III-R of Mental Disorders was published in 1987; and the Clinical Dementia Rating (CDR) was introduced in 1982.
13 Kazunari Ishii et al., 'Decreased Medial Temporal Oxygen Metabolism in Alzheimer's Disease Shown by PET', *Journal of Nuclear Medicine* 37 (1996): 1159; all further references incorporated in the text.
14 Joseph Dumit, *Picturing Personhood: Brain Scans and Biomedical Identity* (Princeton, NJ: Princeton University Press, 2004), 92, 93.
15 Ibid., 95–100.
16 Lisa Cartwright, *Screening the Body: Tracing Medicine's Visual Culture* (Minneapolis: University of Minnesota Press, 1995), xvii; consider also José Van Dijck, *The Transparent Body: A Cultural Analysis of Medical Imaging* (Seattle: University of Washington Press, 2005), 9–14.
17 'Alzheimer's Therapy Hope', *BBC News*, 30 July 2001, http://news.bbc.co.uk/1/hi/health/1446516.stm; consider also the following reports: 'Education Wards Off Mental Decline', *BBC News*, 12 July 1999, http://news.bbc.co.uk/1/hi/health/392161.stm; 'Social Life "May Cut Alzheimer's Risk"', *BBC News*, 25 December 2001, http://news.bbc.co.uk/1/hi/health/1717323.stm, and later: Simon Garfield, 'Memories Are Made of This', *Guardian*, 9 May 2004, https://www.theguardian.com/society/2004/may/09/longtermcare.observermagazine; 'Scan Could Spot Early Alzheimer's', *BBC News*, 27 December 2006, http://news.bbc.co.uk/1/hi/health/6197177.stm.
18 Joyce, *Magnetic Appeal*, 151.
19 Lennard J. Davis, *Obsession: A History* (Chicago: University of Chicago Press, 2008), 23.
20 H. Shinotoh et al., 'Progressive Loss of Cortical Acetylcholinesterase Activity in Association with Cognitive Decline in Alzheimer's Disease: A Positron Emission Tomography Study', *Annals of Neurology* 48 (2000): 194, 199.
21 F. Nobili et al., 'Brain Perfusion Follow-Up in Alzheimer's Patients during Treatment with Acetylcholinesterase Inhibitors', *Journal of Nuclear Medicine* 43 (2002): 983.
22 Ian Hacking, *The Social Construction of What?* (1999; Cambridge, MA: Harvard University Press, 2003), 237 n. 1.
23 Van Dijck, *The Transparent Body*, 6–7.
24 C. Courtney et al., 'Long-Term Donepezil Treatment in 565 Patients with Alzheimer's Disease (AD2000): Randomised Double-Blind Trial', *The Lancet* 363 (2004): 2105–15.
25 Consider, for example, 'Alzheimer's Drug Policy Reviewed', *BBC News*, 23 January 2006, http://news.bbc.co.uk/go/pr/fr/-/1/hi/health/4637484.stm.

26 Rose and Abi-Rached, *Neuro*, 13.
27 Vandenberg, 'What We Can Say', ch. 4.
28 Stephen G. Post, 'The Concept of Alzheimer Disease in a Hypercognitive Society', in *Concepts of Alzheimer Disease: Biological, Clinical, and Cultural Perspectives*, ed. Peter J. Whitehouse, Konrad Maurer and Jesse F. Ballenger (Baltimore: Johns Hopkins University Press, 2000), 245; emphasis original.
29 James Golomb, Alan Kluger and Steven H. Ferris, 'Mild Cognitive Impairment: Historical Development and Summary of Research', *Dialogues in Clinical Neuroscience* 6 (2004): 351.
30 Marco Roth, 'The Rise of the Neuronovel', in *Say What You Mean: The N+1 Anthology*, ed. Christian Lorentzen (2009; London: Notting Hill Editions, 2012), 73–90, https://nplusonemag.com/issue-8/essays/the-rise-of-the-neuronovel/; all references are taken from the online source without page numbers. Roth considered McEwan's *Enduring Love* (1997) as trendsetting.
31 Ian McEwan, *Saturday* (London: Vintage, 2006); references incorporated in the text; Laura Salisbury, 'Narration and Neurology: Ian McEwan's Mother Tongue', *Textual Practice* 24 (2010): 892.
32 Raymond Tallis, *Aping Mankind: Neuromania, Darwinitis and the Misrepresentation of Humanity* (Durham: Acumen, 2011); consider also, as previously referenced: Rose, *The Politics of Life Itself*; and Hartouni, *Cultural Conceptions*.
33 Foley, 'The Experience of Being Demented', 37.
34 These anxieties were also reflected in strengthening caregiver advocacy – which endorsed a robust response to recommendations that pro-cholinergic treatment should be stopped; Jane Kirby, 'New Drug Guidelines Offer Hope to Alzheimer's Sufferers', *Independent*, 7 October 2010, https://www.independent.co.uk/life-style/health-and-families/health-news/new-drug-guidelines-offer-hope-to-alzheimers-sufferers-2099710.html.
35 Roth, 'The Rise of the Neuronovel'.
36 Hans Dieter Mummendey, *Claudia, Alzheimer und ich: Kriminalroman* (Münster: Neues Literaturkontor, 1992); references incorporated in the text.
37 Desai, 'The Narrator Has Alzheimer's'.
38 Peter Brooks, *Reading for the Plot: Design and Intention in Narrative* (1984; Cambridge, MA: Harvard University Press, 1992), 113. My readings of detective fiction also benefitted from thinking with Mieke Bal, *Narratology: Introduction to the Theory of Narrative*, 3rd edn (Toronto: University of Toronto Press, 2009).
39 Brooks, *Reading for the Plot*, 18.
40 Lisa Zunshine, *Why We Read Fiction: Theory of Mind and the Novel* (Columbus: Ohio State University Press, 2006), 121.
41 Michael Dibdin, *The Dying of the Light: A Mystery* (New York: Vintage, 1995); all references incorporated in the text.

42 W. H. Auden, 'The Guilty Vicarage', in *Detective Fiction: A Collection of Critical Essays*, ed. Robin W. Winks (Woodstock: Countryman, 1988), 17.
43 George Grella, 'The Formal Detective Novel', in Winks, *Detective Fiction*, 86.
44 Otis, *Networking*, 174.
45 Alan Palmer, 'Ontologies of Consciousness', in Herman, *The Emergence of Mind*, 278.
46 Zunshine, *Why We Read Fiction*, 129; Palmer, 'Ontologies of Consciousness', 278.
47 Catherine Oppenheimer, 'I Am, Thou Art: Personal Identity in Dementia', in *Dementia: Mind, Meaning, and the Person*, ed. Julian C. Hughes, Stephen J. Louw and Steven R. Sabat (Oxford: Oxford University Press, 2006), 199.
48 Auden, 'The Guilty Vicarage', 21.
49 In 1990, the case of Janet Adkins, who, diagnosed with Alzheimer's disease, committed suicide with the help of Jack Kevorkian, was widely reported in the media; Timothy Egan, ' "Her Mind Was Everything", Dead Woman's Husband Says', *New York Times*, 6 June 1990; Cohen, *No Aging in India*, 58.
50 David C. Thomasma, 'Mercy Killing of Elderly People with Dementia: A Counterproposal', in Binstock, Post and Whitehouse, *Dementia and Aging*, 116.
51 Grella, 'The Formal Detective Novel', 96.
52 Dana Lee Baker, *The Politics of Neurodiversity: Why Public Policy Matters* (Boulder, CO: Lynne Rienner, 2011), 167, 177.
53 Vai Ramanathan, *Alzheimer Discourse: Some Sociolinguistic Dimensions* (Mahwah, NJ: Lawrence Erlbaum Associates, 1997), ch. 1.
54 Baker, *The Politics of Neurodiversity*, 185.
55 Steven R. Sabat, *The Experience of Alzheimer's Disease: Life through a Tangled Veil* (Oxford: Blackwell, 2001), vii.
56 Martin Suter, *Small World* (Zurich: Diogenes, 1999); references incorporated in the text.
57 Miriam Seidler, 'Zwischen Demenz und Freiheit: Überlegungen zum Verhältnis von Altern und Geschlecht in der Gegenwartsliteratur' [Between dementia and freedom: Considerations regarding the relationship between ageing and gender in current literature], in *Graue Theorie: Die Kategorien Alter und Geschlecht im kulturellen Diskurs* [Grey theory: Categories of age and gender in cultural discourse], ed. Heike Hartung et al. (Cologne: Böhlau Verlag, 2007), 208.
58 Consider, for example, M. Hofmann et al., 'Individualisiertes, computergestütztes Gedächtnistraining bei Alzheimer-Patienten' [Computer-assisted individualized memory training in Alzheimer's patients], *Nervenarzt* 66 (1995): 703–7.
59 Brooks, *Reading for the Plot*, 168.
60 Carmelo Aquilina and Julian C. Hughes, 'The Return of the Living Dead: Agency Lost and Found?', in Hughes, Louw and Sabat, *Dementia*, 143–61.
61 Christine Boden, *Who Will I Be When I Die?* (Sydney: HarperCollins, 1998), 54; all further references from this edition are incorporated in the text. Boden's narrative

has recently been reissued by Jessica Kingsley Publishers (under Boden's new surname, Bryden).

62 Ryan, Bannister and Anas, 'The Dementia Narrative'.
63 Aquilina and Hughes, 'The Return of the Living Dead', 145.
64 Kathryn Montgomery Hunter, *Doctors' Stories: The Narrative Structure of Medical Knowledge* (Princeton, NJ: Princeton University Press, 1991), 61.
65 Rita Charon, *Narrative Medicine: Honoring the Stories of Illness* (New York: Oxford University Press, 2006), 132.
66 Malcolm Goldsmith, *Hearing the Voice of People with Dementia: Opportunities and Obstacles* (London: Jessica Kingsley, 1996), 9; emphasis original; Julie Goyder, *We'll Be Married in Fremantle: Alzheimer's Disease and the Everyday Act of Storying* (Fremantle: Fremantle Arts Centre, 2001); consider also Marie A. Mills, *Narrative Identity and Dementia: A Study of Autobiographical Memories and Emotions* (Aldershot: Ashgate, 1998).
67 Brian Gearing, preface to *Dementia Reconsidered: The Person Comes First*, by Tom Kitwood (Maidenhead: Open University Press, 1997), vii.
68 Kitwood, *Dementia Reconsidered*, 7.
69 Christine Bryden, *Dancing with Dementia: My Story of Living Positively with Dementia* (London: Jessica Kingsley, 2005), 141–2.
70 Hawkins, *Reconstructing Illness*, 5, 9, 12.
71 Maurer and Maurer, *Alzheimer*, 17.
72 'Welcome to the Trebus Project', Trebus Project, accessed 25 October 2019, http://www.trebusprojects.org/; Rick Noack, 'A German Nursing Home Tries a Novel Form of Dementia Therapy: Re-creating a Vanished Era for Its Patients', *Washington Post*, 26 December 2017, https://www.washingtonpost.com/news/worldviews/wp/2017/12/26/a-german-nursing-home-tries-a-novel-form-of-dementia-therapy-re-creating-a-vanished-era-for-its-patients/.
73 Julian C. Hughes, Stephen J. Louw and Steven R. Sabat, 'Seeing Whole', in Hughes, Louw and Sabat, *Dementia*, 4.
74 Arthur Frank, *The Wounded Storyteller: Body, Illness, and Ethics* (Chicago: University of Chicago Press, 1995), ch. 5, 97–9.
75 Hawkins, *Reconstructing Illness*, ch. 3; Zimmermann, 'Alzheimer's Disease Metaphors', 74–8.
76 Couturier, *Puzzle*, 72; Zimmermann, *The Poetics and Politics*, 107.
77 Literary scholarship has described a similar, though much earlier, shift in breast cancer life-writing; Couser, *Recovering Bodies*, 63; Lisa Diedrich, *Treatments: Language, Politics, and the Culture of Illness* (Minneapolis: University of Minnesota Press, 2007), 26.
78 Bryden's Alzheimer's disease diagnosis had been amended to that of less aggressively advancing fronto-temporal dementia in 1998.

79 Wendy Mitchell with Anna Wharton, *Somebody I Used to Know* (London: Bloomsbury, 2018), 63, 94; consider also Mitchell's recent 'Alzheimer's in the Workplace', *New York Times*, 12 September 2018; and her presence on Twitter active since March 2015; 'Sharing My Journey', Twitter, accessed 25 October 2019, www.twitter.com/wendypmitchell.

80 Ulrike Draesner, 'Ichs Heimweg macht alles alleine' [I's journey home does everything by itself], in *Es schneit in meinem Kopf: Erzählungen über Alzheimer und Demenz* [It snows in my head: Narratives about Alzheimer's and dementia], ed. Klara Obermüller (Munich: Nagel and Kimche, 2006), 59–81.

81 Kim Howes Zabbia, *Painted Diaries: A Mother and Daughter's Experience through Alzheimer's* (Minneapolis: Fairview, 1996); references incorporated in the text.

82 Vaidehi Ramanathan, 'Alzheimer Pathographies: Glimpses into How People with AD and Their Caregivers Text Themselves', in *Dialogue and Dementia: Cognitive and Communicative Resources for Engagement*, ed. Robert W. Schrauf and Nicole Müller (New York: Psychology, 2014), 253.

83 Jeannette Montgomery Barron, *My Mother's Clothes: An Album of Memories* (New York: Welcome Books, 2009), 94.

84 Consider, for example, Cathy Stein Greenblat, *Alive with Alzheimer's* (Chicago: University of Chicago Press, 2004); Peter Granser, *Alzheimer* (Heidelberg: Kehrer Verlag, 2008); Carlos H. Espinel, 'De Kooning's Late Colours and Forms: Dementia, Creativity, and the Healing Power of Art', *The Lancet* 347 (1996): 1096–8.

85 John Berger, *Ways of Seeing* (1972; London: Penguin, 2008), 8.

86 Woodward, *Aging and Its Discontents*, 196; Simon Biggs, 'Toward Critical Narrativity: Stories of Aging in Contemporary Social Policy', *Journal of Aging Studies* 15 (2001): 303–16.

7 The dichotomy of Alzheimer's disease

1 *Honig im Kopf*, directed by Til Schweiger (2014; Warner Home Video, 2015), DVD; time points indicated in the text.

2 The audience would also remember Hallervorden having received awards for his previous performance as former Olympic medallist Paul Averhoff who, now resident in a nursing home, rebels against the monotony and belittling nature of unsuitable group therapy; *Sein letztes Rennen* [His final race], directed by Kilian Riedhof (2013; Universum Film, 2014), DVD.

3 Medina, 'Alzheimer's Disease', 364.

4 Chivers, *The Silvering Screen*, 60.

5 Among all films starting to show in Germany in 2014, *Honig im Kopf* had attracted most viewers by June 2015; 'Top 100 Deutschland 2014', Inside Kino, accessed 25 October 2019, http://www.insidekino.de/DJahr/D2014.htm. By mid-August 2015, over seventy cinemas across Germany continued to show the movie; 'Honig im Kopf', Kino.de, accessed 20 August 2015, http://www.kino.de/kinofilm/honig-im-kopf/imkino/152963. On 2 February 2015, during peak viewing time, German public television channel ARD broadcasted a talk show, in the *Hart aber fair* (Tough but fair) programme, which asked: 'Alzheimer als Komödie – hilft Lachen gegen die Angst?' (Alzheimer's as comedy – does laughing help against fear?).

6 Joachim Kurz, 'Die vergebliche Gnade des Vergessens' [The futile grace of forgetting], *Kino-Zeit*, 14 January 2015, http://www.kino-zeit.de/blog/b-roll/die-vergebliche-gnade-des-vergessens-die-kino-zeit-de-kolumne.

7 One of the final scenes of *Honig im Kopf* – Amandus dying in the hospital bed in the centre of the Rosenbachs' living room (2:04:18) – I read as alluding to Sieveking's depiction of his own mother's end. The documentary was much less popular than Schweiger's comedy, reaching number 142 in the viewer hit list; 'Top 100 Deutschland 2013', Inside Kino, accessed 25 October 2019, http://www.insidekino.de/DJahr/D2013.htm. I offer a detailed analysis of this documentary in *The Poetics and Politics*, 123–7.

8 Chivers, *The Silvering Screen*.

9 Emily Martin, *Flexible Bodies: Tracking Immunity in American Culture – from the Days of Polio to the Age of AIDS* (Boston: Beacon, 1994), 184–5.

10 Kolata, 'Alzheimer's Researchers Close In on Causes'; cf. ch. 5, 94.

11 Francesco Panza et al., 'Is There Still Any Hope for Amyloid-Based Immunotherapy for Alzheimer's Disease?', *Current Opinion in Psychiatry* 27 (2014): 129.

12 Dale Schenk et al., 'Immunization with Amyloid-β Attenuates Alzheimer-Disease-Like Pathology in the PDAPP Mouse', *Nature* 400 (1999): 173–7; references incorporated in the text.

13 Martin, *Flexible Bodies*, 96, 99, 92, 95.

14 Morris, *Illness and Culture*, 7.

15 Frank, *The Wounded Storyteller*, 6, 7.

16 Basting, 'Looking Back from Loss'.

17 Ballenger, *Self, Senility, and Alzheimer's Disease*, 173, 174.

18 Martina Zimmermann, 'Terry Pratchett's *Living with Alzheimer's* as a Case Study of Late-Life Creativity', in *Creativity in Later Life: Beyond Late Style*, ed. David Amigoni and Gordon McMullan (Abingdon: Routledge, 2019), 198–207.

19 Richard Taylor, *Alzheimer's from the Inside Out* (Baltimore: Health Professions, 2008), 49, 118, 130.

20 Martin, *Flexible Bodies*, 70.

21 Redefinition would not, I believe, have changed the sociocultural perception of Alzheimer's disease, as the representation of multiple sclerosis, a neurodegenerative condition actually defined in immunological terms, suggests. The writer Nancy Mairs openly speaks about 'relentless physical losses'. But it is – as with Parkinson's disease – the maintaining of an intellectually performing self that makes the conceptual difference. In Mairs's words: 'Myelin dissolves, nerves short out, muscle atrophies, but the old brain, riddled now with sclerotic patches, goes on wailing, "I can do it myself!"'; Nancy Mairs, *Waist-High in the World: A Life among the Nondisabled* (Boston: Beacon, 1996), 7, 70.

22 Lawrence M. Fisher, 'Vaccine in Mice Offers Hope in Fight against Alzheimer's', *New York Times*, 8 July 1999.

23 Latour, *Science in Action*, 164.

24 A. J. Bayer et al., 'Evaluation of the Safety and Immunogenicity of Synthetic Abeta42 (AN1792) in Patients with AD', *Neurology* 64 (2005): 94, 99. A further clinical trial was suspended, because several participants developed an unexpected and severe reaction.

25 Petroc Sumner et al., 'The Association between Exaggeration in Health Related Science News and Academic Press Releases: Retrospective Observational Study', *Biomedical Journal* 349 (2014): g7015.

26 Lisa Genova, *Still Alice* (New York: Pocket Books, 2009); references incorporated in the text.

27 Sarah Falcus, 'Storying Alzheimer's Disease in Lisa Genova's *Still Alice*', *Entertext* 12 (2014): 92.

28 Rebecca A. Bitenc, 'Representations of Dementia in Narrative Fiction', in *Knowledge and Pain*, ed. Esther Cohen et al. (Amsterdam: Rodopi, 2012), 320, 318.

29 Block, 'A Place beyond Words'.

30 S. Salloway et al., 'A Phase 2 Multiple Ascending Dose Trial of Bapineuzumab in Mild to Moderate Alzheimer Disease', *Neurology* 73 (2009): 2061.

31 Latour, *Science in Action*, 113–21. Ian Dowbiggin convincingly illustrates the power of the pharmaceutical industry to dictate patient destiny in *The Quest for Mental Health*. On how dementia research teamed up with pharmaceutical companies, see also Shenk, *The Forgetting*, 178–89; Whitehouse, *The Myth of Alzheimer's*, 168–70; Basting, *Forget Memory*, 38; Cornelia Stolze, *Vergiss Alzheimer! Die Wahrheit über eine Krankheit, die keine ist* [Forget Alzheimer's! The truth about a condition that does not exist] (2011; Freiburg im Breisgau: Herder, 2013), 155–206.

32 'Newly Reported Research Advances from the Alzheimers Association International Conference 2012', Canada Newswire, 18 July 2012, http://www.newswire.ca/news-releases/newly-reported-research-advances-from-the-alzheimers-association-international-conference-2012-510488911.html.

33 Panza et al., 'Is There Still Any Hope?', 131.
34 *Still Alice*, directed by Richard Glatzer and Wash Westmoreland (2014; Curzon Atificial Eye, 2015), DVD; time points indicated in the text.
35 Riley II, *Disability and the Media*, 70.
36 'Julianne Moore Winning "Best Actress"', Youtube, 6 March 2015, https://www.youtube.com/watch?v=TzR3CUU5IIU; 'Julianne Moore's Oscars 2015 Acceptance Speech in Full: "I'm Thrilled We Were Able to Shine a Light on Alzheimer's Disease"', *Independent*, 23 February 2015, https://www.independent.co.uk/arts-entertainment/films/oscars/julianne-moore-oscars-2015-acceptance-speech-in-full-im-thrilled-we-were-able-to-shine-a-light-on-10064112.html.
37 Chris Smyth, 'Dementia Clinics Are Swamped by Worried Well', *Times*, 11 March 2015; see also John Sutherland, 'Art Is Shining a Light on Dementia at Last', *Times*, 21 February 2015.
38 Lucy Bolton, '"The Art of Losing"', *Times Higher Education*, 5 March 2015.
39 Baker, *The Politics of Neurodiversity*, 164.
40 John Abraham, *Science, Politics and the Pharmaceutical Industry: Controversy and Bias in Drug Regulation* (London: University College London Press, 1995). This perspective also applies to the fact that pharmacological dementia research continues to appear reluctant to challenge the hypotheses underlying its investigations; Karl Herrup, 'The Case for Rejecting the Amyloid Cascade Hypothesis', *Nature Neuroscience* 18 (2015): 794; [John Hardy and John Mayer for the Models of Dementia: The Good, the Bad and the Future Meeting Attendees], 'The Amyloid Cascade Hypothesis Has Misled the Pharmaceutical Industry', *Biochemical Society Transactions* 39 (2011): 920–3.
41 Eric R. Siemers et al., 'Phase 3 Solanezumab Trials: Secondary Outcomes in Mild Alzheimer's Disease Patients', *Alzheimer's and Dementia* 12 (2016): 110.
42 Consider, for example, James Gallagher, 'The Drug to Slow Alzheimer's?' *BBC News*, 22 July 2015, http://www.bbc.co.uk/news/health-33621109; Panza et al., 'Is There Still Any Hope?' As I am finalizing this manuscript, the press once again gets excited about outcomes related to a further passive immunization compound, aducanumab. It remains to be seen whether the drug can slow or halt the progression of the condition; Tom Whipple, 'Doctors Hail First Drug to "Slow Down" Alzheimer's', *Times*, 22 October 2019, https://www.thetimes.co.uk/article/biogen-new-drug-offers-hope-for-alzheimer-s-sufferers-sh90z3q9h.
43 Latour, *Science in Action*, 114, 115.
44 Claire Hilton, 'Sauerkraut and African Violets: The Art of Old Age Psychiatry' (paper, Medical Humanities and Ageing Workshop, King's College London, 29 June 2015).
45 Bryden, *Dancing with Dementia*, 174.
46 Baker, *The Politics of Neurodiversity*, 35.

47 Consider, for example, Denis Campbell, 'NHS Dementia Plan to Give GPs Cash for Diagnoses Criticised as "Ethical Travesty"', *Guardian*, 22 October 2014, https://www.theguardian.com/society/2014/oct/22/nhs-dementia-diagnoses-gps-patients-criticised.

48 Erving Goffman, *Stigma: Notes on the Management of Spoiled Identity* (1963; London: Penguin, 1990), 31.

49 David Hugendick, 'Ich vergisst sich' [I forgets itself], *Die Zeit*, 25 February 2015.

50 For bibliographies, see Hanna Kappus, *Das Leben ist ein großes: Alzheimer – ein langer Abschied* [Life is a big one: Alzheimer's – a long farewell] (Gütersloh: Gütersloher Verlagshaus, 2012), 159–60; and Erin Y. Sakai, Brian D. Carpenter and Rebecca E. Rieger, '"What's Wrong with Grandma?": Depictions of Alzheimer's Disease in Children's Storybooks', *American Journal of Alzheimer's Disease and Other Dementias* 27 (2012): 589; consider also Falcus and Sako, *Contemporary Narratives of Dementia*, ch. 5.

51 Linda Scacco and Nicole Wong, *Always My Grandpa: A Story for Children about Alzheimer's Disease* (Washington, DC: Magination, 2006), 46, 29.

52 Barbara Schnurbush and Cary Pillo, *Striped Shirts and Flowered Pants: A Story about Alzheimer's Disease for Young Children* (Washington, DC: Magination, 2007), 29 (my counting, beginning with title page).

53 Diedrich, *Treatments*, ch. 3.

54 Mareike Hachemer, *Alzheimer im problemorientierten Bilderbuch: Inhaltliche, künstlerische und sprachliche Aspekte* [Alzheimer's in the problem focused picture book: Contextual, artistic and linguistic aspects] (Norderstedt: Grin Verlag, 2009), 41.

55 Ibid., 19.

56 Frank P. Riga, 'Religion in Children's Literature: Introduction', *Children's Literature Association Quarterly* 14 (1989): 5.

57 Alessandro Borio, *La guardiana di Ulisse: Una malata di Alzheimer, un angelo e una domanda* (Arrone: Edizioni Thyrus, 2006); references incorporated in the text.

58 Cynthia Marshall, 'Reading "The Golden Key": Narrative Strategies of Parable', *Children's Literature Association Quarterly* 14 (1989): 22–5.

59 Atul Gawande, *Being Mortal: Illness, Medicine and What Matters in the End* (2014; London: Profile Books, 2015), 22, 9.

60 Lisa Zunshine, 'Theory of Mind, Social Hierarchy, and the Emergence of Narrative Subjectivity', in Herman, *The Emergence of Mind*, 179, 171.

61 Simon Kemp and Christopher D. B. Burt, 'Memories of Uncertain Origin: Dreamt or Real?', *Memory* 14 (2006): 87–93.

62 Frank P. Riga, 'Mortals Call Their History Fable: Narnia and the Use of Fairy Tale', *Children's Literature Association Quarterly* 14 (1989): 27.

63 Luigi Berzano, preface to Borio, *La guardiana di Ulisse*, 7.

64 A text featuring an angel who peers into the psyche of an ageing woman goes against the grain of the critical reader, but in suspending realistic modes of thinking we may be able 'to apprehend what eludes our conscious grasp'; Teresa Casal, 'Experiential Reading of Jennifer Johnson's *Two Moons*' (paper, Mellon Seminar on Ageing, King's College London, 23 November 2015).

65 Jens Brockmeier, *Beyond the Archive: Memory, Narrative, and the Autobiographical Process* (New York: Oxford University Press, 2015), 258.

66 Douwe Draaisma, *Forgetting: Myths, Perils and Compensations*, trans. Liz Waters (New Haven, CT: Yale University Press, 2015), 175–8; originally published in Dutch in 2010; consider also Basting, *Forget Memory*, 18–19.

67 Brockmeier, *Beyond the Archive*, 20, 17, 21.

68 Andrew O'Hagan, *The Illuminations* (London: Faber and Faber, 2015); references incorporated in the text.

69 Debra Dean, *The Madonnas of Leningrad* (2006; London: Harper Perennial, 2007), 206; Samantha Harvey, *The Wilderness* (2009; London: Vintage, 2010), 276; Emma Healey, *Elizabeth Is Missing* (2014; London: Penguin Books, 2015), 236.

70 Dean, *The Madonnas of Leningrad*, 225. But consider Lucy Burke's perspective on the continued abilities to narrate of Healey's Maud, and the argument made by Falcus and Sako that Healey rather 'draw[s] attention to the social environment which judges and isolates those with dementia'; Lucy Burke, 'Missing Pieces: Trauma, Dementia and the Ethics of Reading in *Elizabeth Is Missing*', in *Dementia and Literature: Interdisciplinary Perspectives*, ed. Tess Maginnes (Abingdon: Routledge, 2018), 88; Falcus and Sako, *Contemporary Narratives of Dementia*, 130.

71 Nicci Gerrard, 'Words Fail Us: Dementia and the Arts', *Guardian*, 19 July 2015, https://www.theguardian.com/culture/2015/jul/19/dementia-and-the-arts-fiction-films-drama-poetry-painting.

72 Salisbury, 'Narration and Neurology', 897.

73 Ian McEwan, *Atonement* (London: Vintage, 2002), 354. Further novels that remove dementia from principally negative notions of memory loss include: Andrea Gerster, *Dazwischen Lili* (Lili inbetween, 2008), Pupi Avati, *Una sconfinata giovinezza* (A boundless youthfulness, 2010), Katharina Hacker, *Die Erdbeeren von Antons Mutter* (The strawberries of Anton's mother, 2010), and Alice Lichtenstein, *Lost* (2010).

74 Janice Turner, 'I Dreamt of Greek Olives as I Fumed in Munich', *Times*, 13 August 2015.

75 Sally Magnusson, *Where Memories Go: Why Dementia Changes Everything* (London: Two Roads, 2014); references incorporated in the text.

76 Elena De Dionigi, *Prima di volare via: Quello che l'Alzheimer non ci può rubare* [Before flying away: What Alzheimer's cannot steal from us] (Magenta: La Memoria del Mondo Libreria Editrice, 2012), 61.

77 Giovanna Venturino, *Il tuo mare di nulla: La mia mamma e l'Alzheimer* [Your sea of nothingness: My mother and Alzheimer's] (Rome: A&B Editrice, 2012), 73; Donatella Di Pietrantonio, *Mia madre è un fiume* [My mother is a river] (Rome: Elliot Edizioni, 2010), 74.

78 Baker, *The Politics of Neurodiversity*, 167–89.

79 Gawande, *Being Mortal*, 193.

80 Callahan, 'Dementia and Appropriate Care', 151.

81 Cécile Huguenin, *Alzheimer mon amour* (Paris: Éditions Héloïse d'Ormesson, 2011); references incorporated in the text. Huguenin's text has been translated into Spanish and Italian in 2013, and into Dutch in 2017.

82 Ngatcha-Ribert, *Alzheimer*, 34–7.

83 'Alzheimer mon amour', Mollat, 30 September 2011, http://www.mollat.com/livres/cecile-huguenin-alzheimer-mon-amour-9782350871707.html.

84 Charlotte L. Clarke et al., *Risk Assessment and Management for Living Well with Dementia* (London: Jessica Kingsley, 2011), 25; all further references incorporated in the text. Also Magnusson refers to *Alice's Adventures* to depict nonsensical experiences regarding the organization of caregiving (287); and Walrath maps her narrative onto Carroll's story to depict the condition itself.

85 Cohen, *No Aging in India*, 233, 267, 292; Sivaramakrishnan, *As the World Ages*, 197; consider also W. Ladson Hinton and Sue Levkoff, 'Constructing Alzheimer's: Narratives of Lost Identities, Confusion and Loneliness in Old Age', *Culture, Medicine and Psychiatry* 23 (1999): 453–75.

86 Jane Wilkinson, 'Remembering Forgetting', *Status Quaestionis* 6 (2014): 111–12. Wilkinson takes the term from Elie Wiesel's 1989 *L'oublié*, translated as *The Forgotten* in 1992, which belongs to the body of memory literature explored in Chapter 4.

87 'Alzheimer: L'éthique en question. Recommendations' [Alzheimer's: Ethics in question. Recommendations], Association Francophone des Droits de l'Homme Âgé, 25 April 2007, http://www.sante.gouv.fr/IMG/pdf/ethique_en_questions.pdf; Department of Health, *Putting People First: A Shared Vision and Commitment to the Transformation of Adult Social Care* (2007) and *Living Well with Dementia: A National Dementia Strategy* (2009). Also note the recent guidelines published by the Department of Health and Social Care, *After a Diagnosis of Dementia: What to Expect from Health and Care Services* (2018).

88 'Philosophie', Demenz Support Stuttgart, last modified 1 January 2019, http://www.demenz-support.de/portraet/philosophie; compare to this the recent agenda

Gemeinsam für Menschen mit Demenz [Together for people with dementia] (Berlin: Bundesministerium für Gesundheit, 2014).

89 Schrauf and Müller, *Dialogue and Dementia*, i.

90 Personal conversation with Kathleen Woodward, Dartmouth College, NH, 13 July 2015; Kathleen Woodward, 'A Public Secret: Assisted Living, Caregivers, Globalization', *International Journal of Ageing and Later Life* 7 (2012): 17–51.

91 De Dionigi, *Prima di volare via*, 22–5, 69; Diane Keaton, *Then Again* (2011; London: Fourth Estate, 2012), 274.

92 Julia Neuberger, *Not Dead Yet: A Manifesto for Old Age* (London: HarperCollins, 2008), 193; also consider Janette Davies, *Living before Dying: Imagining and Remembering Home* (New York: Berghahn, 2017), and my review thereof, Martina Zimmermann, 'Book Review: Living before Dying. Imagining and Remembering Home by Janette Davies', *Times Higher Education*, 8 February 2018.

93 Susan H. McFadden and John T. McFadden, *Aging Together: Dementia, Friendship, and Flourishing Communities* (2011; Baltimore: Johns Hopkins University Press, 2014), 164.

94 'Oral History Meets Dementia: A Staged Reading of the Play Timothy and Mary', Columbia University, 3 September 2014, http://oralhistory.columbia.edu/upcoming-and-past-events/People/oral-history-meets-dementia-a-staged-reading-of-the-play-timothy-and-mary; C. F. Vinci, C. Pontesilli and D. Fo, 'Telling about the Stolen Mind', *Neurological Sciences* 26 (2005): 185–7; Hannah Zeilig, 'The Critical Use of Narrative and Literature in Gerontology', *International Journal of Ageing and Later Life* 6 (2011): 7–37; Hannah J. Roberts and James M. Noble, 'Education Research: Changing Medical Student Perceptions of Dementia. An Arts-Centered Experience', *Neurology* 85 (2015): 739–41.

95 June Andrews, *Dementia: The One-Stop Guide. Practical Advice for Families, Professionals and People Living with Dementia and Alzheimer's Disease* (London: Profile Books, 2015); Peter J. S. Ashley, 'This Is My Life', in *The Law and Ethics of Dementia*, ed. Charles Foster, Jonathan Herring and Israel Doron (Oxford: Hart, 2014), 497–503.

96 Cole and Gadow, *What Does It Mean to Grow Old?*; Gary M. Kenyon, James E. Birren and Johannes J. F. Schroots, eds, *Metaphors of Aging in Science and the Humanities* (New York: Springer, 1991).

97 Eva-Marie Kessler and Clemens Schwender, 'Giving Dementia a Face? The Portrayal of Older People with Dementia in German Weekly News Magazines between the Years 2000 and 2009', *Journals of Gerontology Series B: Psychological Sciences and Social Sciences* 67 (2012): 261–70. But see in contrast: Allison M. Kirkman, 'Dementia in the News: The Media Coverage of Alzheimer's Disease', *Australasian Journal on Ageing* 25 (2006): 74–9.

98　Consider, for example, Hannah Zeilig, 'Late-Life Creativity and the "New Old Age"', archived by King's Digital Lab, October 2018, www.latelifecreativity.org; 'Creative Dementia Arts Network', Creative Dementia Arts Network, last modified 1 January 2019, www.creativedementia.org; Aagje Swinnen, 'Book Club as Intersubjective Space with Transformative Potential' (paper, Dementia and Cultural Narrative Symposium, Aston University, 9 December 2017); and projects described in Basting, *Forget Memory*.

99　Mary Mothersill, 'Old Age', *Proceedings and Addresses of the American Philosophical Association* 73 (1999): 21.

100　Lucy Burke, 'Imagining a Future without Dementia: Fictions of Regeneration and the Crises of Work and Sustainability', *Palgrave Communications* 3 (2018): article 52, p. 7.

101　For a very thoughtful and measured account of the role of new capitalism in the rise of mental health diagnoses made, see Nikolas Rose, *Our Psychiatric Future* (Cambridge: Polity, 2019). Recent evidence suggests that specific lifestyle choices can intensify cognitive decline: stress-related chronic lack of sleep as well as emotional loneliness appears to heighten cognitive deterioration, while robust social networks can enhance cognitive function; Andrew R. Mendelsohn and James W. Larrick, 'Sleep Facilitates Clearance of Metabolites from the Brain: Glymphatic Function in Aging and Neurodegenerative Diseases', *Rejuvenation Research* 16 (2013): 518–23; Jae-Eun Kang et al., 'Amyloid-β Dynamics Are Regulated by Orexin and the Sleep-Wake Cycle', *Science* 326 (2009): 1005–7; Kristin R. Krueger et al., 'Social Engagement and Cognitive Function in Old Age', *Experimental Aging Research* 35 (2009): 45–60.

102　Matthew Thomas, *We Are Not Ourselves* (London: Fourth Estate, 2014); references incorporated in the text.

103　Gerrard, 'Words Fail Us'.

104　Block, 'A Place beyond Words'.

105　Ru, *The Family Novel*, 2.

106　Franzen, *The Corrections*, 630, 641.

107　Block, 'A Place beyond Words'.

108　Woodward, *Aging and Its Discontents*, 32.

109　Michel Foucault, *The Birth of Biopolitics: Lectures at the Collège de France, 1978–1979*, trans. Graham Burchell (2008; Basingstoke: Palgrave Macmillan, 2010), 224.

110　Ru, *The Family Novel*, 119–20.

111　Consider, for example, Cohen, *No Aging in India*, 15; Magnusson, *Where Memories Go*, 305; and Simon Russell Beale's interpretation of Lear's part for the National Theatre in 2014, as referred to in Bate, '"The Infirmity of His Age"'. It is not clear that Lear's madness results from ageing rather than repeated mental trauma and the devastation of his understanding of his position in the world. Performance could readily draw out either interpretation.

112 Elliott, *Concepts of the Self*, 140; Arno Gruen, 'War or Peace? We Cannot Survive with Real-Politik' (acceptance speech, Award Ceremony for the Loviisa Peace Price, Loviisa, 7 August 2010), accessed 27 October 2015, http://www.arnogruen.net/war-or-peace.pdf.

113 For Heike Hartung, the 'dementia narrative' is one of the directions the bildungsroman has taken after its classical period, as it 'define[s] the limits of development by narrating the end of memory and consciousness'; in this she is joined by Rita C. Cavigioli, who places the discussion of senility in Italian family narratives in the context of bildungs- and reifungsroman theories; Hartung, *Ageing, Gender and Illness*, 171, 172; Cavigioli, *Women of a Certain Age*, ch. 5.

114 Foucault, *The Birth of Biopolitics*, lectures 9 and 10, 241, 245.

115 It is tempting to compare Eileen to the 'Man of Property' Soames Forsyte, especially because, like a Forsyte, she has 'pride, dignity, firmness, and persistence'; Ru, *The Family Novel*, 21.

116 Ibid., 32 (apostrophes *sic*).

117 Ibid., 81.

118 Zeilig, 'Dementia as a Cultural Metaphor', 266; Helen Dunmore, '*We Are Not Ourselves* by Matthew Thomas Review – an Extraordinary Portrait of Alzheimer's Disease', *Guardian*, 27 August 2014, https://www.theguardian.com/books/2014/aug/27/we-are-not-ourselves-matthew-thomas-review-novel.

119 Foucault, *The Birth of Biopolitics*, 230.

120 Gerontologist Jan Baars identifies ageing with dementia as 'anti-entrepreneurial ageing'; Jan Baars, 'A Good Life in "Old Age": A Privilege for the Rich?' (paper, Living a Good Life in Older Age Symposium, University of Warwick, 5 July 2018).

121 Morris, *Illness and Culture*, 220–1; Lennard J. Davis, *The End of Normal: Identity in a Biocultural Era* (Ann Arbor: University of Michigan Press, 2013), ch. 1.

122 Gruen, 'War or Peace?', 5.

123 Ibid., 1.

124 Sally Chivers, '*Seeing the Apricot*: A Disability Perspective on Alzheimer's in Lee Chang-dong's *Poetry*', in *Different Bodies: Essays on Disability in Film and Television*, ed. Marja E. Mogk (Jefferson, NC: McFarland, 2013), 72.

125 Similar observations have been made regarding Günter Grass's last novel; Taberner, *Aging and Old-Age Style*, 72.

126 Cathy Caruth, *Unclaimed Experience: Trauma, Narrative, and History* (Baltimore: Johns Hopkins University Press, 1996), 49, 52.

127 McFadden and McFadden, *Aging Together*, 14.

128 James Gallagher, 'Dementia Levels "Are Stabilising"', *BBC News*, 21 August 2015, http://www.bbc.co.uk/news/health-34001144; Steve Illiffe, 'Evidence-Based Medicine and Dementia', *British Journal of General Practice* 65 (2015): 511–12.

129 Woodward, 'A Public Secret', 23, 41.

130 Neuberger, *Not Dead Yet*, 191–214; Davies, *Living before Dying*, 81–104.
131 *Head Full of Honey*, directed by Til Schweiger (Barefoot Films, 2018), released in several European countries including the United Kingdom in spring 2019.

8 Conclusion

1 David Herman, 'Re-minding Modernism', in Herman, *The Emergence of Mind*, 254.
2 Pavel, *The Lives of the Novel*, 279.
3 I borrow the notion of 'cure narrative' from Lennard J. Davis, *Bending over Backwards: Disability, Dismodernism, and Other Difficult Positions* (New York: New York University Press, 2002), 99.
4 Hacking, *Rewriting the Soul*, 239.
5 Brockmeier, *Beyond the Archive*, 20–1.

Bibliography

Abraham, John. *Science, Politics and the Pharmaceutical Industry: Controversy and Bias in Drug Regulation*. London: University College London Press, 1995.

Achenbaum, W. Andrew. *Crossing Frontiers: Gerontology Emerges as a Science*. New York: Cambridge University Press, 1995.

Acocella, Silvia. *Effetto Nordau: Figure della degenerazione nella letteratura Italiana tra Ottocento e Novecento*. Naples: Liguori Editore, 2012.

Adelman, Richard C. 'The Alzheimerization of Aging'. *The Gerontologist* 35 (1995): 526–32.

Alzheimer, Alois. *Der Krieg und die Nerven*. Breslau: Verlag von Preuß und Jünger, 1915. https://digital.staatsbibliothek-berlin.de/werkansicht?PPN=PPN716104202&PHYSID=PHYS_0001&DMDID=DMDLOG_0001.

Alzheimer, Alois. 'Die diagnostischen Schwierigkeiten in der Psychiatrie'. *Zeitschrift für die gesamte Neurologie und Psychiatrie* 1 (1910): 1–19.

Alzheimer, Alois. 'Ein "geborener Verbrecher"'. *Archiv für Psychiatrie und Nervenkrankheiten* 28 (1896): 327–53.

Alzheimer, Alois. 'Neuere Arbeiten über die Dementia senilis und die auf atheromatöser Gefässerkrankung basierenden Gehirnkrankheiten'. *Monatsschrift für Psychiatrie und Neurologie* 3 (1898): 101–15.

Alzheimer, Alois. 'Über die Indikationen für eine künstliche Schwangerschafts-unterbrechung bei Geisteskranken'. *Münchener medizinische Wochenschrift* 54 (1907): 1617–22.

Alzheimer, Alois. 'Über eigenartige Krankheitsfälle des späteren Alters'. *Zeitschrift für die gesamte Neurologie und Psychiatrie* 4 (1911): 356–85.

Alzheimer, Alois. 'Über eine eigenartige Erkrankung der Hirnrinde'. *Allgemeine Zeitschrift für Psychiatrie und Psychisch-gerichtliche Medizin* 64 (1907): 146–8.

Alzheimer, Alois. 'Über einen Fall von spinaler progressiver Muskelatrophie und hinzutretender Erkrankung bulbärer Kerne und der Rinde'. *Archiv für Pychiatrie und Nervenkrankheiten* 23 (1892): 459–85.

Alzheimer, Alois, Rainulf A. Stelzmann, H. Norman Schnitzlein and F. Reed Murtagh. 'An English Translation of Alzheimer's 1907 Paper, "Über eine eigenartige Erkrankung der Hirnrinde"'. *Clinical Anatomy* 8 (1995): 429–31.

Alzheimer's Research UK. 'Support for People Affected by Dementia'. Accessed 25 October 2019. https://www.alzheimersresearchuk.org/about-dementia/helpful-information/support-for-carers/.

Amis, Kingsley. *Ending Up*. 1974. London: Penguin, 2011.

Anderson, P. W. 'More Is Different: Broken Symmetry and the Nature of the Hierarchical Structure of Science'. *Science* 177 (1972): 393–6.

Andrews, June. *Dementia: The One-Stop Guide, Practical Advice for Families, Professionals and People Living with Dementia and Alzheimer's Disease*. London: Profile Books, 2015.

Appignanesi, Lisa. *Losing the Dead: A Family Memoir*. London: Chatto and Windus, 1999.

Appignanesi, Lisa. *Mad, Bad and Sad: A History of Women and the Mind Doctors from 1800 to the Present*. 2008. London: Virago, 2009.

Aquilina, Carmelo, and Julian C. Hughes. 'The Return of the Living Dead: Agency Lost and Found?'. In *Dementia: Mind, Meaning, and the Person*, edited by Julian C. Hughes, Stephen J. Louw and Steven R. Sabat, 143–61. Oxford: Oxford University Press, 2006.

Ariyoshi, Sawako. *The Twilight Years*. Translated by Mildred Tahara. London: Peter Owen, 1984. First published in Japanese in 1972.

Ashley, Peter J. S. 'This Is My Life'. In *The Law and Ethics of Dementia*, edited by Charles Foster, Jonathan Herring and Israel Doron, 497–503. Oxford: Hart, 2014.

Association Francophone des Droits de l'Homme Agé. 'Alzheimer. L'éthique en question. Recommendations'. 25 April 2007. www.sante.gouv.fr/IMG/pdf/ethique_en_questions.pdf.

Auden, W. H. 'The Guilty Vicarage'. In *Detective Fiction: A Collection of Critical Essays*, edited by Robin W. Winks, 15–24. Woodstock: Countryman, 1988.

Avati, Pupi. *Una sconfinata giovinezza*. Milan: Garzanti Libri, 2010.

Baars, Jan. 'A Good Life in "Old Age": A Privilege for the Rich?' Paper presented at the Living a Good Life in Older Age Symposium, University of Warwick, 5 July 2018.

Bailey, Paul. *At The Jerusalem*. 1967. London: Penguin, 1982.

Bailey, Paul. 'Author, Author: Paul Bailey'. *Guardian*, 15 January 2011. https://www.theguardian.com/books/2011/jan/15/paul-bailey-author-author.

Bailey, Paul. 'Contemporary Narrative and the Coming of Ageing'. Literary dialogue with Courttia Newland, Senate House, University of London, 30 October 2013.

Baker, Dana Lee. *The Politics of Neurodiversity: Why Public Policy Matters*. Boulder, CO: Lynne Rienner, 2011.

Bal, Mieke. *Narratology: Introduction to the Theory of Narrative*. 3rd edn. Toronto: University of Toronto Press, 2009.

Ballenger, Jesse F. 'Disappearing in Plain Sight: Public Roles of People with Dementia in the Meaning and Politics of Alzheimer's Disease'. In *The Neurological Patient in History*, edited by L. Stephen Jacyna and Stephen T. Casper, 109–28. Rochester: University of Rochester Press, 2012.

Ballenger, Jesse F. *Self, Senility, and Alzheimer's Disease in Modern America: A History*. Baltimore: Johns Hopkins University Press, 2006.

Barba, Andrés. *Ahora tocad música de baile*. Barcelona: Editorial Anagrama, 2004.

Baréma, Jean. *The Test: Living in the Shadow of Huntington's Disease*. New York: Franklin Square, 2005.

Barlow, A. Ralph. 'Senile Dementia: Metaphor for Our Time: The Effect of Senile Dementia on Relationships with Others Echoes the Sense of Discontinuity in American Society'. *Rhode Island Medical Journal* 66 (1983): 359–60.

Barron, Jeannette Montgomery. *My Mother's Clothes: An Album of Memories*. New York: Welcome Books, 2009.

Basting, Anne Davis. *Forget Memory: Creating Better Lives for People with Dementia*. Baltimore: Johns Hopkins University Press, 2009.

Basting, Anne Davis. 'Looking Back from Loss: Views of the Self in Alzheimer's Disease'. *Journal of Aging Studies* 17 (2003): 87–99.

Bate, Jonathan. '"The Infirmity of His Age": Shakespeare's 400th Anniversary'. *The Lancet* 387 (2016): 1715–16.

Bayer, A. J., R. Bullock, R. W. Jones, D. Wilkinson, K. R. Paterson, L. Jenkins, S. B. Millais and S. Donoghue. 'Evaluation of the Safety and Immunogenicity of Synthetic Abeta42 (AN1792) in Patients with AD'. *Neurology* 64 (2005): 94–101.

Bayley, John. *The Iris Trilogy: A Memoir of Iris Murdoch*. London: Abacus, 2003.

BBC News. 'Alzheimer's Drug Policy Reviewed'. 23 January 2006. http://news.bbc.co.uk/go/pr/fr/-/1/hi/health/4637484.stm.

BBC News. 'Alzheimer's Therapy Hope'. 30 July 2001. http://news.bbc.co.uk/1/hi/health/1446516.stm.

BBC News. 'Education Wards Off Mental Decline'. 12 July 1999. http://news.bbc.co.uk/1/hi/health/392161.stm.

BBC News. 'Scan Could Spot Early Alzheimer's'. 27 December 2006. http://news.bbc.co.uk/1/hi/health/6197177.stm.

BBC News. 'Social Life "May Cut Alzheimer's Risk"'. 25 December 2001. http://news.bbc.co.uk/1/hi/health/1717323.stm.

Becker, Gay. *Disrupted Lives: How People Create Meaning in a Chaotic World*. Berkeley: University of California Press, 1997.

Beckett, Samuel. 'Company'. 1980. In *Company, Ill Seen Ill Said, Worstward Ho, Stirrings Still*, edited by Dirk Van Hulle, 1–42. London: Faber and Faber, 2009.

Beer, Gillian. *Darwin's Plots: Evolutionary Narrative in Darwin, George Eliot and Nineteenth-Century Fiction*. 3rd edn. Cambridge: Cambridge University Press, 2009.

Behuniak, Susan M. 'The Living Dead? The Construction of People with Alzheimer's Disease as Zombies'. *Ageing and Society* 31 (2011): 70–92.

Bellow, Saul. *Mr. Sammler's Planet*. 1969. London: Penguin, 1995.

Berger, John. *Ways of Seeing*. 1972. London: Penguin, 2008.

Berlant, Lauren. 'On the Case'. *Critical Inquiry* 33 (2007): 663–72.

Bernlef, J. [Hendrik Jan Marsman]. *Out of Mind*. Translated by Adrienne Dixon. London: Faber and Faber, 1988. First published in Dutch in 1984.

Berrios, G. E. 'Alzheimer's Disease: A Conceptual History'. *International Journal of Geriatric Psychiatry* 5 (1990): 355–65.

Berrios, G. E., and H. L. Freeman. 'Dementia before the Twentieth Century'. In *Alzheimer and the Dementias*, edited by G. E. Berrios and H. L. Freeman, 9–27. London: Royal Society of Medicine, 1991.

Berrios, G. E., and H. L. Freeman. Introduction to *Alzheimer and the Dementias*, edited by G. E. Berrios and H. L. Freeman, 1–8. London: Royal Society of Medicine, 1991.

Bertman, Stephen. 'The Ashes and the Flame: Passion and Aging in Classical Poetry'. In *Old Age in Greek and Latin Literature*, edited by Thomas M. Falkner and Judith de Luce, 157–71. Albany: State University of New York Press, 1989.

Berzano, Luigi. Preface to *La guardiana di Ulisse: Una malata di Alzheimer, un angelo e una domanda*, by Alessandro Borio, 7–9. Arrone: Edizioni Thyrus, 2006.

Bick, Katherine, Luigi Amaducci and Giancarlo Pepeu, eds. *The Early Story of Alzheimer's Disease: Translation of the Historical Papers by Alois Alzheimer, Oskar Fischer, Francesco Bonfiglio, Emil Kraepelin, Gaetano Perusini*. Padua: Liviana, 1987.

Biggs, Simon. 'Toward Critical Narrativity: Stories of Aging in Contemporary Social Policy'. *Journal of Aging Studies* 15 (2001): 303–16.

Binstock, Robert H., Stephen G. Post and Peter J. Whitehouse. 'The Challenges of Dementia'. In *Dementia and Aging: Ethics, Values, and Policy Choices*, edited by Robert H. Binstock, Stephen G. Post and Peter J. Whitehouse, 1–17. Baltimore: Johns Hopkins University Press, 1992.

Binswanger, Otto Ludwig. 'Die Abgrenzung der allgemeinen progressiven Paralyse'. *Berliner klinische Wochenschrift* 31 (1894): 1180–6.

Bitenc, Rebecca A. 'Representations of Dementia in Narrative Fiction'. In *Knowledge and Pain*, edited by Esther Cohen, Leona Toker, Manuela Consonni and Otniel E. Dror, 305–29. Amsterdam: Rodopi, 2012.

Blessed, G., B. E. Tomlinson and M. Roth. 'The Association between Quantitative Measures of Dementia and of Senile Changes in the Cerebral Grey Matter of Elderly Subjects'. *British Journal of Psychiatry* 114 (1968): 797–811.

Block, Stefan Merrill. 'A Place beyond Words: The Literature of Alzheimer's'. *New Yorker*, 20 August 2014. https://www.newyorker.com/books/page-turner/place-beyond-words-literature-alzheimers.

Blumgart, Herrman L. 'Caring for the Patient'. *New England Journal of Medicine* 270 (1964): 449–56.

Boden, Christine. *Who Will I Be When I Die?* Sydney: HarperCollins, 1998.

Boffey, Philip M. 'Alzheimer's Disease: Families Are Bitter'. *New York Times*, 7 May 1985.

Boller, François. 'History of Dementia'. In *Handbook of Clinical Neurology: Dementias*, edited by Charles Duyckaerts and Irene Litvan, 3–13. New York: Elsevier, 2008.

Boller, François, and Margaret M. Forbes. 'History of Dementia and Dementia in History: An Overview'. *Journal of the Neurological Sciences* 158 (1998): 125–33.

Bolton, Lucy. '"The Art of Losing"'. *Times Higher Education*, 5 March 2015.

Bondareff, William, Janak Raval, Buck Woo, Douglas L. Hauser and Patrick M. Colletti. 'Magnetic Resonance Imaging and the Severity of Dementia in Older Adults'. *Archives of General Psychiatry* 47 (1990): 47–51.

Bonfiglio, Francesco. 'Di speciali reperti in un caso di probabile sifilide cerebrale'. *Rivista Sperimentale di Freniatria e Medicina Legale delle Alienazioni Mentali* 34 (1908): 196–206.

Booth, Charles. *Pauperism: A Picture, and the Endowment of Old Age: An Argument.* London: Macmillan, 1892.

Booth, Wayne. *The Art of Growing Older: Writers on Living and Aging.* 1992. Chicago: University of Chicago Press, 1996.

Borio, Alessandro. *La guardiana di Ulisse: Una malata di Alzheimer, un angelo e una domanda.* Arrone: Edizioni Thyrus, 2006.

Bouman, L. 'Senile Plaques'. *Brain* 57 (1934): 128–42.

Bowen, David M., Carolyn B. Smith, Pamela White and Alan N. Davison. 'Neurotransmitter-Related Enzymes and Indices of Hypoxia in Senile Dementia and Other Abiotrophies'. *Brain* 99 (1976): 459–96.

Boyle, Mary. 'The Non-discovery of Schizophrenia? Kraepelin and Bleuler Reconsidered'. In *Reconstructing Schizophrenia*, edited by Richard P. Bentall, 3–22. London: Routledge, 1990.

Braak, Heiko, and Eva Braak. 'Neuropathological Stageing of Alzheimer-Related Changes'. *Acta Neuropathologica* 82 (1991): 239–59.

Braddon, Mary E. *Lady Audley's Secret.* 1862. London: Penguin, 2012.

Braff, Zach, dir. *Going in Style.* De Line Pictures, 2017.

Brander, Martin S. 'The Scientist and the News Media'. *New England Journal of Medicine* 308 (1983): 1170–3.

Briggs, Julia. Afterword to *There Were No Windows*, by Norah Hoult, 329–41. London: Persephone Books, 2005.

Brockmeier, Jens. *Beyond the Archive: Memory, Narrative, and the Autobiographical Process.* New York: Oxford University Press, 2015.

Brody, Elaine M. '"Women in the Middle" and Family Help to Older People'. *The Gerontologist* 21 (1981): 471–80.

Brody, Howard. *Stories of Sickness.* 2nd edn. New York: Oxford University Press, 2003.

Brooks, Peter. *Reading for the Plot: Design and Intention in Narrative.* 1984. Cambridge, MA: Harvard University Press, 1992.

Brown, Emma. 'Robert N. Butler Dies, "Father of Modern Gerontology" Was 83'. *Washington Post*, 7 July 2010.

Brozan, Nadine. 'Coping with Travail of Alzheimer's Disease'. *New York Times*, 29 November 1982.

Brun, Arne. 'An Overview of Light and Electron Microscopic Changes'. In *Alzheimer's Disease: The Standard Reference*, edited by Barry Reisberg, 37–47. New York: Free Press, 1983.

Bruner, Jerome. *Making Stories: Law, Literature, Life.* Cambridge, MA: Harvard University Press, 2002.

Bryden, Christine. *Dancing with Dementia: My Story of Living Positively with Dementia.* London: Jessica Kingsley, 2005.

Bryden, Christine. *Will I Still Be Me? Finding a Continuing Sense of Self in the Lived Experience of Dementia.* London: Jessica Kingsley, 2018.

Buades, Margarita Retuerto. *Mi vida junto a un enfermo de Alzheimer*. Madrid: La Esfera de los Libros, 2003.

Bundesministerium für Gesundheit. *Gemeinsam für Menschen mit Demenz*. Berlin: Bundesministerium für Gesundheit, 2014.

Burack-Weiss, Ann. *The Caregiver's Tale: Loss and Renewal in Memoirs of Family Life*. New York: Columbia University Press, 2006.

Burke, Lucy. 'Alzheimer's Disease: Personhood and First Person Testimony'. Paper presented at the Inaugural Conference of the Cultural Disability Studies Research Network, Liverpool John Moores University, May 2007. Accessed 22 August 2011. http://www.cdsrn.org.uk/Burke_CDSRN_2007.pdf.

Burke, Lucy. '"The Country of My Disease": Genes and Genealogy in Alzheimer's Life-Writing'. *Journal of Literary and Cultural Disability Studies* 2 (2008): 63–74.

Burke, Lucy. 'Imagining a Future without Dementia: Fictions of Regeneration and the Crises of Work and Sustainability'. *Palgrave Communications* 3 (2018): article 52.

Burke, Lucy. 'Missing Pieces: Trauma, Dementia and the Ethics of Reading in *Elizabeth Is Missing*'. In *Dementia and Literature: Interdisciplinary Perspectives*, edited by Tess Maginess, 88–102. Abingdon: Routledge, 2018.

Burke, Lucy. 'The Poetry of Dementia: Art, Ethics and Alzheimer's Disease in Tony Harrison's *Black Daisies for the Bride*'. *Journal of Literary and Cultural Disability Studies* 1 (2007): 61–73.

Burke, Lucy. 'Thinking about Cognitive Impairment'. *Journal of Literary and Cultural Disability Studies* 2 (2008): i–iv.

Bury, Michael. 'Chronic Illness as Biographical Disruption'. *Sociology of Health and Illness* 4 (1982): 167–82.

Butler, Robert N. 'The Life Review: An Interpretation of Reminiscence in the Aged'. In *New Thoughts on Old Age*, edited by Robert Kastenbaum, 265–80. New York: Springer, 1964.

Butler, Robert N. *Why Survive? Being Old in America*. 1975. Baltimore: Johns Hopkins University Press, 2002.

Butler, Robert N., and Marian Emr. 'An American Perspective'. In *Alzheimer's Disease: The Standard Reference*, edited by Barry Reisberg, 461–4. New York: Free Press, 1983.

Bynum, W. F. *Science and the Practice of Medicine in the Nineteenth Century*. New York: Cambridge University Press, 1994.

Callahan, David. 'Dementia and Appropriate Care: Allocating Scarce Resources'. In *Dementia and Aging: Ethics, Values, and Policy Choices*, edited by Robert H. Binstock, Stephen G. Post and Peter J. Whitehouse, 141–52. Baltimore: Johns Hopkins University Press, 1992.

Campbell, Denis. 'NHS Dementia Plan to Give GPs Cash for Diagnoses Criticised as "Ethical Travesty"'. *Guardian*, 22 October 2014. https://www.theguardian.com/society/2014/oct/22/nhs-dementia-diagnoses-gps-patients-criticised.

Canada Newswire. 'Newly Reported Research Advances from the Alzheimers Association International Conference 2012'. 18 July 2012. http://www.newswire.ca/news-releases/newly-reported-research-advances-from-the-alzheimers-association-international-conference-2012-510488911.html.

Canguilhem, Georges. *The Normal and the Pathological*. 1978. Translated by Carolyn R. Fawcett. New York: Zone Books, 1991. First published in French in 1966.

Carson, Ronald A. Foreword to *What Does It Mean to Grow Old? Reflections from the Humanities*, edited by Thomas R. Cole and Sally A. Gadow, xi–xiv. Durham, NC: Duke University Press, 1986.

Cartwright, Lisa. *Screening the Body: Tracing Medicine's Visual Culture*. Minneapolis: University of Minnesota Press, 1995.

Caruth, Cathy. *Unclaimed Experience: Trauma, Narrative, and History*. Baltimore: Johns Hopkins University Press, 1996.

Casal, Teresa. 'Experiential Reading of Jennifer Johnson's *Two Moons*'. Paper presented at the Mellon Seminar on Ageing, King's College London, 23 November 2015.

Cavigioli, Rita C. *Women of a Certain Age: Contemporary Italian Fictions of Female Aging*. Madison, NJ: Fairleigh Dickinson University Press, 2005.

Charon, Rita. *Narrative Medicine: Honoring the Stories of Illness*. New York: Oxford University Press, 2006.

Chase, Karen. *The Victorians and Old Age*. Oxford: Oxford University Press, 2009.

Chivers, Sally. '*Seeing the Apricot*: A Disability Perspective on Alzheimer's in Lee Chang-dong's *Poetry*'. In *Different Bodies: Essays on Disability in Film and Television*, edited by Marja E. Mogk, 65–74. Jefferson, NC: McFarland, 2013.

Chivers, Sally. *The Silvering Screen: Old Age and Disability in Cinema*. Toronto: University of Toronto Press, 2011.

Clark, M., M. Gosnell and D. Witherspoon. 'A Slow Death of the Mind'. *Newsweek*, 3 December 1984.

Clark, Michael J. '"A Plastic Power Ministering to Organisation": Interpretations of the Mind–Body Relation in Late Nineteenth-Century British Psychiatry'. *Psychological Medicine* 13 (1983): 487–97.

Clark, Michael J. 'The Rejection of Psychological Approaches to Mental Disorder in Late Nineteenth-Century British Psychiatry'. In *Madhouses, Mad-Doctors, and Madmen: The Social History of Psychiatry in the Victorian Era*, edited by Andrew Scull, 271–312. Philadelphia: University of Pennsylvania Press, 1981.

Clark, William R. *Sex and the Origins of Death*. Oxford: Oxford University Press, 1996.

Clarke, Charlotte L., Heather Wilkinson, John Keady and Catherine E. Gibb. *Risk Assessment and Management for Living Well with Dementia*. London: Jessica Kingsley, 2011.

Clarke, Edwin, and L. S. Jacyna. *Nineteenth-Century Origins of Neuroscientific Concepts*. Berkeley: University of California Press, 1987.

Cohen, Donna, and Carl Eisdorfer. *The Loss of Self: A Family Resource for the Care of Alzheimer's Disease and Related Disorders*. New York: W. W. Norton, 1986.

Cohen, Lawrence. *No Aging in India: Alzheimer's, the Bad Family, and Other Modern Things*. 1998. Berkeley: University of California Press, 1999.

Cole, Thomas R. 'The "Enlightened" View of Aging: Victorian Morality in a New Key'. In *What Does It Mean to Grow Old? Reflections from the Humanities*, edited by Thomas R. Cole and Sally A. Gadow, 115–30. Durham, NC: Duke University Press, 1986.

Cole, Thomas R. *The Journey of Life: A Cultural History of Aging in America*. Cambridge: Cambridge University Press, 1992.

Cole, Thomas R. 'Part One: The Tattered Web of Cultural Meanings'. In *What Does It Mean to Grow Old? Reflections from the Humanities*, edited by Thomas R. Cole and Sally A. Gadow, 3–7. Durham, NC: Duke University Press, 1986.

Cole, Thomas R., and Sally A. Gadow, eds. *What Does It Mean to Grow Old? Reflections from the Humanities*. Durham, NC: Duke University Press, 1986.

Collins, Wilkie. *The Woman in White*. 1859. Edited with an introduction and notes by John Sutherland. Oxford: Oxford University Press, 2008.

Columbia University. 'Oral History Meets Dementia: A Staged Reading of the Play Timothy and Mary'. 3 September 2014. http://oralhistory.columbia.edu/upcoming-and-past-events/People/oral-history-meets-dementia-a-staged-reading-of-the-play-timothy-and-mary.

Combe, Kirk, and Kenneth Schmader. 'Shakespeare Teaching Geriatrics: Lear and Prospero as Case Studies in Aged Heterogeneity'. In *Aging and Identity: A Humanities Perspective*, edited by Sara Munson Deats and Lagretta Tallent Lenker, 33–46. Westport, CT: Praeger, 1999.

Connor, Steven. *Postmodernist Culture: An Introduction to Theories of the Contemporary*. 2nd edn. Oxford: Blackwell, 1997.

Conway, Kathlyn. *Illness and the Limits of Expression*. Ann Arbor: University of Michigan Press, 2007.

Cooney, Eleanor. *Death in Slow Motion: A Memoir of a Daughter, Her Mother, and the Beast Called Alzheimer's*. 2003. New York: Harper Perennial, 2004.

Corder, E. H., A. M. Saunders, W. J. Strittmatter, D. E. Schmechel, P. C. Gaskell, G. W. Small, A. D. Roses, J. L. Haines and M. A. Pericak-Vance. 'Gene Dose of Apolipoprotein E Type 4 Allele and the Risk of Alzheimer's Disease in Late Onset Families'. *Science* 261 (1993): 921–3.

Corsellis, J. A. N., and P. H. Evans. 'The Relation of Stenosis of the Extracranial Cerebral Arteries to Mental Disorder and Cerebral Degeneration in Old Age'. *Proceedings of the 5th International Congress of Neuropathology* (1965): 546.

Courtney, C., D. Farrell, R. Gray, R. Hills, L. Lynch, E. Sellwood, S. Edwards, W. Hardyman, J. Raftery, P. Crome, C. Lendon, H. Shaw and P. Bentham. 'Long-Term Donepezil Treatment in 565 Patients with Alzheimer's Disease (AD2000): Randomised Double-Blind Trial'. *The Lancet* 363 (2004): 2105–15.

Couser, G. Thomas. *Recovering Bodies: Illness, Disability, and Life Writing*. Madison: University of Wisconsin Press, 1997.

Couturier, Claude. *Puzzle, Journal d'une Alzheimer*. Paris: Éditions Josette Lyon, 2004.

Creative Dementia Arts Network. 'Creative Dementia Arts Network'. Last modified 1 January 2019. www.creativedementia.org.

Crichton-Browne, James. 'Clinical Lectures on Mental and Cerebral Diseases'. *British Medical Journal* 1 (1874): 601–3.

Crownshaw, Rick. 'Theoretical Anticipations: Memory and Photography in W. G. Sebald's *Austerlitz*'. In *The Politics of Cultural Memory*, edited by Lucy Burke, Simon Faulkner and Jim Aulich, 48–61. Newcastle upon Tyne: Cambridge Scholars, 2010.

Cumming, M. Elaine. 'New Thoughts on the Theory of Disengagement'. In *New Thoughts on Old Age*, edited by Robert Kastenbaum, 3–18. New York: Springer, 1964.

Dames, Nicholas. *Amnesiac Selves: Nostalgia, Forgetting, and British Fiction, 1810–1870*. New York: Oxford University Press, 2001.

Davidson, Ann. *Alzheimer's, a Love Story: One Year in My Husband's Journey*. Secaucus, NJ: Carol Publishing Group, 1997.

Davies, Janette. *Living before Dying: Imagining and Remembering Home*. New York: Berghahn, 2017.

Davies, P., and A. J. F. Maloney. 'Selective Loss of Central Cholinergic Neurons in Alzheimer's Disease'. *The Lancet* 2 (1976): 1403.

Davis, Lennard J. *Bending over Backwards: Disability, Dismodernism, and Other Difficult Positions*. New York: New York University Press, 2002.

Davis, Lennard J. *The End of Normal: Identity in a Biocultural Era*. Ann Arbor: University of Michigan Press, 2013.

Davis, Lennard J. *Enforcing Normalcy: Disability, Deafness, and the Body*. London: Verso, 1995.

Davis, Lennard J. *Obsession: A History*. Chicago: University of Chicago Press, 2008.

Davis, Robert. *My Journey into Alzheimer's Disease: Helpful Insights for Family and Friends, a True Story*. Carol Stream, IL: Tyndale House, 1989.

De Ajuriaguerra, J., and R. Tissot. 'Some Aspects of Psycho-neurologic Disintegration in Senile Dementia'. In *Senile Dementia: Clinical and Therapeutic Aspects*, edited by Ch. Müller and L. Ciompi, 69–79. Bern: Hans Huber, 1968.

De Dionigi, Elena. *Prima di volare via: Quello che l'Alzheimer non ci può rubare*. Magenta: La Memoria del Mondo Libreria Editrice, 2012.

De Morgan, William. *The Old Man's Youth*. London: William Heinemann, 1921.

Dean, Debra. *The Madonnas of Leningrad*. 2006. London: Harper Perennial, 2007.

Deats, Sara Munson. 'The Dialectic of Aging in Shakespeare's *King Lear* and *The Tempest*'. In *Aging and Identity: A Humanities Perspective*, edited by Sara Munson Deats and Lagretta Tallent Lenker, 23–32. Westport, CT: Praeger, 1999.

DeBaggio, Thomas. *Losing My Mind: An Intimate Look at Life with Alzheimer's*. 2002. New York: Free Press, 2003.

DeLillo, Don. *Falling Man*. 2007. London: Picador, 2008.

Demenz Support Stuttgart. 'Philosophie'. Last modified 1 January 2019. http://www.demenz-support.de/portraet/philosophie.

Department of Health. *Living Well with Dementia: A National Dementia Strategy*. London: Department of Health, 2009.

Department of Health. *Putting People First: A Shared Vision and Commitment to the Transformation of Adult Social Care*. London: Department of Health, 2007.

Department of Health and Social Care. *After a Diagnosis of Dementia: What to Expect from Health and Care Services*. London: Department of Health and Social Care, 2018.

Der Spiegel. 'Das Hirn wird brüchig wie ein alter Stiefel'. 19 June 1989.

Desai, Anita. 'The Narrator Has Alzheimer's'. *New York Times*, 17 September 1989.

Devi, Gayatri. *The Spectrum of Hope: An Optimistic and New Approach to Alzheimer's Disease and Other Dementias*. New York: Workman, 2017.

Di Pietrantonio, Donatella. *Mia madre è un fiume*. Rome: Elliot Edizioni, 2010.

Dibdin, Michael. *The Dying of the Light: A Mystery*. 1993. New York: Vintage, 1995.

Diedrich, Lisa. *Treatments: Language, Politics, and the Culture of Illness*. Minneapolis: University of Minnesota Press, 2007.

Douglas, Cristina. 'When Memory Gets Political: Dementia and the Project of Democracy in Post-Communist Romania'. Paper presented at the Dementia and Cultural Narrative Symposium, Aston University, 9 December 2017.

Dowbiggin, Ian. *The Quest for Mental Health: A Tale of Science, Medicine, Scandal, Sorrow, and Mass Society*. New York: Cambridge University Press, 2011.

Draaisma, Douwe. *Forgetting: Myths, Perils and Compensations*. Translated by Liz Waters. New Haven, CT: Yale University Press, 2015. First published in Dutch in 2010.

Draaisma, Douwe. 'A Labyrinth of Tangles: Alzheimer's Disease'. In *Disturbances of the Mind*. Translated by Barbara Fasting, 199–228. New York: Cambridge University Press, 2009. First published in Dutch in 2006.

Drachman, David A. 'The Pharmacological Basis for Cholinergic Investigations of Alzheimer's Disease: Evidence and Implications'. In *Alzheimer's Disease: The Standard Reference*, edited by Barry Reisberg, 340–5. New York: Free Press, 1983.

Draeger, Franziska. 'Da war ich tot'. *Die Zeit*, 27 March 2013.

Draesner, Ulrike. 'Ichs Heimweg macht alles alleine'. In *Es schneit in meinem Kopf: Erzählungen über Alzheimer und Demenz*, edited by Klara Obermüller, 59–81. Munich: Nagel and Kimche, 2006.

Dumit, Joseph. *Picturing Personhood: Brain Scans and Biomedical Identity*. Princeton, NJ: Princeton University Press, 2004.

Dunmore, Helen. Introduction to *Ending Up*, by Kingsley Amis, ix–xiv. London: Penguin, 2011.

Dunmore, Helen. '*We Are Not Ourselves* by Matthew Thomas Review – an Extraordinary Portrait of Alzheimer's Disease'. *Guardian*, 27 August 2014. https://www.theguardian.com/books/2014/aug/27/we-are-not-ourselves-matthew-thomas-review-novel.

Eakin, Paul John. *How Our Lives Become Stories: Making Selves*. Ithaca, NY: Cornell University Press, 1999.

Egan, Timothy. '"Her Mind Was Everything", Dead Woman's Husband Says'. *New York Times*, 6 June 1990.

Ehrenreich, Barbara. *Smile or Die: How Positive Thinking Fooled America and the World*. 2009. London: Granta Books, 2010.

Elliott, Anthony. *Concepts of the Self*. 2nd edn. Cambridge: Polity, 2008.

Ellis, Havelock. *The Criminal*. New York: Scribner and Welford, 1890. https://archive.org/details/criminal00elli.

Ernaux, Annie. *Une femme*. 1987. Lecture accompagnée par Pierre-Louis Fort. Paris: Éditions Gallimard, 2002. English translation as *A Woman's Story*. Translated by Tanya Leslie. New York: Seven Stories Press, 1991.

Esler, Carol Clemeau. 'Horace's Old Girls: Evolution of a Topos'. In *Old Age in Greek and Latin Literature*, edited by Thomas M. Falkner and Judith de Luce, 172–82. Albany: State University of New York Press, 1989.

Espinel, Carlos H. 'De Kooning's Late Colours and Forms: Dementia, Creativity, and the Healing Power of Art'. *The Lancet* 347 (1996): 1096–8.

Esquirol, Jean Étienne. *Mental Maladies: A Treatise on Insanity*. Translated by E. K. Hunt. Philadelphia: Lea and Blanchard, 1845. First published in French in 1838.

Estes, Carroll L., and Elizabeth A. Binney. 'The Biomedicalization of Aging: Dangers and Dilemmas'. *The Gerontologist* 29 (1989): 587–96.

Evans, Martyn, and Jane Macnaughton. 'Should Medical Humanities Be a Multidisciplinary or Interdisciplinary Study?' *Medical Humanities* 30 (2004): 1–4.

Eyre, Richard, dir. *Iris*. 2001. Miramax Home Entertainment, 2002. DVD.

Falcus, Sarah. 'Storying Alzheimer's Disease in Lisa Genova's *Still Alice*'. *Entertext* 12 (2014): 73–94.

Falcus, Sarah, and Katsura Sako. *Contemporary Narratives of Dementia: Ethics, Ageing, Politics*. Abingdon: Routledge, 2019.

Faulkner, William. *The Sound and the Fury*. 1929. London: Vintage, 1995.

Finger, Stanley. 'The Neuropathology of Memory'. In *Origins of Neuroscience: A History of Explorations into Brain Function*, 349–68. New York: Oxford University Press, 1994.

Finley, M. I. Introduction to *Old Age in Greek and Latin Literature*, edited by Thomas M. Falkner and Judith de Luce, 1–20. Albany: State University of New York Press, 1989.

Fisher, Lawrence M. 'Vaccine in Mice Offers Hope in Fight against Alzheimer's'. *New York Times*, 8 July 1999.

Fitzgerald, F. Scott. 'The Curious Case of Benjamin Button'. 1922. In *The Curious Case of Benjamin Button and Six Other Stories*. London: Penguin, 2008.

Fogelman, Dan, dir. *Danny Collins*. Big Indie Pictures, 2015.

Foley, Joseph M. 'The Experience of Being Demented'. In *Dementia and Aging: Ethics, Values, and Policy Choices*, edited by Robert H. Binstock, Stephen G. Post and Peter J. Whitehouse, 30–43. Baltimore: Johns Hopkins University Press, 1992.

Folstein, Marshal F., Susan E. Folstein and Paul R. McHugh. 'Mini-Mental State: A Practical Method for Grading the Cognitive State of Patients for the Clinician'. *Journal of Psychiatric Research* 12 (1975): 189–98.

Fontana, Andrea, and Ronald W. Smith. 'Alzheimer's Disease Victims: The "Unbecoming" of Self and the Normalization of Competence'. *Sociological Perspectives* 32 (1989): 35–46.

Forster, Margaret. *Have the Men Had Enough?* 1989. London: Vintage, 2004.

Fotuhi, Majid, Vladimir Hachinski and Peter J. Whitehouse. 'Changing Perspectives Regarding Late-Life Dementia'. *Nature Reviews Neurology* 5 (2009): 649–58.

Foucault, Michel. *The Birth of Biopolitics: Lectures at the Collège de France, 1978–1979*. 2008. Translated by Graham Burchell. Basingstoke: Palgrave Macmillan, 2010. First published in French in 2004.

Foucault, Michel. *The Birth of the Clinic: An Archaeology of Medical Perception*. 1973. Translated by A. M. Sheridan. London: Routledge, 2005. First published in French in 1963.

Foucault, Michel. *Madness and Civilization: A History of Insanity in the Age of Reason*. 1965. Translated by Richard Howard. London: Vintage, 1988. First published in French in 1961.

Foucault, Michel. *Psychiatric Power: Lectures at the Collège de France, 1973–1974*. 2006. Translated by Graham Burchell. Basingstoke: Palgrave Macmillan, 2008. First published in French in 2003.

Fox, Patrick. 'From Senility to Alzheimer's Disease: The Rise of the Alzheimer's Disease Movement'. *Milbank Quarterly* 67 (1989): 58–102.

Frank, Arthur. *The Wounded Storyteller: Body, Illness, and Ethics*. Chicago: University of Chicago Press, 1995.

Frankfurter Allgemeine Zeitung. 'Alois Alzheimer hat sich nicht geirrt'. 15 April 1998.

Franzen, Jonathan. *The Corrections*. 2001. London: Fourth Estate, 2010.

Freeman, H. L. 'Social Aspects of Alzheimer's Disease'. In *Alzheimer and the Dementias*, edited by G. E. Berrios and H. L. Freeman, 57–68. London: Royal Society of Medicine, 1991.

Friedland, R. P., T. F. Budinger, M. Brant-Zawadzki and W. J. Jagust. 'The Diagnosis of Alzheimer-Type Dementia: A Preliminary Comparison of Positron Emission Tomography and Proton Magnetic Resonance'. *Journal of the American Medical Association* 252 (1984): 2750–2.

Frisch, Max. *Der Mensch erscheint im Holozän*. 1979. Frankfurt am Main: Suhrkamp, 2011. English translation as *Man in the Holocene*. 1980. Translated by Geoffrey Skelton. London: Dalkey Archive, 2007.

Frow, John. *Genre*. Abingdon: Routledge, 2015.

Fuchs, Elinor. *Making an Exit: A Mother–Daughter Drama with Alzheimer's, Machine Tools, and Laughter*. New York: Metropolitan Books, 2005.

Furst, Lilian R. *Between Doctors and Patients: The Changing Balance of Power*. Charlottesville: University Press of Virginia, 1998.

Gadow, Sally A. 'Part Two. Subjectivity: Literature, Imagination, and Frailty. Introduction'. In *What Does It Mean to Grow Old? Reflections from the Humanities*, edited by Thomas R. Cole and Sally A. Gadow, 131–4. Durham, NC: Duke University Press, 1986.

Gallagher, James. 'Dementia Levels "Are Stabilising"'. *BBC News*, 21 August 2015. http://www.bbc.co.uk/news/health-34001144.

Gallagher, James. 'The Drug to Slow Alzheimer's?'. *BBC News*, 22 July 2015. http://www.bbc.co.uk/news/health-33621109.

Gallagher, James. 'Huntington's Breakthrough May Stop Disease'. *BBC News*, 11 December 2017. http://www.bbc.co.uk/news/health-42308341.

Galsworthy, John. *The Forsyte Saga*. 1906–21. Edited with an introduction and notes by Geoffrey Harvey. Oxford: Oxford University Press, 2008.

Gardini, Nicola. *Lo sconosciuto*. Milan: Sironi Editore, 2007.

Garfield, Simon. 'Memories Are Made of This'. *Guardian*, 9 May 2004. https://www.theguardian.com/society/2004/may/09/longtermcare.observermagazine.

Gawande, Atul. *Being Mortal: Illness, Medicine and What Matters in the End*. 2014. London: Profile Books, 2015.

Gearing, Brian. Preface to *Dementia Reconsidered: The Person Comes First*, by Tom Kitwood, vii–viii. Maidenhead: Open University Press, 1997.

Geiger, Arno. *Der alte König in seinem Exil*. Munich: Carl Hanser, 2011. English translation as *The Old King in His Exile*. Translated by Stefan Tobler. Sheffield: And Other Stories, 2017.

Genova, Lisa. *Still Alice*. 2007. New York: Pocket Books, 2009.

George, Daniel R., and Peter J. Whitehouse. 'Marketplace of Memory: What the Brain Fitness Technology Industry Says about Us and How We Can Do Better'. *The Gerontologist* 51 (2011): 590–6.

Gerrard, Nicci. 'Words Fail Us: Dementia and the Arts'. *Guardian*, 19 July 2015. https://www.theguardian.com/culture/2015/jul/19/dementia-and-the-arts-fiction-films-drama-poetry-painting.

Gerster, Andrea. *Dazwischen Lili*. Basel: Lenos Verlag, 2008.

Giles, David, and James C. Jones, dirs. *The Forsyte Saga*. 1967. BBC Worldwide, 2004. DVD.

Gillies, Andrea. *Keeper: Living with Nancy, a Journey into Alzheimer's*. London: Short Books, 2009.

Gilman, Sander L. *The Case of Sigmund Freud: Medicine and Identity at the Fin de Siècle*. Baltimore: Johns Hopkins University Press, 1993.

Gilman, Sander L. *Disease and Representation: Images of Illness from Madness to AIDS*. 1988. Ithaca, NY: Cornell University Press, 1991.

Glatzer, Richard, and Wash Westmoreland, dirs. *Still Alice*. 2014. Curzon Atificial Eye, 2015. DVD.

Glenner, George G., and Caine W. Wong. 'Alzheimer's Disease: Initial Report of the Purification and Characterization of a Novel Cerebrovascular Amyloid Protein'. *Biochemical and Biophysical Research Communications* 120 (1984): 885–90.

Goate, Alison, Marie-Christine Chartier-Harlin, Mike Mullan, Jeremy Brown, Fiona Crawford, Liana Fidani, Luis Giuffra, Andrew Haynes, Nick Irving, Louise James, Rebecca Mant, Phillippa Newton, Karen Rooke, Penelope Roques, Chris Talbot, Margaret Pericak-Vance, Allen Roses, Robert Williamson, Martin Rossor, Mike Owen and John Hardy. 'Segregation of a Missense Mutation in the Amyloid Precursor Protein Gene with Familial Alzheimer's Disease'. *Nature* 349 (1991): 704–6.

Goffman, Erving. *Asylums: Essays on the Social Situation of Mental Patients and Other Inmates*. 1961. London: Penguin, 1991.

Goffman, Erving. *Stigma: Notes on the Management of Spoiled Identity*. 1963. London: Penguin, 1990.

Goldby, Roger, dir. *The Time of Their Lives*. Bright Pictures, 2017.

Goldman, Marlene. *Forgotten: Narratives of Age-Related Dementia and Alzheimer's Disease in Canada*. Montreal: McGill-Queen's University Press, 2017.

Goldsmith, Malcolm. *Hearing the Voice of People with Dementia: Opportunities and Obstacles*. London: Jessica Kingsley, 1996.

Golomb, James, Alan Kluger and Steven H. Ferris. 'Mild Cognitive Impairment: Historical Development and Summary of Research'. *Dialogues in Clinical Neuroscience* 6 (2004): 351–67.

Gordon, Mary. *The Shadow Man: A Daughter's Search for Her Father*. 1996. New York: Vintage, 1997.

Gordon, Sandi. *Parkinson's: A Personal Story of Acceptance*. Boston: Branden, 1992.

Goyder, Julie. *We'll Be Married in Fremantle: Alzheimer's Disease and the Everyday Act of Storying*. Fremantle: Fremantle Arts Centre, 2001.

Graeber, M. B., S. Kösel, E .Grasbon-Frodl, H. J. Möller and P. Mehraein. 'Histopathology and APOE Genotype of the First Alzheimer Disease Patient, Auguste D.' *Neurogenetics* 1 (1998): 223–8.

Granser, Peter. *Alzheimer*. Heidelberg: Kehrer Verlag, 2008.

Grant, Linda. *Remind Me Who I Am, Again*. 1998. London: Granta Books, 1999.

Green, M. A., L. D. Stevenson, J. E. Fonseca and S. B. Wortis. 'Cerebral Biopsy in Patients with Presenile Dementia'. *Diseases of the Nervous System* 13 (1952): 303–7.

Greenblat, Cathy Stein. *Alive with Alzheimer's*. Chicago: University of Chicago Press, 2004.

Greenhalgh, Charlotte. *Aging in Twentieth-Century Britain*. Oakland: University of California Press, 2018.

Greenslade, William. *Degeneration, Culture and the Novel, 1880–1940*. New York: Cambridge University Press, 1994.

Grella, George. 'The Formal Detective Novel'. In *Detective Fiction: A Collection of Critical Essays*, edited by Robin W. Winks, 84–102. Woodstock: Countryman, 1988.

Gruen, Arno. 'War or Peace? We Cannot Survive with Real-Politik'. Acceptance Speech at Award Ceremony for the Loviisa Peace Price, Loviisa, 7 August 2010. Accessed 27 October 2015. http://www.arnogruen.net/war-or-peace.pdf.

Grundke-Iqbal, I., K. Iqbal, M. Quinlan, Y.-C. Tung, M. S. Zaidi and H. M. Wisniewski. 'Microtubule-Associated Protein Tau: A Component of Alzheimer Paired Helical Filaments'. *Journal of Biological Chemistry* 261 (1986): 6084–9.

Grünthal, Ernst. 'Klinisch-anatomisch vergleichende Untersuchungen über den Greisenblödsinn'. *Zeitschrift für die gesamte Neurologie und Psychiatrie* 111 (1927): 763–818.

Grünthal, Ernst. 'Über die Alzheimersche Krankheit: Eine histopathologisch-klinische Studie'. *Zeitschrift für die gesamte Neurologie und Psychiatrie* 101 (1926): 128–57.

Gubrium, Jaber F. 'Structuring and Destructuring the Course of Illness: The Alzheimer's Disease Experience'. *Sociology of Health and Illness* 9 (1987): 1–24.

Guenther, Katja. *Localization and Its Discontents: A Genealogy of Psychoanalysis and the Neuro Disciplines*. Chicago: University of Chicago Press, 2015.

Hachemer, Mareike. *Alzheimer im problemorientierten Bilderbuch: Inhaltliche, künstlerische und sprachliche Aspekte*. Norderstedt: Grin Verlag, 2009.

Hachinski, V. C., N. A. Lassen and J. Marshall. 'Multi-infarct Dementia: A Cause of Mental Deterioration in the Elderly'. *The Lancet* 2 (1974): 207–10.

Hacker, Katharina. *Die Erdbeeren von Antons Mutter*. 2010. Frankfurt am Main: Fischer, 2012.

Hacking, Ian. *Rewriting the Soul: Multiple Personality and the Sciences of Memory*. Princeton, NJ: Princeton University Press, 1995.

Hacking, Ian. *The Social Construction of What?* 1999. Cambridge, MA: Harvard University Press, 2003.

Hake, Egmont. *Regeneration: A Reply to Max Nordau*. Westminster: Archibald Constable, 1895. In *The Fin de Siècle: A Reader in Cultural History, c. 1880–1900*, edited by Sally Ledger and Roger Luckhurst, 17–19. Oxford: Oxford University Press, 2000.

Hall, G. Stanley. *Senescence: The Last Half of Life*. New York: D. Appleton, 1922.

Hardy, John A. 'ApoE, Amyloid, and Alzheimer's Disease'. *Science* 263 (1994): 454.

Hardy, John A., and Gerald A. Higgins. 'Alzheimer's Disease: The Amyloid Cascade Hypothesis'. *Science* 256 (1992): 184–5.

[Hardy, John, and John Mayer for the Models of Dementia: The Good, the Bad and the Future Meeting Attendees]. 'The Amyloid Cascade Hypothesis Has Misled the Pharmaceutical Industry'. *Biochemical Society Transactions* 39 (2011): 920–3.

Hargreaves, Tracy. '"We Other Victorians": Literary Victorian Afterlives'. *Journal of Victorian Culture* 13 (2008): 278–86.

Harlow, John M. *Recovery from the Passage of an Iron Bar through the Head*. Boston: David Clapp and Son, 1869. https://archive.org/details/66210360R.nlm.nih.gov.

Harrison, Tony. *Black Daisies for the Bride*. London: Faber and Faber, 1993.

Hart aber fair. 'Alzheimer als Komödie – hilft Lachen gegen die Angst?'. Aired 2 February 2015, on ARD.

Hartouni, Valerie. *Cultural Conceptions: On Reproductive Technologies and the Remaking of Life*. Minneapolis: University of Minnesota Press, 1997.

Hartung, Heike. *Ageing, Gender and Illness in Anglophone Literature: Narrating Age in the Bildungsroman*. Abingdon: Routledge, 2016.

Harvey, Geoffrey. Introduction to *The Forsyte Saga*, by John Galsworthy, vi–xxi. Oxford: Oxford University Press, 2008.

Harvey, Samantha. *The Wilderness*. 2009. London: Vintage, 2010.

Havemann, Joel. *A Life Shaken: My Encounter with Parkinson's Disease*. Baltimore: Johns Hopkins University Press, 2002.

Hawkins, Anne Hunsaker. *Reconstructing Illness: Studies in Pathography*. 2nd edn. West Lafayette: Purdue University Press, 1999.

Hazell, Kenneth. *Social and Medical Problems of the Elderly*. London: Hutchinson, 1960.

Hazell, Kenneth. *Social and Medical Problems of the Elderly*. London: Hutchinson, 1965.

Hazell, Kenneth. *Social and Medical Problems of the Elderly*. London: Hutchinson, 1973.

Healey, Emma. *Elizabeth Is Missing*. 2014. London: Penguin Books, 2015.

Held, Wolfgang. *Uns hat Gott vergessen*. 2000. Bucha bei Jena: Quartus-Verlag, 2004.

Henderson, Cary Smith. *Partial View: An Alzheimer's Journal*. Dallas, TX: Southern Methodist University Press, 1998.

Hepworth, Mike. *Stories of Ageing*. Buckingham: Open University Press, 2000.

Herman, David. Introduction to *The Emergence of Mind: Representations of Consciousness in Narrative Discourse in English*, edited by David Herman, 1–40. Lincoln: University of Nebraska Press, 2011.

Herman, David. 'Re-minding Modernism'. In *The Emergence of Mind: Representations of Consciousness in Narrative Discourse in English*, edited by David Herman, 243–72. Lincoln: University of Nebraska Press, 2011.

Herrup, Karl. 'The Case for Rejecting the Amyloid Cascade Hypothesis'. *Nature Neuroscience* 18 (2015): 794–9.

Hilton, Claire. 'Sauerkraut and African Violets: The Art of Old Age Psychiatry'. Paper presented at the Medical Humanities and Ageing Workshop, King's College London, 29 June 2015.

Hinton, W. Ladson, and Sue Levkoff. 'Constructing Alzheimer's: Narratives of Lost Identities, Confusion and Loneliness in Old Age'. *Culture, Medicine and Psychiatry* 23 (1999): 453–75.

Hoff, P. 'Alzheimer and His Time'. In *Alzheimer and the Dementias*, edited by G. E. Berrios and H. L. Freeman, 29–55. London: Royal Society of Medicine, 1991.

Hofmann, M., C. Hock, A. Kühler and F. Müller-Spahn. 'Individualisiertes, computergestütztes Gedächtnistraining bei Alzheimer-Patienten'. *Nervenarzt* 66 (1995): 703–7.

Holderman, Bill, dir. *Book Club*. June Pictures, 2018.

Honel, Rosalie Walsh. 'Alzheimer's Disease'. *New England Journal of Medicine* 309 (1983): 1524.

Honel, Rosalie Walsh. *Journey with Grandpa: Our Family's Struggle with Alzheimer's Disease*. Baltimore: Johns Hopkins University Press, 1988.

Horton, Richard. 'Editorial. Shakespeare: The Bard at the Bedside'. *The Lancet* 387 (2016): 1693.

Hoult, Norah. *There Were No Windows*. 1944. London: Persephone Books, 2005.

Howell, Trevor H. *Our Advancing Years: An Essay on Modern Problems of Old Age*. London: Phoenix House, 1976.

Hubble, Nick, and Philip Tew. *Ageing, Narrative and Identity: New Qualitative Social Research*. Basingstoke: Palgrave Macmillan, 2013.

Hugendick, David. 'Ich vergisst sich'. *Die Zeit*, 25 February 2015.

Hughes, Julian C., Stephen J. Louw and Steven R. Sabat. 'Seeing Whole'. In *Dementia: Mind, Meaning, and the Person*, edited by Julian C. Hughes, Stephen J. Louw and Steven R. Sabat, 1–39. Oxford: Oxford University Press, 2006.

Huguenin, Cécile. *Alzheimer mon amour*. Paris: Éditions Héloïse d'Ormesson, 2011.

Hunter, Kathryn Montgomery. *Doctors' Stories: The Narrative Structure of Medical Knowledge*. Princeton, NJ: Princeton University Press, 1991.

Huyssen, Andreas. *Present Pasts: Urban Palimpsests and the Politics of Memory*. Stanford, CA: Stanford University Press, 2003.

Hydén, Lars-Christer, Hilde Lindemann and Jens Brockmeier. Introduction to *Beyond Loss: Dementia, Identity, Personhood*, edited by Lars-Christer Hydén, Hilde Lindemann and Jens Brockmeier, 1–7. New York: Oxford University Press, 2014.

Hyman, Bradley T. 'Alzheimer's Disease or Alzheimer's Diseases? Clues from Molecular Epidemiology'. *Annals of Neurology* 40 (1996): 135–6.

Ignatieff, Michael. *Scar Tissue*. 1993. London: Vintage, 1994.

Illiffe, Steve. 'Evidence-Based Medicine and Dementia'. *British Journal of General Practice* 65 (2015): 511–12.

Independent. 'Julianne Moore's Oscars 2015 Acceptance Speech in Full: "I'm Thrilled We Were Able to Shine a Light on Alzheimer's Disease"'. 23 February 2015. https://www.independent.co.uk/arts-entertainment/films/oscars/julianne-moore-oscars-2015-acceptance-speech-in-full-im-thrilled-we-were-able-to-shine-a-light-on-10064112.html.

Ingram, Jay. *The End of Memory: A Natural History of Aging and Alzheimer's*. 2014. London: Rider Books, 2016.

Inoue, Yasushi. *Chronicle of My Mother*. 1982. Translated by Jean O. Moy. New York: Kodansha International, 1985. First published in Japanese in 1975.

Inside Kino. 'Top 100 Deutschland 2013'. Accessed 25 October 2019. www.insidekino.de/DJahr/D2013.htm.

Inside Kino. 'Top 100 Deutschland 2014'. Accessed 25 October 2019. www.insidekino.de/DJahr/D2014.htm.

Irigaray, Luce. *Le Langage des Déments*. The Hague: Mouton, 1973.

Ishii, Kazunari, Hajime Kitagaki, Michio Kono and Etsuro Mori. 'Decreased Medial Temporal Oxygen Metabolism in Alzheimer's Disease Shown by PET'. *Journal of Nuclear Medicine* 37 (1996): 1159–65.

Jackson, John Hughlings. 'On Affections of Speech from Disease of the Brain'. *Brain* 1 (1879): 304–30.
Jackson, John Hughlings. *On Convulsive Seizures*. London: British Medical Association, 1890.
Jens, Tilman. *Demenz: Abschied von meinem Vater*. Gütersloh: Gütersloher Verlagshaus, 2009.
Jordanova, Ludmilla J. Introduction to *Languages of Nature: Critical Essays on Science and Literature*, edited by Ludmilla J. Jordanova, 15–47. London: Free Association Books, 1986.
Joyce, Kelly A. *Magnetic Appeal: MRI and the Myth of Transparency*. Ithaca, NY: Cornell University Press, 2008.
Joyner, Michael J., Laszlo G. Boros and Gregory Fink. 'Biological Reductionism versus Redundancy in a Degenerate World'. *Perspectives in Biology and Medicine* 61 (2018): 517–26.
Kang, Jae-Eun, Miranda M. Lim, Randall J. Bateman, James J. Lee, Liam P. Smyth, John R. Cirrito, Nobuhiro Fujiki, Seiji Nishion and David M. Holtzman. 'Amyloid-β Dynamics Are Regulated by Orexin and the Sleep-Wake Cycle'. *Science* 326 (2009): 1005–7.
Kang, Jie, Hans-Georg Lemaire, Axel Unterbeck, J. Michael Salbaum, Colin L. Masters, Karl-Heinz Grzeschik, Gerd Multhaup, Konrad Beyreuther and Benno Müller-Hill. 'The Precursor of Alzheimer's Disease Amyloid A4 Protein Resembles a Cell-Surface Receptor'. *Nature* 325 (1987): 733–6.
Kappus, Hanna. *Das Leben ist ein großes: Alzheimer – ein langer Abschied*. Gütersloh: Gütersloher Verlagshaus, 2012.
Katz, Stephen. *Disciplining Old Age: The Formation of Gerontological Knowledge*. Charlottesville: University Press of Virginia, 1996.
Katz, Stephen, and Barbara L. Marshall. 'Tracked and Fit: FitBits, Brain Games, and the Quantified Aging Body'. *Journal of Aging Studies* 45 (2018): 63–8.
Katzman, Robert. 'Editorial: The Prevalence and Malignancy of Alzheimer Disease: A Major Killer'. *Archives of Neurology* 33 (1976): 217–18.
Keaton, Diane. *Then Again*. 2011. London: Fourth Estate, 2012.
Kemp, Simon, and Christopher D. B. Burt. 'Memories of Uncertain Origin: Dreamt or Real?' *Memory* 14 (2006): 87–93.
Kennaway, James. 'Singing the Body Electric: Nervous Music and Sexuality in Fin-de-Siècle Literature'. In *Neurology and Literature, 1860–1920*, edited by Anne Stiles, 141–60. Basingstoke: Palgrave Macmillan, 2007.
Kennedy, Meegan. *Revising the Clinic: Vision and Representation in Victorian Medical Narrative and the Novel*. Columbus: Ohio State University Press, 2010.
Kenyon, Gary M., James E. Birren and Johannes J. F. Schroots, eds. *Metaphors of Aging in Science and the Humanities*. New York: Springer, 1991.
Kessler, Eva-Marie, and Clemens Schwender. 'Giving Dementia a Face? The Portrayal of Older People with Dementia in German Weekly News Magazines between the Years

2000 and 2009'. *Journals of Gerontology Series B: Psychological Sciences and Social Sciences* 67 (2012): 261–70.

Kidd, Michael. 'Paired Helical Filaments in Electron Microscopy of Alzheimer's Disease'. *Nature* 197 (1963): 192–3.

King, Jeanette. *Discourses of Ageing in Fiction and Feminism: The Invisible Woman*. Basingstoke: Palgrave Macmillan, 2013.

King, Nicola. *Memory, Narrative, Identity: Remembering the Self*. Edinburgh: Edinburgh University Press, 2000.

King's Fund. *Living Well into Old Age: Applying Principles of Good Practice to Services for People with Dementia*. London: King's Fund, 1986.

Kino.de. 'Honig im Kopf'. Accessed 20 August 2015. http://www.kino.de/kinofilm/honig-im-kopf/imkino/152963.

Kirby, Jane. 'New Drug Guidelines Offer Hope to Alzheimer's Sufferers'. *Independent*, 7 October 2010. https://www.independent.co.uk/life-style/health-and-families/health-news/new-drug-guidelines-offer-hope-to-alzheimers-sufferers-2099710.html.

Kirkman, Allison M. 'Dementia in the News: The Media Coverage of Alzheimer's Disease'. *Australasian Journal on Ageing* 25 (2006): 74–9.

Kitwood, Tom. *Dementia Reconsidered: The Person Comes First*. Maidenhead: Open University Press, 1997.

Kleinman, Arthur. 'Caregiving: The Odyssey of Becoming More Human'. *The Lancet* 373 (2009): 292–3.

Kleinman, Arthur. *The Illness Narratives: Suffering, Healing, and the Human Condition*. New York: Basic Books, 1988.

Kolata, Gina. 'Alzheimer's Researchers Close In on Causes'. *New York Times*, 26 February 1991.

Konek, Carol Wolfe. *Daddyboy: A Family's Struggle with Alzheimer's*. 1991. Saint Paul: Graywolf, 1992.

Kosik, Kenneth S., Catherine L. Joachim and Dennis J. Selkoe. 'Microtubule Associated Protein Tau (τ) Is a Major Antigenic Component of Paired Helical Filaments in Alzheimer Disease'. *Proceedings of the National Academy of Sciences* 83 (1986): 4044–8.

Kotulak, Ronald. 'Stalking "Demons of the Mind"'. *Chicago Tribune*, 4 June 1980.

Kraepelin, Emil. 'Altersblödsinn'. In *Einführung in die psychiatrische Klinik: Dreissig Vorlesungen*, 234–44. Leipzig: Johann Ambrosius Barth, 1901. https://archive.org/details/einfhrungindiep01kraegoog.

Kraepelin, Emil. 'Das senile and präsenile Irresein'. In *Psychiatrie: Ein Lehrbuch für Studierende und Ärzte*, 533–632. Leipzig: Johann Ambrosius Barth, 1909–10.

Krueger, Kristin R., Robert S. Wilson, Julia M. Kamenetsky, Lisa L. Barnes, Julia L. Bienias and David A. Bennett. 'Social Engagement and Cognitive Function in Old Age'. *Experimental Aging Research* 35 (2009): 45–60.

Kuhn, Thomas S. *The Structure of Scientific Revolutions*. 4th edn. Chicago: University of Chicago Press, 2012.

Kurz, Joachim. 'Die vergebliche Gnade des Vergessens'. *Kino-Zeit*, 14 January 2015. http://www.kino-zeit.de/blog/b-roll/die-vergebliche-gnade-des-vergessens-die-kino-zeit-de-kolumne.

Laborde, Françoise. *Pourquoi ma mère me rend folle*. Paris: Flammarion, 2002.

Lafora, Gonzalo R. 'Beitrag zur Kenntnis der Alzheimerschen Krankheit oder präsenilen Demenz mit Herdsymptomen'. *Zeitschrift für die gesamte Neurologie und Psychiatrie* 6 (1911): 15–20.

Laing, R. D. *The Divided Self: An Existential Study in Sanity and Madness*. 1960. London: Penguin, 2010.

Laing, R. D. *The Politics of Experience* and *The Bird of Paradise*. 1967. Harmondsworth: Penguin, 1987.

Lane, Joan. *A Social History of Medicine: Health, Healing and Disease in England, 1750–1950*. London: Routledge, 2001.

Latour, Bruno. *Science in Action: How to Follow Scientists and Engineers through Society*. Cambridge, MA: Harvard University Press, 1987.

Latour, Bruno, and Steve Woolgar. *Laboratory Life: The Construction of Scientific Facts*. 1979. Princeton, NJ: Princeton University Press, 1986.

Levin, Sidney. 'Depression in the Aged: The Importance of External Factors'. In *New Thoughts on Old Age*, edited by Robert Kastenbaum, 179–85. New York: Springer, 1964.

Lewis, Jane, and Barbara Meredith. *Daughters Who Care: Daughters Caring for Mothers at Home*. London: Routledge, 1988.

Library of Congress. 'Project on the Decade of the Brain'. Last modified 1 March 2000. http://www.loc.gov/loc/brain/.

Lichtenstein, Alice. *Lost*. New York: Scribner, 2010.

Lindbergh, Reeve. *No More Words: A Journal of My Mother, Anne Morrow Lindbergh*. 2001. New York: Touchstone, 2002.

Littlefield, Melissa M. 'Matter for Thought: The Psychon in Neurology, Psychology and American Culture, 1927–1943'. In *Neurology and Modernity: A Cultural History of Nervous Systems, 1800–1950*, edited by Laura Salisbury and Andrew Shail, 267–86. Basingstoke: Palgrave Macmillan, 2010.

Lloyd, Phyllida, dir. *The Iron Lady*. 2011. Twentieth Century Fox Home Entertainment, 2012. DVD.

Lock, Margaret. *The Alzheimer Conundrum: Entanglements of Dementia and Aging*. Princeton, NJ: Princeton University Press, 2013.

Lodge, David. *Consciousness and the Novel: Connected Essays*. Cambridge, MA: Harvard University Press, 2002.

Luckhurst, Roger, and Josephine McDonagh. Introduction to *Transactions and Encounters: Science and Culture in the Nineteenth Century*, edited by Roger Luckhurst and Josephine McDonagh, 1–15. Manchester: Manchester University Press, 2002.

Lynch, Michael. 'Discipline and the Material Form of Images: An Analysis of Scientific Visibility'. *Social Studies of Science* 15 (1985): 37–66.

Macchi, G., C. Brahe and M. Pomponi. 'Alois Alzheimer and Gaetano Perusini: Should Man Divide What Fate United?'. *European Journal of Neurology* 4 (1997): 210–13.

Mace, Nancy L., and Peter V. Rabins. *The 36-Hour Day: A Family Guide to Caring for Persons with Alzheimer Disease, Related Dementing Illnesses, and Memory Loss in Later Life*. Baltimore: Johns Hopkins University Press, 1981.

Mace, Nancy L., and Peter V. Rabins. *The 36-Hour Day: A Family Guide to Caring for Persons with Alzheimer Disease, Related Dementing Illnesses, and Memory Loss in Later Life*. 4th edn. New York: Wellness Central, 1999.

Mace, Nancy L., and Peter V. Rabins. *The 36-Hour Day: A Family Guide to Caring for People Who Have Alzheimer Disease, Related Dementias, and Memory Loss*. 5th edn. Baltimore: Johns Hopkins University Press, 2011.

Magnusson, Sally. *Where Memories Go: Why Dementia Changes Everything*. London: Two Roads, 2014.

Mairs, Nancy. *Waist-High in the World: A Life among the Nondisabled*. Boston: Beacon, 1996.

Mann, Thomas. *Buddenbrooks*. 1901. Stuttgart: Deutscher Bücherbund, n.d.

Manthorpe, Jill. 'A Child's Eye View: Dementia in Children's Literature'. *British Journal of Social Work* 35 (2005): 305–20.

Marrot, Harold V. *The Life and Letters of John Galsworthy*. London: William Heinemann, 1935.

Marshall, Cynthia. 'Reading "The Golden Key": Narrative Strategies of Parable'. *Children's Literature Association Quarterly* 14 (1989): 22–5.

Martin, Emily. *Flexible Bodies: Tracking Immunity in American Culture – from the Days of Polio to the Age of AIDS*. Boston: Beacon, 1994.

Martin, Richard J., and Stephen G. Post. 'Human Dignity, Dementia, and the Moral Basis of Caregiving'. In *Dementia and Aging: Ethics, Values, and Policy Choices*, edited by Robert H. Binstock, Stephen G. Post and Peter J. Whitehouse, 55–68. Baltimore: Johns Hopkins University Press, 1992.

Matus, Jill. 'Emergent Theories of Victorian Mind Shock: From War and Railway Accident to Nerves, Electricity and Emotion'. In *Neurology and Literature, 1860–1920*, edited by Anne Stiles, 163–83. Basingstoke: Palgrave Macmillan, 2007.

Maurer, Konrad, and Ulrike Maurer. *Alzheimer: Das Leben eines Arztes und die Karriere einer Krankheit*. Munich: Piper, 1998. English translation as *Alzheimer: The Life of a Physician and the Career of a Disease*. Translated by Neil Levi with Alistair Burns. New York: Columbia University Press, 2003.

May, William F. 'The Virtues and Vices of the Elderly'. In *What Does It Mean to Grow Old? Reflections from the Humanities*, edited by Thomas R. Cole and Sally A. Gadow, 41–61. Durham, NC: Duke University Press, 1986.

Mazzarello, Paolo. *Golgi: A Biography of the Founder of Modern Neuroscience*. Translated by Aldo Badiani and Henry A. Buchtel. New York: Oxford University Press, 2010.

McDonagh, Patrick. *Idiocy: A Cultural History*. Liverpool: Liverpool University Press, 2008.

McEwan, Ian. *Atonement*. 2001. London: Vintage, 2002.

McEwan, Ian. *Saturday*. 2005. London: Vintage, 2006.

McFadden, Susan H., and John T. McFadden. *Aging Together: Dementia, Friendship, and Flourishing Communities*. 2011. Baltimore: Johns Hopkins University Press, 2014.

McGowin, Diana Friel. *Living in the Labyrinth: A Personal Journey through the Maze of Alzheimer's*. 1993. New York: Dell, 1994.

McHaffie, Hazel. *Right to Die*. Edinburgh: Luath, 2008.

McMenemey, W. H. 'Alzheimer's Disease: A Report of Six Cases'. *Journal of Neurology and Psychiatry* 3 (1940): 211–40.

Medina, Raquel. 'Alzheimer's Disease, a Shifting Paradigm in Spanish Film: *¿Y tú quién eres?* and *Amanecer de un sueño*'. *Hispanic Research Journal* 14 (2013): 356–72.

Medina, Raquel. *Cinematic Representations of Alzheimer's Disease*. London: Palgrave Macmillan, 2018.

Medina, Raquel. 'From the Medicalisation of Dementia to the Politics of Memory and Identity in Three Spanish Documentary Films: *Bicicleta, cullera, poma, Las voces de la memoria* and *Bucarest: la memòria perduda*'. *Ageing and Society* 34 (2014): 1688–710.

Mendelsohn, Andrew R., and James W. Larrick. 'Sleep Facilitates Clearance of Metabolites from the Brain: Glymphatic Function in Aging and Neurodegenerative Diseases'. *Rejuvenation Research* 16 (2013): 518–23.

Metchnikoff, Elie. 'Old Age'. In *Annual Report of the Smithsonian Institution for the Year Ending June 30, 1904*, 533–50. Washington, DC: Smithsonian Institution, 1903–4.

Meyer, Lawrence. 'A Cure Is Sought for Disease of Aged'. *Washington Post*, 8 January 1982.

Meyer, Lawrence. 'A Family Stranger'. *Washington Post*, 9 January 1982.

Meyers, Nancy, dir. *The Intern*. Waverly Films, 2015.

Micale, Mark S. 'Medical and Literary Discourses of Trauma in the Age of the American Civil War'. In *Neurology and Literature, 1860–1920*, edited by Anne Stiles, 184–206. Basingstoke: Palgrave Macmillan, 2007.

Mills, Marie A. *Narrative Identity and Dementia: A Study of Autobiographical Memories and Emotions*. Aldershot: Ashgate, 1998.

Mingazzini, G. 'On Aphasia due to Atrophy of the Cerebral Convolutions'. *Brain* 36 (1914): 493–524.

Mitchell, Wendy. 'Alzheimer's in the Workplace'. *New York Times*, 12 September 2018.

Mitchell, Wendy. 'Sharing My Journey'. Twitter. Accessed 25 October 2019. www.twitter.com/wendypmitchell.

Mitchell, Wendy, with Anna Wharton. *Somebody I Used to Know*. London: Bloomsbury, 2018.

Mobley, Tracy. *Young Hope: The Broken Road*. Denver, CO: Outskirts, 2007.

Mollat. 'Alzheimer mon amour'. 30 September 2011. www.mollat.com/livres/cecile-huguenin-alzheimer-mon-amour-9782350871707.html.

Moncrieff, Joanna. 'The Politics of a New Mental Health Act'. *British Journal of Psychiatry* 183 (2003): 8–9.

Monicelli, Mario, dir. *Speriamo che sia femmina*. 1986. Sony Pictures Home Entertainment, 2012. DVD.

Moog, Ferdinand P., and Daniel Schäfer. 'Aspekte der Altersdemenz im antiken Rom. Literarische Fiktion und faktische Lebenswirklichkeit'. *Sudhoffs Archiv* 91 (2007): 73–81.

Morel, Bénédict-Auguste. *Traité des Maladies Mentales*. Paris: Librairie Victor Masson, 1860.

Morris, David B. *Illness and Culture in the Postmodern Age*. Berkeley: University of California Press, 1998.

Mothersill, Mary. 'Old Age'. *Proceedings and Addresses of the American Philosphical Association* 73 (1999): 9–23.

Moylan, Philippa. 'The Nervous Economies of John Galsworthy's Forsyte Chronicles'. *English Literature in Transition, 1880–1920* 54 (2011): 56–78.

Mulisch, Harry. *The Assault*. Translated by Claire N. White. New York: Pantheon, 1985. First published in Dutch in 1982.

Müller, Ch. Foreword to *Senile Dementia: Clinical and Therapeutic Aspects*, edited by Ch. Müller and L. Ciompi, 11–12. Bern: Hans Huber, 1968.

Müller, Ch., and L. Ciompi, eds. *Senile Dementia: Clinical and Therapeutic Aspects*. Bern: Hans Huber, 1968.

Mummendey, Hans Dieter. *Claudia, Alzheimer und ich: Kriminalroman*. Münster: Neues Literaturkontor, 1992.

Murray, Stuart. *Representing Autism: Culture, Narrative, Fascination*. Liverpool: Liverpool University Press, 2008.

Nalbantian, Suzanne. *Memory in Literature: From Rousseau to Neuroscience*. Basingstoke: Palgrave Macmillan, 2003.

Neuberger, Julia. *Not Dead Yet: A Manifesto for Old Age*. London: HarperCollins, 2008.

New York Times. 'Notable Books of the Year'. 3 December 1989.

Ngatcha-Ribert, Laëtitia. *Alzheimer: la construction sociale d'une maladie*. Paris: Dunod, 2012.

Noack, Rick. 'A German Nursing Home Tries a Novel Form of Dementia Therapy: Re-creating a Vanished Era for Its Patients'. *Washington Post*, 26 December 2017. https://www.washingtonpost.com/news/worldviews/wp/2017/12/26/a-german-nursing-home-tries-a-novel-form-of-dementia-therapy-re-creating-a-vanished-era-for-its-patients/.

Nobili, F., M. Koulibaly, P. Vitali, O. Migneco, G. Mariani, K. Ebmeier, A. Pupi, P. H. Robert, G. Rodriguez and J. Darcourt. 'Brain Perfusion Follow-Up in Alzheimer's Patients during Treatment with Acetylcholinesterase Inhibitors'. *Journal of Nuclear Medicine* 43 (2002): 983–90.

Nye, Robert A. *Crime, Madness, and Politics in Modern France: The Medical Concept of National Decline*. Princeton, NJ: Princeton University Press, 1984.
O'Hagan, Andrew. *The Illuminations*. London: Faber and Faber, 2015.
Omran, Abdel R. 'The Epidemiologic Transition: A Theory of the Epidemiology of Population Change'. *Milbank Quarterly* 83 (2005): 731–57. Reissue of 49 (1971): 509–38.
O'Neill, Desmond. 'Ageing with Style'. *The Lancet* 387 (2016): 639.
Oppenheimer, Catherine. 'I Am, Thou Art: Personal Identity in Dementia'. In *Dementia: Mind, Meaning, and the Person*, edited by Julian C. Hughes, Stephen J. Louw and Steven R. Sabat, 193–203. New York: Oxford University Press, 2006.
O'Rourke, Norm. 'Alzheimer's Disease as a Metaphor for Contemporary Fears of Aging'. *Journal of the American Geriatrics Society* 44 (1996): 220–1.
Orr, David M. R., and Yugin Teo. 'Carers' Responses to Shifting Identity in Dementia in *Iris* and *Away from Her*: Cultivating Stability or Embracing Change?' *Medical Humanities* 41 (2015): 81–5.
Orton, Samuel T. 'A Study of the Satellite Cells in Fifty Selected Cases of Mental Disease'. *Brain* 36 (1914): 525–42.
Otis, Laura. *Membranes: Metaphors of Invasion in Nineteenth-Century Literature, Science, and Politics*. Baltimore: Johns Hopkins University Press, 1999.
Otis, Laura. *Networking: Communicating with Bodies and Machines in the Nineteenth Century*. Ann Arbor: University of Michigan Press, 2011.
Otis, Laura. *Organic Memory: History and the Body in the Late Nineteenth and Early Twentieth Centuries*. Lincoln: University of Nebraska Press, 1994.
Oxford English Dictionary Online. Oxford: Oxford University Press, June 2014.
Palmer, Alan. 'Ontologies of Consciousness'. In *The Emergence of Mind: Representations of Consciousness in Narrative Discourse in English*, edited by David Herman, 273–97. Lincoln: University of Nebraska Press, 2011.
Panza, Francesco, Giancarlo Logroscino, Bruno P. Imbimbo and Vincenzo Solfrizzi. 'Is There Still Any Hope for Amyloid-Based Immunotherapy for Alzheimer's Disease?' *Current Opinion in Psychiatry* 27 (2014): 128–37.
Pavel, Thomas G. *The Lives of the Novel: A History*. Princeton, NJ: Princeton University Press, 2013. First published in French in 2003.
Perry, E. K., B. E. Tomlinson, G. Blessed, K. Bergmann, P. H. Gibson and R. H. Perry. 'Correlation of Cholinergic Abnormalities with Senile Plaques and Mental Test Scores in Senile Dementia'. *British Medical Journal* 2 (1978): 1457–9.
Perusini, Gaetano. 'Über klinisch und histologisch eigenartige psychische Erkrankungen des späteren Lebensalters'. In *Histopathologische Arbeiten über die Großhirnrinde unter besonderer Berücksichtigung der pathologischen Anatomie der Geisteskrankheiten*, vol. 3, edited by Franz Nissl and Alois Alzheimer, 297–352. Jena: Gustav Fischer, 1910.
Petrie, A. A. W. 'Differential Diagnosis of Organic and Functional Nervous Disorders'. *British Medical Journal* 2 (1934): 503–6.

Philibert, Nicolas. *La moindre des choses*. La Sept Cinéma, 1996. English subtitles in *Every Little Thing*. La Sept Cinéma, 1997.

Pick, Daniel. *Faces of Degeneration: A European Disorder, c. 1848–c. 1918*. Cambridge: Cambridge University Press, 1989.

Pierce, Charles P. *Hard to Forget: An Alzheimer's Story*. New York: Random House, 2000.

Pinel, Philippe. *Nosographie philosophique ou méthode de l'analyse appliquée à la médecine*. Vol. 3. 6th edn. Paris: Brosson, 1818. https://archive.org/details/nosographiephilo03pine.

Pollen, Daniel A. *Hannah's Heirs: The Quest for the Genetic Origins of Alzheimer's Disease*. Exp. edn. New York: Oxford University Press, 1996.

Polley, Sarah, dir. *Away from Her*. 2006. Ascot Elite Home Entertainment, 2008. DVD.

Post, Stephen G. 'The Concept of Alzheimer Disease in a Hypercognitive Society'. In *Concepts of Alzheimer Disease: Biological, Clinical, and Cultural Perspectives*, edited by Peter J. Whitehouse, Konrad Maurer and Jesse F. Ballenger, 245–56. Baltimore: Johns Hopkins University Press, 2000.

Prichard, James C. *A Treatise on Insanity and Other Disorders Affecting the Mind*. Philadelphia: Haswell, Barrington and Haswell, 1837. https://wellcomelibrary.org/item/b21007597#?c=0&m=0&s=0&cv=0&z=-1.0448%2C-0.0907%2C3.0895%2C1.8137.

Pym, Barbara. *Quartet in Autumn*. 1977. London: Bello, 2013.

Rabins, Peter V. Foreword to *Journey with Grandpa: Our Family's Struggle with Alzheimer's Disease*, by Rosalie Walsh Honel, ix–x. Baltimore: Johns Hopkins University Press, 1988.

Ramanathan, Vai. *Alzheimer Discourse: Some Sociolinguistic Dimensions*. Mahwah, NJ: Lawrence Erlbaum Associates, 1997.

Ramanathan, Vaidehi. 'Alzheimer Pathographies: Glimpses into How People with AD and Their Caregivers Text Themselves'. In *Dialogue and Dementia: Cognitive and Communicative Resources for Engagement*, edited by Robert W. Schrauf and Nicole Müller, 245–61. New York: Psychology, 2014.

Rasool, C. G., C. N. Svendsen and Dennis J. Selkoe. 'Neurofibrillary Degeneration of Cholinergic and Noncholinergic Neurons of the Basal Forebrain in Alzheimer's Disease'. *Annals of Neurology* 20 (1986): 482–8.

Reisberg, Barry, ed. *Alzheimer's Disease: The Standard Reference*. New York: Free Press, 1983.

Reisberg, Barry. 'Clinical Presentation, Diagnosis, and Symptomatology of Age-Associated Cognitive Decline and Alzheimer's Disease'. In *Alzheimer's Disease: The Standard Reference*, edited by Barry Reisberg, 173–87. New York: Free Press, 1983.

Resnais, Alain, dir. *Hiroshima mon amour*. 1959. Nouveaux Pictures, 2004. DVD.

Reynolds, Andrew S. *The Third Lens: Metaphor and the Creation of Modern Cell Biology*. Chicago: University of Chicago Press, 2018.

Ribot, Théodule A. *The Diseases of Memory*. Translated by J. Fitzgerald. New York: Humboldt, 1883.

Richardson, Alan. *British Romanticism and the Science of the Mind*. Cambridge: Cambridge University Press, 2001.

Riedhof, Kilian, dir. *Sein letztes Rennen*. 2013. Universum Film, 2014. DVD.

Riga, Frank P. 'Mortals Call Their History Fable: Narnia and the Use of Fairy Tale'. *Children's Literature Association Quarterly* 14 (1989): 26–30.

Riga, Frank P. 'Religion in Children's Literature: Introduction'. *Children's Literature Association Quarterly* 14 (1989): 4–5.

Riley II, Charles A. *Disability and the Media: Prescriptions for Change*. Lebanon: University Press of New England, 2005.

Roberts, Hannah J., and James M. Noble. 'Education Research: Changing Medical Student Perceptions of Dementia. An Arts-Centered Experience'. *Neurology* 85 (2015): 739–41.

Roberts, Louis. 'Portrayal of the Elderly in Classical Greek and Roman Literature'. In *Perceptions of Aging in Literature: A Cross-Cultural Study*, edited by Prisca von Dorotka Bagnell and Patricia Spencer Soper, 17–33. New York: Greenwood, 1989.

Robertson, Ann. 'The Politics of Alzheimer's Disease: A Case Study in Apocalyptic Demography'. *International Journal of Health Services* 20 (1990): 429–42.

Rogers, S. L., M. R. Farlow, R. S. Doody, R. Mohs, L. T. Friedhoff and the Donepezil Study Group. 'A 24-Week, Double-Blind, Placebo-Controlled Trial of Donepezil in Patients with Alzheimer's Disease'. *Neurology* 50 (1998): 136–45.

Rooke, Constance. 'Old Age in Contemporary Fiction: A New Paradigm of Hope'. In *Handbook of the Humanities and Aging*, edited by Thomas R. Cole, David D. Van Tassel and Robert Kastenbaum, 241–57. New York: Springer, 1992.

Rose, Charles L. 'Social Correlates of Longevity'. In *New Thoughts on Old Age*, edited by Robert Kastenbaum, 75–91. New York: Springer, 1964.

Rose, Larry. *Show Me the Way to Go Home*. Forest Knolls, CA: Elder Books, 1996.

Rose, Nikolas. *Our Psychiatric Future*. Cambridge: Polity, 2019.

Rose, Nikolas. *The Politics of Life Itself: Biomedicine, Power, and Subjectivity in the Twenty-First Century*. Princeton, NJ: Princeton University Press, 2007.

Rose, Nikolas, and Joelle M. Abi-Rached. *Neuro: The New Brain Sciences and the Management of the Mind*. Princeton, NJ: Princeton University Press, 2013.

Rota, Nucci A. *La bimbamamma: Cosa vuol dire convivere con l'Alzheimer. Il diario di una figlia*. Naples: Iuppiter Edizioni, 2009.

Roth, Marco. 'The Rise of the Neuronovel'. 2009. In *Say What You Mean: The N+1 Anthology*, edited by Christian Lorentzen, 73–90. London: Notting Hill Editions, 2012. https://nplusonemag.com/issue-8/essays/the-rise-of-the-neuronovel/.

Rousseau, G. S. 'The Discourses of Literature and Medicine: Theory and Practice (1)'. In *Enlightenment Borders: Pre- and Post-Modern Discourses, Medical, Scientific*, 2–25. Manchester: Manchester University Press, 1991.

Ru, Yi-Ling. *The Family Novel: Toward a Generic Definition*. New York: Peter Lang, 1992.

Rubin, Lillian B. *Tangled Lives: Daughters, Mothers, and the Crucible of Aging*. Boston: Beacon, 2000.

Russell, Charlie, dir. *Terry Pratchett: Living with Alzheimer's*. Aired 4 February 2009, on BBC Two. IWC Media, 2009. DVD. https://www.youtube.com/watch?v=KmejLjxFmCQ and https://www.youtube.com/watch?v=tTgqocgY5Ww.

Ryan, Ellen Bouchard, Karen A. Bannister and Ann P. Anas. 'The Dementia Narrative: Writing to Reclaim Social Identity'. *Journal of Aging Studies* 23 (2009): 145–57.

Sabat, Steven R. *The Experience of Alzheimer's Disease: Life through a Tangled Veil*. Oxford: Blackwell, 2001.

Sacks, Oliver. *The Man Who Mistook His Wife for a Hat*. 1985. London: Picador, 2007.

Sáez, Flora. 'Mi marido tiene Alzheimer'. *El Mundo*, 22 September 2002. https://www.elmundo.es/cronica/2002/362/1032772217.html.

Sahyouni, Ronald, Aradhana Verma and Jefferson Chen. *Alzheimer's Disease Decoded: The History, Present, and Future of Alzheimer's Disease and Dementia*. New Jersey: World Scientific, 2017.

Sakai, Erin Y., Brian D. Carpenter and Rebecca E. Rieger. ' "What's Wrong with Grandma?": Depictions of Alzheimer's Disease in Children's Storybooks'. *American Journal of Alzheimer's Disease and Other Dementias* 27 (2012): 584–91.

Salisbury, Laura. 'Narration and Neurology: Ian McEwan's Mother Tongue'. *Textual Practice* 24 (2010): 883–912.

Salisbury, Laura, and Andrew Shail. Introduction to *Neurology and Modernity: A Cultural History of Nervous Systems, 1800–1950*, edited by Laura Salisbury and Andrew Shail, 1–40. Basingstoke: Palgrave Macmillan, 2010.

Salloway, S., R. Sperling, S. Gilman, N. C. Fox, K. Blennow, M. Raskind, M. Sabbagh, L. S. Honig, R. Doody, C. H. Van Dyck, R. Mulnard, J. Barakos, K. M. Gregg, E. Liu, I. Lieberburg, D. Schenk, R. Black and M. Grundman, for the Bapineuzumab 201 Clinical Trial Investigators. 'A Phase 2 Multiple Ascending Dose Trial of Bapineuzumab in Mild to Moderate Alzheimer Disease'. *Neurology* 73 (2009): 2061–70.

Saunders, Barry F. *CT Suite: The Work of Diagnosis in the Age of Noninvasive Cutting*. Durham, NC: Duke University Press, 2008.

Scacco, Linda, and Nicole Wong. *Always My Grandpa: A Story for Children about Alzheimer's Disease*. Washington, DC: Magination, 2006.

Scarry, Elaine. *The Body in Pain: The Making and Unmaking of the World*. New York: Oxford University Press, 1985.

Schäubli-Meyer, Ruth. *Alzheimer: Wie will ich noch leben – wie sterben*. 2008. Zurich: Oesch Verlag, 2010.

Schenk, Dale, Robin Barbour, Whitney Dunn, Grace Gordon, Henry Grajeda, Teresa Guido, Kang Hu, Jiping Huang, Kelly Johnson-Wood, Karen Khan, Dora Kholodenko, Mike Lee, Zhenmei Liao, Ivan Lieberburg, Ruth Motter, Linda Mutter, Ferdie Soriano, George Shopp, Nicki Vasquez, Christopher Vandevert, Shannan Walker, Mark Wogulis, Ted Yednock, Dora Games and Peter Seubert. 'Immunization with Amyloid-β Attenuates Alzheimer-Disease-Like Pathology in the PDAPP Mouse'. *Nature* 400 (1999): 173–7.

Schlich, Thomas. 'Farmer to Industrialist: Lister's Antisepsis and the Making of Modern Surgery in Germany'. *Notes and Records of the Royal Society* 67 (2013): 245–60.

Schnurbush, Barbara, and Cary Pillo. *Striped Shirts and Flowered Pants: A Story about Alzheimer's Disease for Young Children*. Washington, DC: Magination, 2007.

Schottky, Johannes. 'Über präsenile Verblödungen'. *Zeitschrift für die gesamte Neurologie und Psychiatrie* 140 (1932): 333–97.

Schrauf, Robert W., and Nicole Müller, eds. *Dialogue and Dementia: Cognitive and Communicative Resources for Engagement*. New York: Psychology, 2014.

Schweiger, Til, dir. *Honig im Kopf*. 2014. Warner Home Video, 2015. DVD. English remake as *Head Full of Honey*. Directed by Til Schweiger. Barefoot Films, 2018.

Seab, J. P., W. J. Jagust, S. T. Wong, M. S. Roos, B. R. Reed and T. F. Budinger. 'Quantitative NMR Measurements of Hippocampal Atrophy in Alzheimer's Disease'. *Magnetic Resonance in Medicine* 8 (1988): 200–8.

Sebald, W. G. *Austerlitz*. 2001. Frankfurt am Main: Fischer, 2013. English translation as *Austerlitz*. 2001. Translated by Anthea Bell. London: Penguin, 2008.

Segers, Kurt. 'Degenerative Dementias and Their Medical Care in the Movies'. *Alzheimer Disease and Associated Disorders* 21 (2007): 55–9.

Seidler, Miriam. 'Zwischen Demenz und Freiheit: Überlegungen zum Verhältnis von Altern und Geschlecht in der Gegenwartsliteratur'. In *Graue Theorie: Die Kategorien Alter und Geschlecht im kulturellen Diskurs*, edited by Heike Hartung, Dorothea Reinmuth, Christiane Streubel and Angelika Uhlmann, 195–212. Cologne: Böhlau Verlag, 2007.

Selkoe, Dennis J. 'The Deposition of Amyloid Proteins in the Aging Mammalian Brain: Implications for Alzheimer's Disease'. *Annals of Medicine* 21 (1989): 73–6.

Selkoe, Dennis J. 'Missense on the Membrane'. *Nature* 375 (1995): 734–5.

Shakespeare, William. *King Lear*. 1606. Edited by Grace Ioppolo. New York: W. W. Norton, 2008.

Sheldon, Joseph H. *The Social Medicine of Old Age: Report of an Inquiry in Wolverhampton*. London: Nuffield Foundation, 1948.

Shenk, David. *The Forgetting: Alzheimer's: Portrait of an Epidemic*. 2001. New York: Anchor Books, 2003.

Shinotoh, H., H. Namba, K. Fukushi, S. Nagatsuka, N. Tanaka, A. Aotsuka, T. Ota, S. Tanada and T. Irie. 'Progressive Loss of Cortical Acetylcholinesterase Activity in Association with Cognitive Decline in Alzheimer's Disease: A Positron Emission Tomography Study'. *Annals of Neurology* 48 (2000): 194–200.

Shuttleworth, Sally. *George Eliot and Nineteenth-Century Science: The Make-Believe of a Beginning*. Cambridge: Cambridge University Press, 1984.

Shuttleworth, Sally. '"The Malady of Thought": Embodied Memory in Victorian Psychology and the Novel'. In *Memory and Memorials: From the French Revolution to World War One*, edited by Matthew Campbell, Jacqueline M. Labbe and Sally Shuttleworth, 46–59. New Brunswick, NJ: Transaction, 2004.

Shuttleworth, Sally. *The Mind of the Child: Child Development in Literature, Science, and Medicine, 1840–1900*. Oxford: Oxford University Press, 2010.

Siemers, Eric R., Karen L. Sundell, Christopher Carlson, Michael Case, Gopalan Sethuraman, Hong Liu-Seifert, Sherie A. Dowsett, Michael J. Pontecorvo, Robert A. Dean and Ronald Demattos. 'Phase 3 Solanezumab Trials: Secondary Outcomes in Mild Alzheimer's Disease Patients'. *Alzheimer's and Dementia* 12 (2016): 110–20.

Sieveking, David, dir. *Vergiss mein nicht: Wie meine Mutter ihr Gedächtnis verlor und meine Eltern die Liebe neu entdeckten*. Farbfilm Home Entertainment, 2013. DVD.

Sieveking, David. *Vergiss mein nicht: Wie meine Mutter ihr Gedächtnis verlor und ich meine Eltern neu entdeckte*. 2012. Freiburg im Breisgau: Herder, 2013.

Sim, Myre, and W. Thomas Smith. 'Alzheimer's Disease Confirmed by Cerebral Biopsy: A Therapeutic Trial with Cortisone and ACTH'. *Journal of Mental Science* 101 (1955): 604–9.

Sim, Myre, Eric Turner and W. Thomas Smith. 'Cerebral Biopsy in the Investigation of Presenile Dementia: I. Clinical Aspects, *British Journal of Psychiatry* 112 (1966): 119–25.

Sivaramakrishnan, Kavita. *As the World Ages: Rethinking a Demographic Crisis*. Cambridge, MA: Harvard University Press, 2018.

Small, Helen. *The Long Life*. Oxford: Oxford University Press, 2007.

Small, Helen. 'The Unquiet Limit: Old Age and Memory in Victorian Narrative'. In *Memory and Memorials: From the French Revolution to World War One*, edited by Matthew Campbell, Jacqueline M. Labbe and Sally Shuttleworth, 60–79. New Brunswick, NJ: Transaction, 2004.

Smith, F. B. 'Health'. In *The Working Class in England, 1875–1914*, edited by John Benson, 36–62. Beckenham: Croom Helm, 1985.

Smith, F. B. 'Old Age'. In *The People's Health, 1830–1910*, 316–413. 1979. London: Weidenfeld and Nicolson, 1990.

Smith, W. Thomas, Eric Turner and Myre Sim. 'Cerebral Biopsy in the Investigation of Presenile Dementia: II. Pathological Aspects'. *British Journal of Psychiatry* 112 (1966): 127–33.

Smyth, Chris. 'Dementia Clinics Are Swamped by Worried Well'. *Times*, 11 March 2015.

Snowdon, David A. *Aging with Grace: The Nun Study and the Science of Old Age: How We Can All Live Longer, Healthier and More Vital Lives*. 2001. London: Fourth Estate, 2002.

Snowdon, David A., Susan J. Kemper, James A. Mortimer, Lydia H. Greiner, David R. Wekstein and William R. Markesbery. 'Linguistic Ability in Early Life and Cognitive Function and Alzheimer's Disease in Late Life: Findings from the Nun Study'. *Journal of the American Medical Association* 275 (1996): 528–32.

Snyder, Lisa. *Speaking Our Minds: What It's Like to Have Alzheimer's*. Rev. edn. Baltimore: Health Professions, 2009.

Society for Neuroscience. 'Alzheimer's Disease Information Page'. Last modified 27 March 2019. https://www.ninds.nih.gov/Disorders/All-Disorders/Alzheimers-Disease-Information-Page.

Solomon, Miriam. 'Epistemological Reflections on the Art of Medicine and Narrative Medicine'. *Perspectives in Biology and Medicine* 51 (2008): 406–17.

Sontag, Susan. *Illness as Metaphor* and *AIDS and Its Metaphors*. 1977 and 1988. London: Penguin, 2002.

Soto, Ana M., and Carlos Sonnenschein. 'Reductionism, Organicism, and Causality in the Biomedical Sciences: A Critique'. *Perspectives in Biology and Medicine* 61 (2018): 489–502.

Spijker, Jeroen, and John MacInnes. 'Population Ageing: The Timebomb That Isn't?' *Biomedical Journal* 347 (2013): f6598.

Spinozzi, Paola. 'Representing and Narrativizing Science'. In *Discourses and Narrations in the Biosciences*, edited by Paola Spinozzi and Brian Hurwitz, 31–60. Göttingen: V&R Unipress, 2011.

Spohr, Betty Baker, with Jean Valens Bullard. *Catch a Falling Star: Living with Alzheimer's*. Seattle: Storm Peak, 1995.

St. George-Hyslop, Peter H., Rudolph E. Tanzi, Ronald J. Polinsky, Jonathan L. Haines, Linda Nee, Paul C. Watkins, Richard H. Myers, Robert G. Feldman, Daniel Pollen, David Drachman, John Growdon, Amalia Bruni, Jean-François Foncin, Denise Salmon, Peter Frommelt, Luigi Amaducci, Sandro Sorbi, Silva Piacentini, Gordon D. Stewart, Wendy J. Hobbs, P. Michael Conneally and James F. Gusella. 'The Genetic Defect Causing Familial Alzheimer's Disease Maps on Chromosome 21'. *Science* 235 (1987): 885–90.

Stevens, Earl Eugene. 'John Galsworthy: An Annotated Bibliography of Writings about Him. Supplement I'. *English Literature in Transition, 1880–1920* 7 (1964): 93–110.

Stiles, Anne. Introduction to *Neurology and Literature, 1860–1920*, edited by Anne Stiles, 1–23. Basingstoke: Palgrave Macmillan, 2007.

Stolze, Cornelia. *Vergiss Alzheimer! Die Wahrheit über eine Krankheit, die keine ist*. 2011. Freiburg im Breisgau: Herder, 2013.

Sulaiman, Sandy. *Learning to Live with Huntington's Disease: One Family's Story*. London: Jessica Kingsley, 2007.

Sumner, Petroc, Solveiga Vivian-Griffiths, Jacky Boivin, Andy Williams, Christos A. Venetis, Aimée Davies, Jack Ogden, Leanne Whelan, Bethan Hughes, Bethan Dalton, Fred Boy and Christopher D. Chambers. 'The Association between Exaggeration in Health Related Science News and Academic Press Releases: Retrospective Observational Study'. *Biomedical Journal* 349 (2014): g7015.

Suter, Martin. *Small World*. 1997. Zurich: Diogenes, 1999. English translation as *Small World*. Translated by Sandra Harper. London: Harvill, 2001.

Sutherland, John. 'Art Is Shining a Light on Dementia at Last'. *Times*, 21 February 2015.

Swinnen, Aagje. 'Book Club as Intersubjective Space with Transformative Potential'. Paper presented at the Dementia and Cultural Narrative Symposium, Aston University, 9 December 2017.

Symes, Peter, dir. *Black Daisies for the Bride*. Aired 30 June 1993, on BBC Two. https://www.youtube.com/watch?v=c8YxHk7yMo8 and https://www.youtube.com/watch?v=6Z2r1brcBrY.

Taberner, Stuart. *Aging and Old-Age Style in Günter Grass, Ruth Klüger, Christa Wolf, and Martin Walser: The Mannerism of a Late Period*. Rochester: Camden House, 2013.

Tallis, Raymond. *Aping Mankind: Neuromania, Darwinitis and the Misrepresentation of Humanity*. Durham: Acumen, 2011.

Tanzi, Rudolph E., and Ann B. Parson. *Decoding Darkness: The Search for the Genetic Causes of Alzheimer's Disease*. Cambridge, MA: Perseus, 2000.

Tappen, Ruth M. 'Awareness of Alzheimer Patients'. *American Journal of Public Health* 78 (1988): 987–8.

Taunton, Nina. *Fictions of Old Age in Early Modern Literature and Culture*. New York: Routledge, 2007.

Taylor, Elizabeth. *Mrs Palfrey at the Claremont*. 1971. Introduced by Paul Bailey. 1982. London: Virago, 2012.

Taylor, Jenny Bourne, and Sally Shuttleworth. Introduction to *Embodied Selves: An Anthology of Psychological Texts, 1830–1890*, edited by Jenny Bourne Taylor and Sally Shuttleworth, xiii–xviii. Oxford: Clarendon, 1998.

Taylor, Richard. *Alzheimer's from the Inside Out*. 2007. Baltimore: Health Professions, 2008.

Thane, Pat. 'Geriatrics'. In *Companion Encyclopedia of the History of Medicine*, edited by W. F. Bynum and Roy Porter, 1092–115. Abingdon: Routledge, 1993.

Thane, Pat. *Old Age in English History: Past Experiences, Present Issues*. Oxford: Oxford University Press, 2000.

Thomas, Dylan. 'Do Not Go Gentle into That Good Night'. 1937. In *Selected Poems*, edited with an introduction by Walford Davies, 100–1. London: Penguin, 2000.

Thomas, Matthew. *We Are Not Ourselves*. London: Fourth Estate, 2014.

Thomasma, David C. 'Mercy Killing of Elderly People with Dementia: A Counterproposal'. In *Dementia and Aging: Ethics, Values, and Policy Choices*, edited by Robert H. Binstock, Stephen G. Post and Peter J. Whitehouse, 101–17. Baltimore: Johns Hopkins University Press, 1992.

Tinniswood, Peter. '"Company" by Samuel Beckett'. *Times*, 26 June 1980.

Tissot, R. 'Discussion Remarks on Senile Disintegration'. In *Senile Dementia: Clinical and Therapeutic Aspects*, edited by Ch. Müller and L. Ciompi, 83–4. Bern: Hans Huber, 1968.

Toledano, Phillip. *Days with My Father*. San Francisco: Chronicle Books, 2010.

Tomlinson, B. E., G. Blessed and M. Roth. 'Observations on the Brains of Demented Old People'. *Journal of the Neurological Sciences* 11 (1970): 205–42.

Toombs, S. Kay. 'The Meaning of Illness: A Phenomenological Approach to the Patient–Physician Relationship'. *Journal of Medicine and Philosophy* 12 (1987): 219–40.

Torack, Richard M. 'The Early History of Senile Dementia'. In *Alzheimer's Disease: The Standard Reference*, edited by Barry Reisberg, 23–8. New York: Free Press, 1983.

Trebus Project. 'Welcome to the Trebus Project'. Accessed 25 October 2019. www.trebusprojects.org/.

Trevor, William. *The Boarding House*. 1965. London: Penguin, 1968.

Turner, Janice. 'I Dreamt of Greek Olives as I Fumed in Munich'. *Times*, 13 August 2015.

Turteltaub, Jon, dir. *Last Vegas*. CBS Films, 2013.

Updike, John. *The Poorhouse Fair*. 1958. London: Penguin, 2006.

Van Dijck, José. *Imagenation: Popular Images of Genetics*. Basingstoke: Palgrave Macmillan, 1998.

Van Dijck, José. *The Transparent Body: A Cultural Analysis of Medical Imaging*. Seattle: University of Washington Press, 2005.

Vanden Bosch, James, dir. *My Mother, My Father*. 1984. Concord Media, n.d. DVD.

Vandenberg, Ann E. 'What We Can Say about Cognition in Aging: Arguments for and against Cognitive Health Promotion'. PhD thesis, Emory University, 2012.

Vassilas, Christopher A. 'Dementia and Literature'. *Advances in Psychiatric Treatment* 9 (2003): 439–45.

Venturino, Giovanna. *Il tuo mare di nulla: La mia mamma e l'Alzheimer*. Rome: A&B Editrice, 2012.

Villa, J. L., and L. Ciompi. 'Therapeutic Problems of Senile Dementia'. In *Senile Dementia: Clinical and Therapeutic Aspects*, edited by Ch. Müller and L. Ciompi, 107–49. Bern: Hans Huber, 1968.

Vinci, C. F., C. Pontesilli and D. Fo. 'Telling about the Stolen Mind'. *Neurological Sciences* 26 (2005): 185–7.

Wall, Frank. *Where Did Mary Go?* Amherst: Prometheus Books, 1996.

Walrath, Dana. *Aliceheimer's: Alzheimer's through the Looking Glass*. University Park: Pennsylvania State University Press, 2016.

Wardle, Irving. 'A Worthy Guest'. *Times*, 12 June 1974.

Warren, Marjorie. 'Care of the Chronic Aged Sick'. *The Lancet* 1 (1946): 841–3.

Weindling, Paul. *Health, Race and German Politics between National Unification and Nazism, 1870–1945*. 1989. Cambridge: Cambridge University Press, 1993.

Weiner, Jonathan. *His Brother's Keeper: One Family's Journey to the Edge of Medicine*. 2004. New York: Ecco, 2005.

Whipple, Tom. 'Doctors Hail First Drug to "Slow Down" Alzheimer's'. *Times*, 22 October 2019. https://www.thetimes.co.uk/article/biogen-new-drug-offers-hope-for-alzheimer-s-sufferers-sh90z3q9h.

Whitehead, Anne, and Angela Woods. Introduction to *The Edinburgh Companion to the Critical Medical Humanities*, edited by Anne Whitehead and Angela Woods, 1–31. Edinburgh: Edinburgh University Press, 2016.

Whitehouse, Peter J., with Daniel George. *The Myth of Alzheimer's: What You Aren't Being Told about Today's Most Dreaded Diagnosis*. New York: St. Martin's Griffin, 2008.

Whitehouse, Peter J., Donald L. Price, Robert G. Struble, Arthur W. Clark, Joseph T. Coyle and Mahlon R. DeLong. 'Alzheimer's Disease and Senile Dementia: Loss of Neurons in the Basal Forebrain'. *Science* 215 (1982): 1237–9.

Wiesel, Elie. *L'oublié*. Paris: Éditions du Seuil, 1989. English translation as *The Forgotten*. Translated by Stephen Becker. New York: Schocken Books, 1992.

Wildgen, Michelle. *You're Not You*. 2006. New York: Picador, 2007.

Wilkinson, Jane. 'Remembering Forgetting'. *Status Quaestionis* 6 (2014): 103–21.

Wilks, Samuel. 'Clinical Notes on Atrophy of the Brain'. *Journal of Mental Science* 10 (1864): 381–92.

Wilson, Andy, dir. *The Complete Forsyte Saga*. 2002. ITV Studios Home Entertainment, 2003. DVD.

Wisniewski, Henryk M. 'Neuritic (Senile) and Amyloid Plaques'. In *Alzheimer's Disease: The Standard Reference*, edited by Barry Reisberg, 57–61. New York: Free Press, 1983.

Woods, Angela. 'The Limits of Narrative: Provocations for the Medical Humanities'. *Medical Humanities* 37 (2011): 73–8.

Woodward, Kathleen. *Aging and Its Discontents: Freud and Other Fictions*. Bloomington: Indiana University Press, 1991.

Woodward, Kathleen. 'A Public Secret: Assisted Living, Caregivers, Globalization'. *International Journal of Ageing and Later Life* 7 (2012): 17–51.

Wuest, J., P. K. Ericson and P. N. Stern. 'Becoming Strangers: The Changing Family Caregiving Relationship in Alzheimer's Disease'. *Journal of Advanced Nursing* 20 (1994): 437–43.

Yahnke, Robert E. 'Old Age and Loss in Feature-Length Films'. *The Gerontologist* 43 (2003): 426–8.

Young, Robert M. *Mind, Brain and Adaptation in the Nineteenth Century: Cerebral Localization and Its Biological Context from Gall to Ferrier*. Oxford: Clarendon, 1970.

Youtube. 'Julianne Moore Winning "Best Actress"'. 6 March 2015. www.youtube.com/watch?v=TzR3CUU51IU.

Zabbia, Kim Howes. *Painted Diaries: A Mother and Daughter's Experience through Alzheimer's*. Minneapolis: Fairview, 1996.

Zeilig, Hannah. 'The Critical Use of Narrative and Literature in Gerontology'. *International Journal of Ageing and Later Life* 6 (2011): 7–37.

Zeilig, Hannah. 'Dementia as a Cultural Metaphor'. *The Gerontologist* 54 (2014): 258–67.

Zeilig, Hannah. 'Gaps and Spaces: Representations of Dementia in Contemporary British Poetry'. *Dementia* 13 (2014): 160–75.

Zeilig, Hannah. 'Late-Life Creativity and the "New Old Age"'. Archived by King's Digital Lab. October 2018. www.latelifecreativity.org.

Zilberman, Yaron, dir. *A Late Quartet*. Opening Night Productions, 2012.

Zimmermann, Martina. 'Alzheimer's Disease Metaphors as Mirror and Lens to the Stigma of Dementia'. *Literature and Medicine* 35 (2017): 71–97. https://muse.jhu.edu/article/659107.

Zimmermann, Martina. 'Book Review: Living before Dying. Imagining and Remembering Home by Janette Davies'. *Times Higher Education*, 8 February 2018.

Zimmermann, Martina. 'Deliver Us from Evil: Carer Burden in Alzheimer's Disease'. *Medical Humanities* 36 (2010): 101–7.

Zimmermann, Martina. 'Dementia in Life Writing: Our Health Care System in the Words of the Sufferer'. *Neurological Sciences* 32 (2011): 1233–8.

Zimmermann, Martina. '"Journeys" in the Life-Writing of Adult-Child Dementia Caregivers'. *Journal of Medical Humanities* 34 (2013): 385–97.

Zimmermann, Martina. *The Poetics and Politics of Alzheimer's Disease Life-Writing*. Basingstoke: Palgrave Macmillan, 2017. https://www.palgrave.com/gb/book/9783319443874.

Zimmermann, Martina. 'Terry Pratchett's *Living with Alzheimer's* as a Case Study of Late-Life Creativity'. In *Creativity in Later Life: Beyond Late Style*, edited by David Amigoni and Gordon McMullan, 198–207. Abingdon: Routledge, 2019.

Zunshine, Lisa. 'Theory of Mind, Social Hierarchy, and the Emergence of Narrative Subjectivity'. In *The Emergence of Mind: Representations of Consciousness in Narrative Discourse in English*, edited by David Herman, 161–86. Lincoln: University of Nebraska Press, 2011.

Zunshine, Lisa. *Why We Read Fiction: Theory of Mind and the Novel*. Columbus: Ohio State University Press, 2006.

Zweers, Alexander. 'The Narrator's Position in Selected Novels by J. Bernlef'. *Canadian Journal of Netherlandic Studies* 19 (1998): 35–40.

Index

absence 49, 50, 63, 76, 79, 101, 112, 141, 142, 152, 161, 165
acetylcholine 89, 189 n.74
acetylcholinesterase inhibitors 189 n.74
  See also donepezil; tacrine; therapy, pro-cholinergic; treatment, pro-cholinergic
Acocella, Silvia 34
activist 102, 137
ADRDA
  See Alzheimer's disease, and Related Disorders Association
aducanumab 215 n.42
advocacy 16, 18, 69, 73, 209 n.34
Africa 14, 151
age 27, 38, 55–60, 65, 141, 144
  middle 41, 45, 80, 141, 157, 166–7
  of onset 35, 39–40, 44, 47, 124, 127, 157, 188 n.65, 204 n.47
ageing
  as challenge 13, 50, 51
  and demographic shifts 15, 17, 61, 79, 104
  failed 98–9
  and gain 5, 162
  healthy 18, 98–9, 134, 193 n.4
  as involution 28, 37, 43
  and life-experience 65, 162
  and loss 5, 6, 8, 16, 23, 28–9, 33, 105, 142
  process 6, 114
  as regression 38, 41, 43
ageism 5, 9, 58, 80
  See also Butler, Robert N.
Aging with Grace
  See Snowdon, David A.
Ahora tocad música de baile
  See Barba, Andrés
Alice in Wonderland 150, 218 n.84
Aliceheimer's
  See Walrath, Dana
ALS
  See amyotrophic lateral sclerosis
Alzheimer, Alois 16, 32, 34, 36, 40–1, 56, 58, 87, 93, 109, 126
Alzheimer's Association 99, 106, 138, 140
Alzheimer's disease
  vs. amyotrophic lateral sclerosis 13, 90–1
  cause of 60, 70, 88, 94, 96, 100
  as construct 15
  early-onset 35, 88, 94, 98, 104, 141, 154, 157
  as experience (See also experience, of dementia) 154
  familial 94–5, 98
  vs. Huntington's disease 95–7
  as killer (See also Katzman, Robert) 59–60, 70–1, 78, 82, 94, 95, 102, 120, 136, 157
  late-onset 35, 104
  vs. multiple sclerosis 214 n.21
  vs. Parkinson's disease 13, 91–2
  popularization of 12, 143
  presenile vs. senile 35, 37, 38, 42, 44, 45, 60
  and Related Disorders Association 111, 198 n.12
Alzheimer's from the Inside Out
  See Taylor, Richard
Alzheimer mon amour
  See Huguenin, Cécile
America, American 14, 17, 47, 55–6, 59, 65, 71, 75, 79, 90, 93, 96, 100, 102, 114, 143, 149, 155, 158
  See also United States
Amis, Kingsley 46, 65
amnesia 75, 79, 115
  See also trauma
amyloid
  burden 99
  fragments 95
  hypothesis 94, 135
  as infectious agent 135–6

inoculants 139
   plaques 37, 39, 44, 69, 87–8, 93, 94, 136
   precursor protein 94–5
amyotrophic lateral sclerosis 13, 91,
      202 n.17
   See also motor neuron disease
aphasia 12, 32
   See also Freud, Sigmund; Jackson, John
      Hughlings
Appignanesi, Lisa 100–1, 104
aricept
   See donepezil
Aristophanes 6
Ariyoshi, Sawako 68, 71, 72, 198 n.13
arteriosclerotic 35, 189 n.69
   See also lesion
Asia 14
   See also India; Japan; South Korea
The Assault
   See Mulisch, Harry
asylum 35, 48–9, 192 n.108
Atonement
   See McEwan, Ian
atrophy 18, 34, 43–5, 56, 88, 188 n.60
At The Jerusalem
   See Bailey, Paul
Auguste D. 8, 33, 34–7, 40, 41, 47, 58,
      126, 157
Austerlitz
   See Sebald, W. G.
Australia, Australian 14, 124, 125
Avati, Pupi 72
Away from Her
   See Polley, Sarah

Baars, Jan 221 n.120
Bailey, Paul 17, 46, 60–6, 76, 79–80,
      90, 165
Baker, Dana Lee 121–2, 141, 142
Ballenger, Jesse F. 3, 74, 87, 137
bapineuzumab 139–40
Barba, Andrés 1, 4
Baréma, Jean 97
Barlow, A. Ralph 79
Barron, Jeannette Montgomery 128
basal forebrain
   See nucleus, basalis of Meynert
Bayley, John 17, 70, 106, 154
Beale, Simon Russell 220 n.111
Beckett, Samuel 74–5, 165

Beer, Gillian 31
Bellow, Saul 78, 220 n.54
Berger, John 129
Bernlef, J. 17, 74–83, 116, 121
Berrios, G. E. 109, 189 n.71
bildungsroman 80, 155, 158, 221 n.113
   See also reifungsroman
La bimbamamma
   See Rota, Nucci A.
Binswanger, Otto Ludwig 187 n.53
biomarker 2, 96
   See also diagnosis
biomedical model 5
   and literary scholarship 3, 177 n.8
biomedicalization 87, 100, 115, 163–4
biopsychosocial approach 16, 124
   See also Kitwood, Tom; whole-person
      approach
Bitenc, Rebecca A. 139
Black Daisies for the Bride
   See Harrison, Tony
Blessed Dementia Scale 111
Block, Stefan Merrill 12, 139, 155–6
The Boarding House
   See Trevor, William
Boden, Christine
   See Bryden, Christine
body 1, 8, 36, 44, 56, 59, 62, 64–5, 75, 77,
      90, 91–2, 95–6, 102, 110, 113, 128,
      136–8, 150, 153, 156, 161
Bolton, Lucy 141
Bonfiglio, Francesco 37–8, 41
Booth, Charles 26, 183 n.10
Booth, Wayne 105
Borio, Alessandro 18, 142–7, 148, 150,
      153, 154, 160–1
Boyle, Mary 39
Braddon, Mary E. 49, 193 n.116
Bradford dementia group 126
Braff, Zach 135
brain
   ageing 99
   atrophy (*See* atrophy)
   basal fore (*See* nucleus, basalis of
      Meynert)
   cortex 89–90
   decade of the 109
   examination 42
   hippocampus 89–90, 112
   imaging (*See* imaging)

as organ of mind 7
reserve (*See also* cognitive, reserve) 98
Briggs, Julia 47, 48, 192 n.98
Britain, British 13–14, 17, 18, 23, 27, 30–3, 35, 51, 57, 65, 95, 101, 113, 114, 127
   Geriatrics Society 57
Broca, Paul 185 n.36
Brockmeier, Jens 146
Brody, Elaine M. 68
Brooks, Peter 116
Brun, Arne 88
Bryden, Christine 123–9, 137, 142, 153, 211 n.78
Buades, Margarita Retuerto 105–6, 123
Buddenbrooks 34, 155
Burack-Weiss, Ann 105
burden 2, 17, 19, 45, 61, 67–8, 70–2, 99, 107, 135, 156
Burke, Lucy 82, 154, 217 n.70
Bush, George W. 109
Butler, Robert N. 55, 58, 61, 64, 65, 69, 80, 82

Canada, Canadian 3, 100, 198 n.12
cancer 118, 136, 142, 211 n.77
Canguilhem, Georges 56
care
   vs. cure 18, 60, 120, 141, 149
   patient-centred 121–2, 126
   See also healthcare
caregiver
   adult-child 13, 68, 72, 100–7, 116, 147, 148
   as archivist/detective 100
   burden (*See also* burden) 17, 68, 71–2
   burnout 102
   as companion 152
   experience 105, 156, 161
   guides 67–73
   life-writing (*See also* life-writing) 4, 13, 129, 147, 149
   professional 139, 151–2
   spouse/partner as 13, 73, 105–6
Cartwright, Lisa 112
Caruth, Cathy 161
case
   clinical 7, 34–6, 42, 44
   report 10
   study 33, 36–7, 41, 188 n.58
Catch a Falling Star

   See Spohr, Betty Baker
Cavigioli, Rita C. 221 n.113
CDR
   See Clinical Dementia Rating
cell
   brain (*See also* neuron) 32, 87, 94, 112, 113
   death (*See also* atrophy; loss, of neurons; neuron) 2, 5, 32, 34, 37, 89–90, 94
   and identity 1, 5
centenarianism 27
Cephalus 65
cerebrospinal fluid 96
Charon, Rita 124
child
   addressed by dementia narratives 142–3
   as image of underachievement (*See also* phylogeny) 28
   vs. madman (*See also* Foucault, Michel) 49
   perspective of the 72, 142, 160–1, 165
childhood 59, 72, 77, 158, 160
   second 6–7, 28
Chivers, Sally 134, 161
cholinergic 89, 90, 93, 111
chromosomes 94, 97
Chronicle of My Mother
   See Inoue, Yasushi
class 23–25, 49, 157–8
Claudia, Alzheimer und ich
   See Mummendey, Hans Dieter
Clinical Dementia Rating 111, 208 n.12
clinical trial 113, 139, 205 n.53, 214 n.24
   See also drugs
cognition 1, 3, 18, 95, 110, 114, 163, 167
cognitive
   decline 98, 105, 114, 141, 164, 220 n.101
   health 99, 113
   health industry (*See* industry, cognitive health)
   performance 13, 15, 16, 98, 111–14, 140
   reserve 98–9, 100
   testing 113, 134, 138, 139
Cohen, Lawrence 69–70
Cole, Thomas, R. 26
Collins, Wilkie 49
communication 43, 65, 88–91, 93, 102, 107, 124, 136, 162, 165, 202 n.17
   See also gossip; network

Company
  See Beckett, Samuel
computer tomography 110
Connor, Steven 76
consciousness 12, 51, 61, 74, 81, 91, 115, 221 n.113
Conway, Kathlyn 104
Cooney, Eleanor 102–5, 123
The Corrections
  See Franzen, Jonathan
Couturier, Claude 1, 4, 127, 153, 166
creativity 9, 128, 145, 153
Crichton-Browne, James 34
CSF
  See cerebrospinal fluid
CT
  See computer tomography
Cullen, William 185 n.27
cure 17, 18, 95–7, 109, 136, 138–42, 161, 166
  vs. care 18, 60, 120, 141, 149
  narrative (*See also* Davis, Lennard J.) 166
'The Curious Case of Benjamin Button'
  See Fitzgerald, F. Scott

Daddyboy
  See Konek, Carol Wolfe
Dames, Nicholas 8, 29–30, 146
Dancing with Dementia
  See Bryden, Christine
Darwin, Charles 25, 31, 183 n.8, 185 n.32
Darwin, Erasmus 7
Davidson, Ann 73
Davis, Lennard J. 113, 222 n.3
Davis, Robert 82, 124–6, 137
Days with My Father
  See Toledano, Phillip
De Dionigi, Elena 148–9, 152
Dean, Debra 147
death
  cell (*See also* loss, of neurons) 2, 5, 32, 34, 37, 89–90, 94
  living (*See also* living dead) 19, 71, 102, 115
  of self 19, 104, 154, 161
  social 17, 92–3, 100, 114, 123
Death in Slow Motion
  See Cooney, Eleanor
DeBaggio, Thomas 76, 78–9

Decoding Darkness
  See Tanzi, Rudolph E.
degeneration 15–16, 19, 34, 36–8, 43, 56–7, 155, 160, 164, 185 n.30
  cellular (*See also* cell, death) 40–1, 44, 87–8, 93, 113
  post-Darwinian concepts of 23–8, 30–1, 33–4, 49–50, 72, 97
  vs. regeneration 189–90 n.76
  See also Morel, Bénédicte-Auguste; Nordau, Max
DeLillo, Don 147
delusion
  See hallucinations
dementia
  as age-related decline 31
  as death sentence 79, 140
  as disorder of cognition 3, 164
  as dispossession 19, 26, 61–3, 65, 76, 92, 142, 155
  fronto-temporal 211 n.78
  as global trope 79
  as narrative device 80, 106
  as organic disease 3, 4, 8, 15
  as postmodern condition 74
  and post-war trauma (*See also* trauma) 17, 74–5, 78–9, 100–1
  as second childhood (*See* childhood, second)
  as torture of the mind 78
  vascular 15, 39, 44, 101, 189 n.69
  See also feeble-mindedness; idiocy
Demenz
  See Jens, Tilman
De Morgan, William 45
depression 28, 40, 57, 60, 62–3, 102
Der alte König in seinem Exil
  See Geiger, Arno
detention 48–9
Deutsch, Helene 45–6
Devi, Gayatri 99
Di Pietrantonio, Donatella 149, 151
diagnosis 2, 18, 33, 42–3, 45, 46–7, 58, 65–66, 69, 97, 111, 114, 124, 128, 139, 140–2, 159
  differential 37, 191 n.84
  Ex vivo 44
  In vivo 44, 107, 110
diary 72, 106, 127–8, 166
Dibdin, Michael 18, 117–22, 129, 165

Dickens, Charles 30
Diedrich, Lisa 143
disability 140
disappearance 36, 43, 80, 128
discourse
  caregiver 2
  counter- 4, 60, 107, 125
  materialist 32
disease
  hypotheses 2, 88, 140
  label 97
disengagement 195 n.37
disintegration 34, 36, 50, 59, 65, 104, 133, 157
dispossession 19, 26, 61–3, 65, 76, 92, 142, 155
documentary 14, 197 n.12
  Black Daisies for the Bride (*See also* Harrison, Tony) 81
  Living with Alzheimer's (*See also* Pratchett, Terry) 137
  Vergiss mein nicht (*See also* Sieveking, David) 134, 213 n.7
donepezil 111, 113
dopamine 92, 203 n.20
Drachman, David A. 90, 93, 101
Draesner, Ulrike 127
drama 12, 34, 81, 104, 121
  See also play
drugs 107, 109–11, 113, 138–9, 141
  See also acetylcholinesterase inhibitors; donepezil; tacrine; therapy, pro-cholinergic; treatment, pro-cholinergic
DS
  See Blessed Dementia Scale
Dumit, Joseph 112
The Dying of the Light
  See Dibdin, Michael

Eakin, Paul John 81
Ehrenreich, Barbara 104
Electron Microscopy 88, 93, 94
Eliot, George 184 n.18
Elizabeth Is Missing
  See Healey, Emma
Elliott, Anthony 158, 199 n.37
Ellis, Havelock 28
empathy 64, 76, 79, 142, 148, 153, 160–3, 167

encephalography 43
  See also imaging
Ending Up
  See Amis, Kingsley
Entartung
  See Nordau, Max
Ernaux, Annie 100
Esquirol, Jean Étienne 31, 33, 35
Europe 13, 31, 34, 44, 57, 59, 74, 78, 100, 194 n.9
experience 4, 5, 69, 103, 105, 136, 139, 156
  of ageing 9, 154
  of dementia 71, 75, 78, 80, 127, 154
  See also caregiver; patient
Eyre, Richard 106, 166

Falcus, Sarah 139, 207 n.70
Falling Man
  See DeLillo, Don
Faulkner, William 47, 165
feeble-mindedness 30, 47, 49
Ferrier, David 186–7 n.36
film
  See documentary; television
Fitzgerald, F. Scott 65
Fogelman, Dan 134
Foley, Joseph M. 75, 81, 115
Folstein, M. F. 111
forgetting 17, 75, 79, 89, 101, 104, 129, 146–7, 153, 161
Forster, Margaret 68, 72
Forsyte, Timothy 16, 23–9, 30–1, 33–4, 38, 41, 50, 55, 62, 64, 146, 158, 165
  See also Galsworthy, John
Foucault, Michel 7, 37, 96, 110, 157–8, 160
fragmentation 24, 74, 136–7, 183 n.5, 199 n.37
  See also Beckett, Samuel; Bernlef, J.; genes, genetics
France, French 13–14, 19, 30, 31, 33, 34, 68, 75, 100, 148–54, 161, 185 n.32, 186 n.40, 189 n.76, 191 n.84, 192 nn.108, 111
Frank, Arthur 126, 136–7
Franzen, Jonathan 92, 156
freedom 5, 47, 61, 78, 134, 157
Freeman, H. L. 109
Freud, Sigmund 16, 32, 40, 47, 51, 189 nn.71, 73
Frisch, Max 74

Fritsch, Gustav 185–6 n.36
frontal lobe disorder 15, 185 n.33
Frow, John 12
Fuchs, Elinor 128

Gadny, Faith 17, 46, 60–6, 79, 90, 165
   See also Bailey, Paul
Gadow, Sally A. 60
Gage, Phineas 185 n.33
gain 5, 154, 162, 189 n.74
   See also growth
Gall, Franz Joseph 7, 185 n.36
Galsworthy, John 16, 23–9, 30–4, 38, 41, 50–1, 55, 70, 79, 103, 155
Galt, John 30
Gardini, Nicola 149
Gaskell, Elizabeth 30
Gawande, Atul 144, 149, 162
gaze 4, 7, 43, 80, 96, 110, 145
   See also Foucault, Michel; visibility
GDS
   See global deterioration scale
Geiger, Arno 128
gender 9, 14, 30, 77, 80
genes, genetics 17, 87, 88, 94–7, 100, 104, 107, 110, 129, 204 n.38
   See also presenilin
Genova, Lisa 138–9, 165
genre 10, 12–13, 115, 116–18, 123, 143, 155, 156
geriatrics 16, 56–58, 194 nn.8, 9
German, Germany 13–14, 18, 33, 34, 40, 42, 75, 77–78, 133–4, 152, 186 n.46, 189–90 n.76, 192 nn.108, 111, 213 n.5
gerontology 16, 56–7, 60, 194 nn.8, 9
Gerrard, Nicci 147, 155
Gillies, Andrea 70, 102, 206 n.82
Gilman, Sander L. 189 n.73
Glatzer, Richard 166
global deterioration scale 77
glutamate 189 n.74
Goffman, Erving 63
Goldby, Roger 134
Goldman, Marlene 3, 177 n.8, 197–8 n.12
Goldsmith, Malcolm 124–5
Golgi, Camillo 8, 35
gossip 24, 184 n.18
Goyder, Julie 125
La guardiana di Ulisse
   See Borio, Alessandro
Grant, Linda 101
Grass, Günter 79, 221 n.125
Great War
   See War, Great
Greenslade, William 24, 192 n.111
Griesinger, Wilhelm 33, 186 n.44
growth 5, 31, 33, 48, 57, 126, 146, 162, 164
Gruen, Arno 160–1
Grünthal, Ernst 42–3

Hacking, Ian 97, 111, 113, 166
Haeckel, Ernst 28
Hall, G. Stanley 45
Hallervorden, Dieter 133–4, 212 n.2
hallucinations 35, 39, 78
Hannah's Heirs
   See Pollen, Daniel A.
Hard to Forget
   See Pierce, Charles P.
Hardy, John A. 94
Harrison, Tony 81–2
Harry Potter 161
Hartung, Heike 221 n.113
Harvey, Geoffrey 26
Harvey, Samantha 147
Havemann, Joel 91–2
Have the Men Had Enough?
   See Forster, Margaret
Hawkins, Anne Hunsaker 97, 126
Hayworth, Rita 70, 198 n.22
Hazell, Kenneth 57–8, 60, 62–3, 65, 66
Healey, Emma 147, 217 n.70
health
   brain 98–9
   humanities (*See also* medical humanities; narrative, medicine) 2–3, 11, 73, 153, 164
   public 17, 24, 98–9, 113, 140
healthcare
   approaches (*See also* whole-person approach) 149, 153, 161
   campaigns 60, 70
   caregiver-centred 65, 76
   in the community 61, 121
   low cost 162
   person-centred 152
   politics 55, 68, 189–90 n.76
Held, Wolfgang 127

Henderson, Cary Smith 74, 193 n.120
Herman, David 12
Hiroshima mon amour 161
His Brother's Keeper
  See Weiner, Jonathan
Hitzig, Eduard 185–6 n.36
Holderman, Bill 134
Holocaust 17, 78–9
home
  care 68, 148
  nursing 64, 68, 102, 122–3, 145, 151, 156, 212 n.2
  See also institutionalization
Homer 6, 144, 146
Honel, Rosalie Walsh 17, 67–73, 105, 197 n.12
Honig im Kopf
  See Schweiger, Til
  See also Hallervorden, Dieter
Horace 6
hospice 121, 152
hospital 30, 49, 64–5, 68, 77, 81, 118, 213 n.7
Hoult, Norah 16, 45–51, 55, 76, 165
Howe, Samuel G. 185 n.30
Howell, Trevor H. 58
Howland, Alice 138–42
  See also Genova, Lisa
Hubble, Nick 47
Huguenin, Cécile 19, 148–54, 161
humanities
  See health, humanities; medical humanities; narrative, medicine
Hunter, Kathryn Montgomery 124
Huntington's disease 13, 15, 95, 97, 114, 205 n.53
Huyssen, Andreas 101
Hyman, Bradley T. 105
hypercognitive society 114
  See also Post, Stephen G.
Hysteria 40

identity
  loss of 5, 25, 47, 90
idiocy 8
  See also feeble-mindedness; retardation
Ignatieff, Michael 90–1
Il tuo mare di nulla
  See Venturino, Giovanna
The Illuminations
  See O'Hagan, Andrew
imaging 9, 18, 43–4, 107, 109–10, 113–14, 128
  See also computer tomography; encephalography; magnetic resonance imaging; positron emission tomography; single photon emission computed tomography; X-ray technology
immunization 135–42
incontinence 58, 133
India 14, 151
  See also Asia
individuality 11, 18, 51, 60, 97, 125, 136, 157, 160, 164, 166
  See also personhood
industrialization 14, 48
industry
  cognitive health 18, 99
  pharmaceutical 136, 140, 214 n.31
infantilization 42
Inoue, Yasushi 197 n.2, 198 n.13
insanity 31, 45, 46, 185 n.35
institution 28, 48–9, 64–5
institutionalization 49, 64–5
involution 16, 28, 37, 38, 41–3, 67, 93
  See also regression
Irigaray, Luce 77
Iris
  film (See Eyre, Richard)
  memoir (See Bayley, John)
The Iron Lady
  See Lloyd, Phyllida
Italy, Italian 13, 14, 18, 34, 144, 148, 160, 185 n.32, 189 n.76, 197 n.9, 205–6 n.64, 218 n.81, 221 n.113
Ivory, Marcia 46, 58, 65, 195 n.37
  See also Pym, Barbara

Jackson, John Hughlings 31–2
Japan, Japanese 68, 112, 161, 197 n.2, 198 n.13
  See also Asia
Jens, Tilman 206 n.82
Johann F. 41, 47
journalism
  See mass media; newspapers; print media
Journey with Grandpa
  See Honel, Rosalie Walsh

Joyce, Kelly A. 110
Juvenal 6

Katz, Stephen 194 n.7
Katzman, Robert 59–60, 69, 71
Keeper
    See Gillies, Andrea
King Lear 7, 47, 157
King, Nicola 101
Kitwood, Tom 125, 126
Klein, Maarten 74–81
    See also Bernlef, J.
Kleinman, Arthur 73
Konek, Carol Wolfe 73
Kraepelin, Emil 37, 38–41, 42, 49, 87, 93, 164
Kuhn, Thomas S. 189 n.67

Laborde, Françoise 150
Lady Audley's Secret
    See Braddon, Mary E.
Lafora, Gonzalo R. 42
Laing, R. D. 64
La moindre des choses
    See Philibert, Nicolas
A Late Quartet
    See Zilberman, Yaron
Latour, Bruno 9–10, 95, 110, 138, 140, 141
Leary, Edmund 154–61
    See also Thomas, Matthew
Lesion 7, 8, 31, 32, 39–40, 44, 77, 89, 110–11
Lewandowsky, Max 187 n.56
Lewy bodies 15
    See also Parkinson's disease
life review 61, 126
    See also Butler, Robert N.
A Life Shaken
    See Havemann, Joel
life-writing
    caregiver 4, 13, 129, 147, 149
    patient 4, 11, 115, 123, 128, 153
    See also diary; memoir
Lindbergh, Reeve 128
living dead 47, 55, 62, 70, 73, 101, 103, 123, 147, 184 n.18
    See also death, living; walking dead
Living in the Labyrinth
    See McGowin, Diana Friel
Living with Alzheimer's
    See Pratchett, Terry
Lloyd, Phyllida 154
L'oublié
    See Wiesel, Elie
Lock, Margaret 94
Lombroso, Cesare 189–90 n.76
loneliness 60–5, 220 n.101
Lo sconosciuto
    See Gardini, Nicola
Losing My Mind
    See DeBaggio, Thomas
Losing the Dead
    See Appignanesi, Lisa
loss
    of awareness 5, 141
    of control 5, 92, 149
    of identity 5, 25, 47, 90
    of language 5, 47, 76
    of memory 1, 5, 58, 80, 114, 142, 185 n.27
    of neurons (*See also* cell, death) 44, 90
    of self 71, 139, 147
A Love Story
    See Davidson, Ann
Lunacy Act 49
Lynch, Michael 36

Mace, Nancy L. 67–9, 71–2
madness 7, 8, 64, 188 n.65, 192 n.108, 220 n.111
The Madonnas of Leningrad
    See Dean, Debra
magnetic resonance imaging 110
    See also imaging
Magnusson, Sally 148, 151, 153, 218 n.84
Mairs, Nancy 214 n.21
Making an Exit
    See Fuchs, Elinor
management 17, 51, 60, 152
marketplace 17, 60, 88, 100, 107, 135, 139, 140
Marshall, Cynthia 144
Martin, Emily 136
mass media 4, 94, 107, 112–13, 115, 129, 134, 136, 138, 140
materialist 7, 26, 28, 32, 51, 56, 125, 164, 189 n.71
Maudsley, Henry 189 n.76
McDonagh, Patrick 30, 47
McEwan, Ian 114, 148

McGowin, Diana Friel 82, 124, 125, 127
McHaffie, Hazel 91
MCI
   See mild cognitive impairment
McMenemey, William H. 43, 59
medical humanities 2, 3, 124
   See also health, humanities; narrative, medicine
Medina, Raquel 14, 134
melancholia 7, 185 n.35
memoir 13, 67, 97, 100, 102, 104, 127, 128
memory
   archival notions of 101
   crisis 146
   politics of 14, 100
   sciences of 7–8
menopause 65
Der Mensch erscheint im Holozän
   See Frisch, Max
mental illness 6, 7, 46, 51, 59
Mental Treatment Act 49
Merlo, Maria 144–6, 149, 153, 165
   See also Borio, Alessandro
metaphor 5–6, 11, 27, 28, 41, 60, 70, 78, 79, 90, 91, 95, 110, 164
   See also myth
Metchnikoff, Elie 56, 193 n.4
Meyers, Nancy 134
Mia madre è un fiume
   See Di Pietrantonio, Donatella
middle age
   See age, middle
Middlemarch
   See Eliot, George
mild cognitive impairment 114
mind
   failing 7, 15, 81, 115, 166
   and modernism 12, 16, 50–1, 80, 116, 165
   See also consciousness
Mini-Mental State Examination 111
Mitchell, Wendy 127, 151
Mi vida junto a un enfermo de Alzheimer
   See Buades, Margarita Retuerto
MMSE
   See Mini-Mental State Examination
Mobley, Tracy 204 n.47
Monicelli, Mario 68
Moore, Julianne 140–1
morbidity 29

Morel, Bénédicte-Auguste 31, 49, 189 n.76
morphology 8, 37, 188 n.60
   See also imaging; lesion
Morris, David B. 88, 136
mortality 29, 55, 65
Mothersill, Mary 154
motor neuron disease 13, 91
   See also amyotrophic lateral sclerosis
Mr. Sammler's Planet
   See Bellow, Saul
MRI
   See magnetic resonance imaging
Mrs Palfrey at the Claremont
   See Taylor, Elizabeth
Mulisch, Harry 78–9
multiple sclerosis 189 n.73, 214 n.21
Mummendey, Hans Dieter 115–17, 122
Murdoch, Iris 17, 47, 70, 106
mutation 94, 98
My Journey into Alzheimer's Disease
   See Davis, Robert
myth 24, 25, 27, 31, 50, 113, 126

narrative
   capabilities 17, 117
   caregiver 17, 105, 129, 154
   chaos 126
   chronological 61, 105, 137, 156
   device 80, 106, 139
   (of) dispossession 76, 142
   form 23, 26, 149, 151, 155–6
   illness 13, 97
   medicine 153
   patient 18, 82, 125, 137
   perspective 151
   triumphalist 79
Nascher, Ignatz 56
National Dementia Strategy in England 152
National Health Service 57, 121, 127
National Institute for Health and Care Excellence 113
National Institute on Aging 17, 55
network 63, 90, 92, 93, 105, 118, 119, 129
neurocognitive reserve
   See cognitive, reserve
neurodegeneration 87, 164
neurodiversity 121
neurology 4, 32, 36, 40, 138
neuron 1, 13, 44, 88, 89–90, 91, 95, 127

See also cell
neuropathological features 87, 88, 93, 136, 164
neurotransmitter 89, 189 n.74, 203 n.20
  See also acetylcholine; dopamine; glutamate
newspapers 105, 197 n.12, 204 n.38
  See also print media
NHS
  See National Health Service
NIA
  See National Institute on Aging
NICE
  See National Institute for Health and Care Excellence
Night of the Living Dead
  See Romero, George
No Aging in India
  See Cohen, Lawrence
No More Words
  See Lindbergh, Reeve
Nordau, Max 31, 41, 189 n.76
norm 24, 112, 149, 151
nucleus 35, 37–8, 41, 94
  basalis of Meynert 90
Nun Study 98
  See also Snowdon, David A.

O'Hagan, Andrew 146–7
Odyssey 146
  See also Borio, Alessandro
The Old Man's Youth
  See De Morgan, William
ontogeny 28
Otis, Laura 34, 90, 93, 119
Out of Mind
  See Bernlef, J.

Painted Diaries
  See Zabbia, Kim Howes
paired helical filaments
  See tangles
Palmer, Alan 119
Parkinson's disease 13, 15, 91–2, 156, 203 n.20, 214 n.21
Partial View
  See Henderson, Cary Smith
passivity 2, 19, 50, 153, 164
pathogenesis 109, 137

pathology 6, 7, 31, 38, 44, 47, 79, 88, 89, 136, 186 n.44
pathway 9, 91, 100
  See also projection
patient
  as child (*See* childhood, second)
  infantilization of the 42
  invisible 24
  life-writing (*See* life-writing, patient)
  as living dead (*See* living dead)
  as living in the past 47–8, 156
  as passive 17, 18, 70, 123, 128, 136, 164, 165
  poster 70
  as stranger 80, 184 n.18
  as unreliable narrator 115, 117–19
Pavel, Thomas G. 50, 165
performance 9, 13, 15–16, 18, 98, 109–15, 124, 140, 166, 183 n.10, 220 n.111
Perowne, Henry 114
  See also McEwan, Ian
personhood 1, 3, 11, 16, 76, 96, 115, 124, 126, 143, 150, 164, 165, 166
Perusini, Gaetano 37, 38, 41
PET
  See positron emission tomography
Philibert, Nicolas 152
phylogeny 28
physiology 32, 40
Pick, Arnold 189 n.71
Pierce, Charles P. 96–7
Pinel, Philippe 31, 33, 73
plaques 37, 39, 44, 69, 87–8, 93, 94
Plato 65
play 121
  See also drama
plot 12, 79, 80, 95, 115, 119, 122, 124, 126, 144–5, 156, 165
poetry 12, 81
Pollen, Daniel A. 95–7
Polley, Sarah 154
The Poorhouse Fair
  See Updike, John
popular
  press (*See also* mass media) 5, 60, 78, 80
  science 10, 88, 93, 94, 95, 100, 115, 128
positron emission tomography 110
Post, Stephen G. 114
postmodernity 74

poverty 24, 30, 48, 57
power relations 63, 66, 67
Pratchett, Terry 137
prediction
  vs. prevention 16, 98, 167
presenile 104, 187 n.53, 188 n.65
  vs. senile (*See* Alzheimer's disease, presenile vs. senile)
presenilin 95, 204 n.39
Prichard, James Cowles 31
Prima di volare via
  See De Dionigi, Elena
print media 70, 74
  See also mass media; newspapers
prison 49, 63, 75
pro-cholinergic drugs
  See drug
productivity 26, 45, 114, 154, 157, 161
projection 90, 91
  See also network; pathway
psychiatry 38, 40, 44, 56, 58, 165
psycholinguistic 76–7, 121
psychology 16, 29, 30, 32, 36, 125, 126, 146, 160
Puzzle, Journal d'une Alzheimer
  See Couturier, Claude
Pyke, Marigold 46, 65
  See also Amis, Kingsley
Pym, Barbara 46, 58, 65, 195 n.37

Quartet in Autumn
  See Pym, Barbara
Quirk, Anne 146
  See also O'Hagan, Andrew

Rabins, Peter V. 67–9, 71–2
Ramanathan, Vaidehi 128
Reagan, Nancy 104
Reagan, Ronald 70, 197 n.12
realism
  new 64
  social 29, 50
reductionism 2, 3, 110, 114
regression 38, 41, 42, 43, 79, 147
reifungsroman 221 n.113
Reisberg, Barry 77, 87, 105
Remind Me Who I Am, Again
  See Grant, Linda
retardation 8

  See also Faulkner, William; feeble-mindedness; idiocy
Reynolds, Andrew S. 11
Ribot, Théodule A. 186 n.40
Riga, Frank P. 143
Right to Die
  See McHaffie, Hazel
Romero, George 70
Rose, Larry 71, 82, 124
Rose, Nikolas 2, 220 n.101
Rota, Nucci A. 149, 206 n.82
Roth, Marco 114
Rousseau, G. S. 12
Ru, Yi-Ling 155, 159
Russell, Charlie 137
  See also Pratchett, Terry

Sabat, Steven R. 122
Sacks, Oliver 4, 40
Sako, Katsura 217 n.70
Salisbury, Laura 148
Saturday
  See McEwan, Ian
Scarry, Elaine 28
Scar Tissue
  See Ignatieff, Michael
Schäubli-Meyer, Ruth 150
Schlich, Thomas 186 n.46
Schottky, Johannes 42, 43
Schweiger, Til 18, 133–5, 142, 162
Sebald, W. G. 77–8
Seidler, Miriam 122, 144
Selkoe, Dennis J. 93, 94–5, 135, 139
senescence 44, 57, 183 n.9
Senescence: The Last Half of Life
  See Hall, G. Stanley
senile
  vs. presenile (*See* Alzheimer's disease, presenile vs. senile)
senility 8, 34, 47, 56, 65, 66, 69, 102, 106, 221 n.113
Senium praecox 38–9
Shakespeare, William 7, 47
  See also *King Lear*
Sheldon, Joseph A. 57
Show Me the Way to Go Home
  See Rose, Larry
Shuttleworth, Sally 8, 184 n.20
Sieveking, David 134, 206 n.82, 213 n.7

See also *Vergiss mein nicht*
silence 89, 91, 184 n.18
single photon emission computed tomography 110
Small, Helen 7, 9, 27, 62, 65, 200 n.54
Small World
   See Suter, Martin
Smith, F. B. 29
Snowdon, David A. 98, 105
   See also Nun Study
Social and Medical Problems of the Elderly
   See Hazell, Kenneth
The Social Medicine of Old Age
   See Sheldon, Joseph A.
sociolinguistic 121–2, 152
solanezumab 140–1
Somebody I Used to Know
   See Mitchell, Wendy
Sontag, Susan 5
The Sound and the Fury
   See Faulkner, William
South Korea 161
   See also Asia
Spain, Spanish 14, 34, 105, 205–6 n.64, 218 n.81
SPECT
   See single photon emission computed tomography
spectrum disorder 99
Speriamo che sia femmina
   See Monicelli, Mario
Spohr, Betty Baker 73
spontaneity 18, 135–6
stigma 49, 69, 99, 114, 125
Still Alice
   film (*See also* Glatzer, Richard) 140, 141, 142, 154, 157, 166
   novel (*See also* Genova, Lisa) 138–9, 142, 157
The Story of Forgetting
   See Block, Stefan Merrill
stream-of-consciousness 50
Strepsiades 6
suicide 91, 119, 121, 139
Suter, Martin 122

tacrine 111
tangles 39, 44, 69, 87–8, 93
Tanzi, Rudolph E. 18, 93, 95, 96, 100, 106, 107, 139
tau hypothesis 93
Taylor, Elizabeth 46, 65
Taylor, Richard 137
television 81, 134, 137, 161, 197 n.12, 213 n.5
Temple, Claire 16, 45–50, 55, 165
   See also Hoult, Nora
Tew, Philip 47
The Test
   See Baréma, Jean
Thane, Pat 6
Thatcher, Margaret 154
   See also Lloyd, Phyllida
The 36-Hour Day 67–73, 123, 199 n.28
   See also Mace, Nancy L.; Rabins, Peter V.
The Forsyte Saga 16, 23–9, 159, 160
   televised 23, 184 n.17
   See also Galsworthy, John
theory of mind 119, 122
   See also Zunshine, Lisa
therapy
   blocking of amyloid as 94, 135–7
   pro-cholinergic (*See also* treatment, pro-cholinergic) 111
There Were No Windows
   See Hoult, Norah
Thomas, Dylan 118
Thomas, Matthew 19, 154–62
tissue 9, 16, 34–44, 56, 59, 88, 97, 110
Toledano, Phillip 128
Townsend, Peter 194 n.13
trauma 30, 75, 78–9, 101, 147, 220 n.111
Travis, Rosemary
   See also Dibdin, Michael
treatment
   inoculant as 139
   pro-cholinergic (*See also* therapy, pro-cholinergic) 111, 113, 209 n.34
Trebus Project
   See Bradford dementia group
Trevor, William 46
Turteltaub, Jon 134
The Twilight Years
   See Ariyoshi, Sawako

United States 13, 31, 56–7, 68, 74, 111, 185 nn.30, 32, 192 n.108, 201 n.73
   See also America
Uns hat Gott vergessen

See Held, Wolfgang
Updike, John 66

Vagrancy Laws 49
Van Dijck, José 95, 110
vascular dementia
　See dementia, vascular
Venturino, Giovanna 149
Vergiss mein nicht
　film 134, 213 n.7
　memoir 206 n.82
　See also Sieveking, David
victim 70, 75, 78, 82, 94, 99, 102, 119, 136, 140, 142
Victorian 23, 25–7, 29, 30, 32, 41, 47–8, 50, 68
Virchow, Rudolf 188 n.60
visibility 7, 25, 36, 39, 141, 166
voice 2, 24, 50, 72, 73, 77, 104, 106, 123, 124, 127, 152, 165
Von Krafft-Ebing, Richard 189 n.76

walking dead 141
Wall, Frank 73
Walrath, Dana 104, 218 n.84
War
　Boer 24, 26, 30
　Cold 101
　Great 30, 49
　World I 29, 48
　World II 29, 48, 76, 136, 161
Warren, Marjorie 57
We Are Not Ourselves
　See Thomas, Matthew
Weiner, Jonathan 91
welfare state 57
Wernicke, Carl 185–6 n.36
Western society 3, 5, 9, 66, 74
Westphal, Carl 33, 186 n.44

Where Did Mary Go?
　See Wall, Frank
Where Memories Go
　See Magnusson, Sally
Whitehouse, Peter J. 90, 92, 99
whole-person approach 4, 106, 119, 125, 126
Who Will I Be When I Die?
　See Bryden, Christine
Why Survive?
　See Butler, Robert N.
Wiesel, Elie 218 n.86
The Wilderness
　See Harvey, Samantha
Wildgen, Michelle 91
Wilks, Samuel 34
Will I Still Be Me?
　See Bryden, Christine
wisdom 62, 92, 128, 147, 153, 162
A Woman's Story
　See Ernaux, Annie
The Woman in White
　See Collins, Wilkie
Woodward, Kathleen 63, 79, 129, 152, 164

Xenophon 6
X-ray technology 43, 110
　See also imaging

You're Not You
　See Wildgen, Michelle
Young Hope
　See Mobley, Tracy

Zabbia, Kim Howes 127–8
Zilberman, Yaron 92
Zunshine, Lisa 116
Zweers, Alexander 81

www.ingramcontent.com/pod-product-compliance
Lightning Source LLC
Chambersburg PA
CBHW072129290426
44111CB00012B/1837